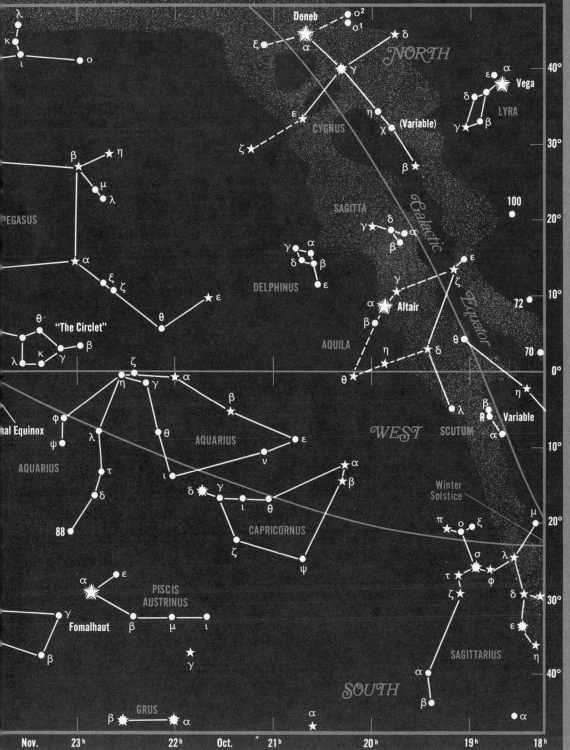

Fall Stars

C0-ATD-624

an introduction to
ASTRONOMY
SECOND EDITION

an introduction to
ASTRONOMY
SECOND EDITION

CHARLES M. HUFFER

PROFESSOR EMERITUS
*University of Wisconsin and
University of California, San Diego*

FREDERICK E. TRINKLEIN

*Nassau Community College
Garden City, N. Y.*

MARK BUNGE

San Jose City College

HOLT, RINEHART AND WINSTON, INC.

New York   Chicago   San Francisco   Toronto   London

Cover photograph courtesy of the Hale Observatories.
Copyright by the California Institute of Technology
and the Carnegie Institute of Washington.

Copyright © 1973, 1967 by Holt, Rinehart and Winston, Inc.
All Rights Reserved
Library of Congress Catalog Card Number: 72-89039
ISBN: 0-03-084685-4
Printed in the United States of America
3 4 5 6 7 8 9 071 1 2 3 4 5 6 7 8 9

preface

This text has been written to provide the college student in his freshman or sophomore year with a non-mathematical treatment of introductory astronomy. It was thought necessary in writing the first edition to include a few simple mathematical developments. The mathematics in this edition has been further reduced due to modern improvements in distance determination by the use of lasers and close-up photographs of the moon from orbiting vehicles. Computation of orbits are now made by high-speed electronic computers, too.

Astronomy has become more popular as a science which fulfills the requirement for a general college degree. It is for these students that this text is designed.

It is the aim of this book to give not merely the results of the studies of astronomy and all its branches. It is our desire to inform the student about the developments that have led to that knowledge. To give the bare facts about the solar system and its members, the stars and their distribution in space, and the universe and its galaxies is not enough. The men behind the important discoveries, their thinking and reasoning in stating their theories, and the modern theories of the extent and composition of the universe and its evolution are more important.

Included in this book is a discussion of quasi-stellar radio and non-radio sources (quasars) and of the more exciting discovery of neutron stars (pulsars)— those old stars that have collapsed until they are nothing but collections of neutrons and are rotating at high speeds, in some cases many times per second.

Our students have told us that they are most interested in the evolution of the sun and its effect on the earth. We discuss the changing ideas about evolution in this text.

The old idea of an astronomer as a bearded man wearing a skullcap and peering through his telescope has given way to a more factual view of an astronomer as a scientist who uses his telescope to collect light which is then analyzed by the physical apparatus needed to investigate the composition, temperature, and motions of stars and stellar systems.

The science of astronomy has changed so rapidly that it is almost impossible to keep up to date, particularly in a printed book. The authors of this text depend on current scientific journals and on personal contacts with the men who are carrying on astronomical research. It is hoped that the reader will be inspired by the discussions here and will look for and understand the numerous articles that appear nearly every day in the newspapers and more technical journals.

Alpine, California
December, 1972

C.M.H.
F.E.T.
M.M.B.

contents

an introduction to
ASTRONOMY
SECOND EDITION

1 the science of astronomy

"That's one small step for man, one giant leap for mankind!" These were the first words spoken from the surface of the moon. The speaker was Neil A. Armstrong; the time was Sunday, July 20, 1969, at 10:56:20 P.M. (Eastern Daylight Time). The words were relayed to earth for millions of persons watching the landing by television to hear.

1.1 BEGINNINGS

Man has watched the sun rise and set for thousands of years. He has seen the day fade into night and has wondered about the stars. He has watched the growth of plants and trees during spring and summer and has seen them become dormant in autumn and winter. Perhaps early man wondered if there was any relation between such changes and certain changes in the positions of celestial objects—the sun, the moon, and the stars.

It was important that early man should know how to predict the times of the seasons because the survival of his crops and herds depended on them. In other words, he needed a calendar. The calendar was therefore one of the first contributions of astronomy.

Some historians believe that the first calendar was developed in ancient Egypt. The Egyptians and other ancient people had watched the sun for hundreds of years. The Egyptians recognized that the point on their horizon where the sun rises or sets moves from day to day. In the spring the sunrise and sunset points move north along the horizon until they reach a northern limit at the time when daylight lasts longest. Then they move south until they arrive at a southern limit when the period of daylight is shortest. By counting the number

1

of sunrises or sunsets from either one of these two points until it reached that point the next time, the Egyptians found the number to be about 365. So they introduced a calendar containing 365 days.

Important observations of the sky were also made by people in other parts of the ancient world, especially in India, China, and Mesopotamia. These early observers realized that it was easier to describe the location of a particular object in the night sky if the stars were divided into recognizable groups. These apparent groupings are called *constellations* (Fig. 1-1). Ancient carvings of such constellations have been discovered in Egypt and Mesopotamia and even in Europe and Central America. They were named for mythical heroes and animals. The constellation names we use today probably originated in the Euphrates Valley and were transmitted from there to Greece, where the names were translated into Latin, in which form they appear in modern tables. (See Table 3-1 in Chapter 3.) Some of the more recently listed constellations were named for various instruments.

FIG. 1-1 A long-exposure photograph of the constellation Orion. The central figure consists of three bright stars that form the diagonal "Belt" and three fainter stars that form the vertical "Sword." The bright star Rigel is at the lower right. Other objects can be identified from the star maps. *(Photograph from the Hale Observatories)*

The stars in the constellations are called *fixed stars*. Actually they only seem to remain fixed with respect to each other because of their great distances from the earth. In reality they are all moving through space with different velocities.

The study of the constellations became particularly important when ships first sailed beyond the sight of land. This required a method of determining direction and location. Navigation thus became another important early contribution of astronomy.

Some starlike objects in the sky change their positions in relation to the fixed stars. Some of these objects are called *planets*, from the Greek word meaning "wanderers." At times other starlike objects, often with tails, appear and move among the stars. These are called *comets*, from the Greek word meaning "long hair."

The ancient astronomers listed seven planets. The sun and the moon were among them because they could be seen to change their positions in relation to the stars. The other five were Mercury, Venus, Mars, Jupiter, and Saturn. Venus is the brightest of the five as seen from the earth. These seven bodies had been so carefully studied by the ancient astronomers that it was possible to compute their positions in advance and predict with considerable accuracy where they would be found in the sky.

1.2 ASTRONOMICAL INSTRUMENTS

Pretelescope Instruments

Before the invention of the telescope, the direction of the sun was observed by watching the shadow cast by the gnomon, a vertical column of stone (see Fig. 9-1). Positions of the moon, planets, and stars were determined by sightings made with instruments like those shown in Fig. 1-2. These instruments were used by the 16th-century astronomer, Tycho Brahe (1546–1601). The lower drawing shows Tycho's sextant. This instrument consisted of a sighting arrangement by which the observer could measure the angle between the object and the horizontal plane, which is perpendicular to the plumb bob (see Fig. 1-3). This angle is now called *altitude*. The entire sextant could be turned also to determine the angle between the direction of the object and a mark on the horizon. This angle, measured from the north, is now called *azimuth*.

With the more accurate wall quadrant (upper drawing, Fig. 1-2) the altitude of an object in a fixed direction—north, for example—could be found by sighting through a hole in the wall and reading the graduated arc shown in the figure. Tycho invented a special device that made it possible to read the angle on this arc with an accuracy of one minute of arc. The wall paintings in the background show the interior of Tycho's observatory, Uraniborg, on the island of Hveen, off the coast of Denmark.

Tycho's observations with these instruments were so accurate that his meas-

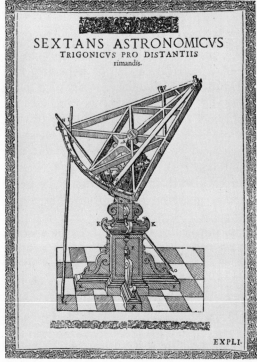

FIG. 1-2 Tycho's wall quadrant (top) and sextant (bottom) were used for accurate measures of star and planet positions before the invention of the telescope. (*New York Public Library; The Granger Collection*)

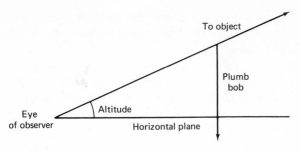

FIG. 1-3 Method of observing altitude.

ures of the positions of Mars, in particular, were used by Johannes Kepler (1571–1630), who discovered that the planets move in ellipses around the sun.[1]

Telescopes

The telescope was invented in Holland by Hans Lippershey (?–1619?) in the early part of the 17th century. Its use by Galileo Galilei (1564–1642) had a profound influence on the development of astronomy. It permitted Galileo to see objects invisible to the unaided eye (Fig. 1-4). He discovered the seas and craters

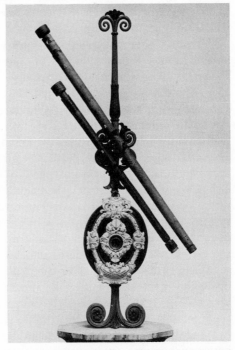

FIG. 1-4 Galileo demonstrating his first telescope in 1609 from the tower of St. Mark's Cathedral in Venice, Italy. At the right are shown two of the telescopes used by Galileo. (Bausch & Lomb; Alinari-Art Reference Bureau)

[1] The astrolabe, the armillary sphere, the quadrant, the compass, and the German torquetum were also used before the invention of the telescope.

on the moon, the spots on the sun, the moons of Jupiter, and the fact that Venus shows changes in apparent shape, as the moon does. He also confirmed the belief that the Milky Way was composed of stars.

Later, telescopes were mounted on instruments similar to those used by Tycho. This made it possible to locate planets and comets with greater precision. These observations led the mathematical astronomers to their investigations of celestial motions and to the statement of fundamental laws. The most famous of these men was Sir Isaac Newton (1642–1727).

Galileo's telescopes were made with lenses. Such lens-type instruments are called *refracting telescopes,* or *refractors.* An *objective lens* collects the light from an object and brings it to a focus. The observer looks through a second lens, called the *eyepiece,* which enlarges the image formed by the objective lens. The world's largest refractor, 40 inches in diameter, was erected at the Yerkes Observatory in Wisconsin in 1895 (see Fig. 8-6).

The *reflecting telescope,* or *reflector,* was invented by Newton. In the reflector the image is formed by a concave mirror placed at the lower end of the instrument. The image is therefore formed at the top. As in the refracting telescope, an eyepiece may be used to magnify the image. However, the eyepiece can be replaced by a photographic film, which makes a permanent record. It may also be replaced by other instruments, such as the spectrograph or the photocell, which analyze the light directly.

The first large reflecting telescopes were the 60-inch and 100-inch instruments at the Mount Wilson Observatory and the 200-inch Hale telescope on Mount Palomar, all in California. Other observatories with large reflectors include the Lick Observatory (Fig. 1-5) and the American National Observatory on Kitt Peak near Tucson, Arizona. Other large telescopes have been built by several European nations. There are also two newer large telescopes in Chile, built for and operated by United States astronomers.

Radio telescopes were developed after Karl Jansky (1905–1950) accidentally discovered radio waves from space in 1931. He had built a receiver at the Bell Telephone Laboratories on the New Jersey coast to investigate the causes of radio static interference. His receiver was placed on a circular track and could be rotated (see Fig. 1-6). To his surprise, he found that the maximum static noises were coming from a particular area in the sky that moved across the receiver at the rate of the daily motion of the stars. This area is located in the Milky Way.

Jansky's discovery led to the building of radio telescopes, which consist of dish-shaped receivers much like the mirrors of reflecting telescopes—only much larger (Fig. 1-7). Surveys of the sky at radio frequencies have been made with these telescopes. In addition to radio signals from stars and the Milky Way, the sun and Jupiter have been found to emit this type of radiation. Radio telescopes have now become very important astronomical instruments.

Two previously unknown types of celestial objects have been discovered with radio telescopes. One type is called *quasi-stellar radio sources,* popularly called *quasars.* Discovered in 1960, their nature is still under discussion. Are they

FIG. 1-5 The 120-inch reflecting telescope of the Lick Observatory, University of California, located on Mount Hamilton, near San Jose, California. (Lick Observatory photograph)

very distant objects, as some astronomers think, or are they relatively nearby? They seem to be moving away from us at velocities approaching the velocity of light. They may be the most distant objects known in the universe.

Another type of radio source, called *pulsars*, was discovered in 1967. These

FIG. 1-6   Karl Jansky with the rotating antenna he used to discover radio waves coming from space. Jansky's investigations during the 1930s into the causes of strange noise in telephone equipment resulted in the discovery of radiation noise from the center of the Milky Way and gave the world a new science—radio astronomy. *(Bell Laboratories)*

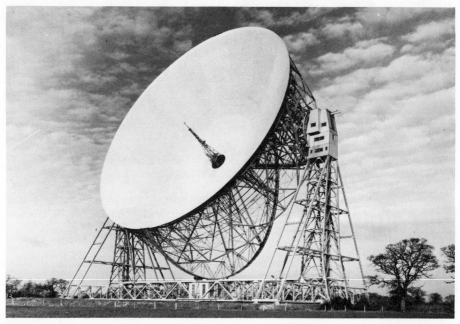

FIG. 1-7   The radio-telescope at Jodrell Bank, near Manchester, England. The dish is 250 feet in diameter. The radio receiver is at the focal point of the parabolic reflector. *(British Information Services)*

sources emit radio waves that suddenly change in intensity at regular intervals of 3.75 seconds or less. In spite of very careful search with reflecting telescopes for over a year, no visible star could be found that was a pulsar. Finally, three astronomers pinpointed one with the 36-inch telescope of the University of Arizona. It is the brighter of two stars in the Crab Nebula (Fig. 17-8). This nebula is an expanding mass of gas, probably ejected from a supernova first seen in A.D. 1054.

1.3 ASTRONOMY AND OTHER SCIENCES

Astronomers rely on other sciences for the tools and methods they need to investigate the universe. They use chemistry and physics for the study of the composition, nature, and motion of objects in space. Biology is needed as space probes gather samples of possible living material from the moon, Mars, and other bodies. Geology is very important. Obviously, a student of astronomy must plan to get a wide background in other sciences.

The original science was probably astronomy. Now research workers in other sciences look increasingly to astronomy for new ideas in their investigations. For example, chemists and physicists use astronomical information about the behavior of atoms in the unusual conditions of interplanetary space to form a better picture of the nature of matter. Many theoretical scientists use the motions of distant stars and star systems to test their theories and decide whether they are universally true.

An excellent example of this interdependence on a worldwide scale was the International Geophysical Year, or IGY, of 1957–1958, an organized program involving all branches of science to study the earth and its environment. Scientists from 66 countries cooperated in this investigation with all the facilities and ingenuity the sciences of that time could provide. Astronomy played an important role. The dates were chosen to coincide with the times when an unusual amount of sunspot activity and accompanying phenomena were predicted. Solar explosions have an important effect on the earth and its surroundings.

1.4 ASTRONOMY IN THE SPACE AGE

The impetus of the IGY and the subsequent achievements of space science have resulted in great progress in the study of the universe. The first man-made satellites orbited the earth, and they were soon followed by vehicles that went into orbit around the sun. Soon thereafter man himself piloted earth satellites.

Most earth satellites are equipped with instruments for the study of particular phenomena, such as cosmic radiation, cloud formations and storms on the earth, and interplanetary particles. The Van Allen radiation belts around the earth were discovered in 1958 by observations made from American satellites.

When objects were sent up that could orbit the earth, men realized that it might be possible to go also to the moon and beyond. Attempts were made to construct vehicles capable of escaping the gravitational pull of the earth. These vehicles are called *space probes*. The first probes did not reach the moon and went into orbit around the earth. Then on January 2, 1959, the Russians launched Luna I, which passed within about 4000 miles of the moon and went into orbit around the sun.

Finally, the U.S.S.R. succeeded in sending Luna III around the moon. It was launched on October 4, 1959, and sent back photographs of the far side. Closeup photographs of the near side were first obtained by the American probe, Ranger 7, just before it crash-landed on the lunar surface on July 31, 1964. These photographs showed the surface to be pitted with small craters, some as small as 3 feet in diameter (Figs. 1-8 and 1-9).

Soft landings on the moon by unmanned probes began two years later in 1966. The American Lunar Orbiters photographed all parts of the moon from a

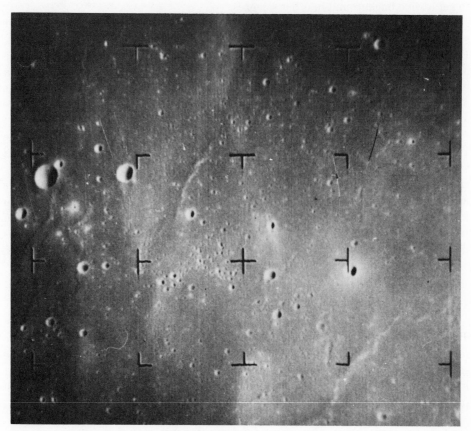

FIG. 1-8   Photograph taken by Ranger VII spacecraft prior to impact on the moon on July 31, 1964. Altitude above the moon, 85 miles. The photo covers an area 8 miles on a side; it shows craters as small as 500 feet in diameter. (NASA)

distance of a few miles to survey it for possible landing sites. At about the same time, the Surveyors landed on the surface. They found that the surface was rigid enough to support the weight of any future landing module; discovered that the dust layer was only a few inches thick; dug into the surface; and, with special equipment, analyzed the surface material. They determined that it was not very different from that of the earth's crust.

When the Surveyors showed that the surface of the moon could support landing craft, the final decision was made to go ahead with preparations to land men on the moon. Until this time prominent scientists disagreed as to whether or not the surface could support the weight. Some thought that any craft that attempted to land would sink beneath the surface.

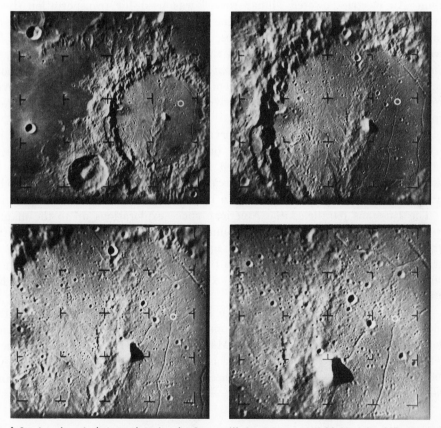

FIG. 1-9 A series of photographs taken by Ranger IX. Impact was at 6:08:20 A.M., PST, March 24, 1965. The white circle marks the point of impact in the crater Alphonsus. Top left: Altitude above lunar surface, 266 miles; $3^m\ 02^s$ before impact; area covered, 126 by 133 miles. Top right: Altitude 141 miles; $1^m\ 33^s$ before impact; area covered, 67.5 by 62 miles. Bottom left: Altitude 95.5 miles; $1^m\ 45^s$ before impact; area covered, 46 by 43 miles. Bottom right: Altitude 65.4 miles; 48.9 seconds before impact; area covered, 31.6 by 28.5 miles. The floor of Alphonsus shows intricate pattern of ridges, rills, and craters. The central peak is 3000 feet high. (NASA)

On July 20, 1969, the actual landing took place from Apollo 11. Neil Armstrong and Edwin Aldrin landed the lunar module and walked on the surface for a couple of hours. They took off again, using the powered portion of the landing module, docked with the command module, which had been piloted by Michael Collins, and were retrieved from the Pacific Ocean. After being flown to the Space Center in Houston, Texas, they were kept in quarantine for several weeks to be sure they had not brought back any unknown infectious virus, and then were released with clean bills of health.

This mission, which was so successfully completed, was the culmination of very careful planning and experiments between 1962 and 1969.

The Apollo 11 mission was followed by other successful missions. Apollo 12 landed in Oceanus Procellarum and Apollo 14 in the Fra Mauro highlands, both near the lunar equator. Apollo 15 landed in the north near the Hadley Rille at the foot of the Apennines, a high range of mountains bordering the Mare Imbrium. An electric-powered "lunar rover" was used for transportation during surface explorations. On each mission, the astronauts who landed took color photographs, collected soil and rock samples, and carried out experiments. They left several scientific instruments, some of them powered by an atomic reactor, for continuing investigations. The astronaut in the command module took photographs and performed other experiments. The electric-powered "lunar rover" first used in the Apollo 15 mission proved successful in extending the range of exploration.

Meanwhile, the orbiting solar observatory and the orbiting geophysical observatory were successfully put into orbit. Later, OAO-1, the first Orbiting Astronomical Observatory, was orbited but was rendered ineffective when the batteries failed to operate. OAO-2 was entirely successful after being launched on December 7, 1968, and according to the latest reports was still in operation.

The Russians paralleled the American space explorations up to the manned landings on the surface. They landed unmanned spacecraft, including one with a surface "dune buggy," which scooped up rock and soil samples that were returned to the spacecraft and finally to the earth. One lunar rover was remotely controlled by a crew on the earth as it explored the surface and transmitted data to the earth.

Both Russians and Americans have sent space probes to Venus and Mars. There are some differences in the announced results, which have not been completely explained. Mariner 9 reached Mars in November, 1971. The first photos were not very satisfactory because of a huge dust storm on the planet. However, the spaceship successfully photographed the two satellites of Mars, Deimos and Phobos.

In this book we can give you only an introduction to astronomy in an elementary form, hoping that you will be inspired to learn more details on your own or to go on to advanced study. Perhaps some of you will become astronauts or research astronomers at large observatories, or teachers in schools, colleges, or universities. The future of astronomy depends on such workers.

In astronomy, as in other sciences, the search for knowledge is never finished. Let us hope that it will be increasingly possible for scientists from many countries to cooperate and to coordinate their efforts as man pushes toward an ever-better understanding of the universe.

QUESTIONS AND PROBLEMS

Group A

1. Name some constellations that you knew before studying astronomy in a class.
2. (a) How can you tell which object in the sky is a planet? What planets can you identify? (b) Which are visible now? *Answers:* (a) By its motion; sometimes by its brightness. (b) (See a current sky map.)
3. How was the calendar developed by the use of astronomy?
4. What space satellites have you seen? What lunar landings have you seen on television?
5. List the books, magazine articles, and newspaper reports about space that you have read. Bring to class any articles of interest.

Group B

6. Explain how you can measure the altitude of a star.
7. How does the telescope help in observing the moon and the planets? How could you look at the sun?
8. Why do you think mathematics, physics, and chemistry are important in the study of astronomy?
9. If there is a radio telescope in your vicinity, plan to visit the observatory where it is located.

2 celestial coordinates

Many people think of an astronomer as an old man in a skullcap who sits and looks at the stars through a telescope. They probably get this idea from pictures of Galileo or Copernicus studying the stars. But modern astronomers do not work that way. They have sophisticated instruments attached to their telescopes and rarely look through them at the objects they are studying. A visit to an observatory would more likely show the astronomer guiding on a star with an auxiliary telescope or letting an electric eye do his guiding for him. He might be taking a photograph of a narrow band of light (a spectrum) or taking measures of the changing brightness of a star with a photoelectric cell. In the latter case, an automatic device would be tracing a line on a moving chart while the astronomer kept watch to see that everything was working properly.

If a visitor to an observatory tried to find out what the telescope was pointed at by sighting along its edge, he might not see any stars at all in that area. In this case, the telescope would probably be collecting and measuring light from an object not visible to the naked eye. To find such an object, the astronomer must be able to pinpoint it by knowing its exact location and finding it by means of scales on the telescope.

One hundred years ago astronomers spent most of their observing time looking through their telescopes. They counted the stars and looked for new faint stars and other objects not visible to the naked eye and located them accurately. Even before 1900 the astronomer was not merely looking at the stars; he was trying to find out how far away they are, how fast they are moving, or what chemical elements they are made of. Today, he is more interested in determining the nature of all kinds of celestial objects, their temperatures, and the conditions of the atoms and molecules that compose them. He is trying to formulate an acceptable theory of the evolution of stars—what has happened to them in the past and what they will be like billions of years in the future. To do this, he must have a good background in mathematics, chemistry, and physics. Some astronomers have their PhD's in physics.

The story is told of an internationally known European theoretical astronomer who was aboard a ship crossing the Atlantic Ocean. One night some passengers asked him to point out a few constellations. He had to confess that he did not know any of them; he had done all his theoretical research without observing the stars himself.

Astronomers are not the only scientists who sometimes limit their interests to a very narrow field. The well-known writer and conservationist Joseph Wood Krutch deplored the fact that some biologists in pursuing their specialty lose sight of the natural beauty around them. He cites the famous biologist who complacently admitted that he recognized only a dozen plants by sight, and the specialist in water beetles who dozed throughout a trek into the wilderness of the Baja California peninsula. This young man ignored the magnificent scenery and sprang to life only when the trail crossed a puddle, a rare occurrence in that desert country.

Unlike the narrow specialists, most students who take a course in astronomy want to enjoy the beauty of the night sky and see the objects they are studying in class. If they travel to wide-open spaces where the sky is clear and where there are no bright lights or city smog to interfere, they like to be able to identify the constellations, name the bright stars, and find out where to locate the planets.

In order to identify objects in the sky, we need a system like that on a terrestrial globe or map; we need a star map or a reference table as the student learns the fundamental systems for locating objects in the sky. The earlier in the course he learns these systems, the more pleasure he will have later on, especially when he can trace the motions of the moon and the planets on his star maps.

2.1 COORDINATE SYSTEMS

As a convenience for locating objects in the sky and especially for transmitting information about their positions to other people, it is desirable to set up one or more systems of coordinates. A *coordinate system* is a set of numbers that may be used to locate an object. It must have initial reference points from which measurements are made. For example, a straight line has only one dimension: length. If it has an initial point, such as the end of a ruler, only one coordinate is needed to describe the location of any point on the line. A straight line is said to be one-dimensional.

A flat surface is two-dimensional, having length and width. Thus two coordinates are required to locate a point on a surface. An address in a city requires a coordinate system of two dimensions: the street and the number of a house on the street.

A solid, such as a cube, has three dimensions: length, width, and height.

Points on the surface of a sphere can be located by the use of three coordinates, but it is easier to use a two-coordinate system employing circles and points. Such a system is called a *spherical coordinate system*. Several such coordinate systems have been established for the location of celestial objects on the apparent sphere of the sky.

2.2 ANGULAR MEASURE ON A SPHERE

The *sphere* is a surface every point of which is equally distant from a point inside, called the *center*. The *radius* of the sphere is any straight line from the center to the surface or its length; the *diameter* is a straight line from one part of the surface through the center to the surface on the opposite side. Therefore, the diameter is twice the radius in length; or $d = 2r$.

If a plane is passed through the center of a sphere, it will cut a circle on the surface of the sphere where the plane and the surface intersect. This, the largest circle that can be drawn on the sphere, is called a *great circle*. Since any number of planes can pass through the center of a sphere, there are any number of great circles on a sphere. A plane that does not pass through the center of the sphere intersects the surface in a circle smaller than a great circle, and is called a *small circle* (see Fig. 2-1).

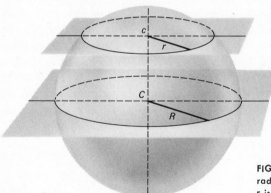

FIG. 2-1 Construction of great circle with radius *R* and small circle with radius *r*, where *r* is less than *R*.

For example, the equator of the earth is a great circle. Parallels of latitude are small circles parallel to the equator. The equator has two poles: the north pole and the south pole. Each is 90 degrees from all points on the equator; that is, the poles are equidistant from the equator.

A great circle has the following properties:

1. Its center is located at the center of the sphere.
2. It divides the sphere into two equal parts, called hemispheres.
3. Each point on the great circle has a corresponding point on the opposite side of the sphere, which also lies on the great circle.

4. It has two poles that are 90° (equidistant) from all parts of the circle.
5. All great circles drawn through the poles of a great circle are perpendicular to it.
6. Only one great circle can be drawn through any two points on a sphere, unless they are 180° apart.
7. An infinite number of great circles can be drawn on a sphere.
8. The shortest distance between two points on the surface of a sphere is measured along the arc of a great circle.

Angles on a sphere can be measured in two ways. For example, given two points A and B on the surface of a sphere (Fig. 2-2). Connect each point to the center, C, by a radius. The angle between A and B is the angle formed at the center by the two radii. Or, if a great circle is drawn through A and B (property 6), the angle between A and B is equal to the length of the arc of the great circle measured in degrees.

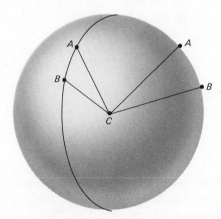

FIG. 2-2　Measurement of angles on a sphere.

2.3 TERRESTRIAL LATITUDE AND LONGITUDE

A familiar set of spherical coordinates is that used to describe positions of points on the earth. The fundamental great circle is the earth's equator with two poles 90° away. Secondary great circles perpendicular to the equator and passing through the poles are called *meridians*. The angular distance north or south from the equator to any point, measured along a meridian, is called *latitude*.

The second coordinate is *longitude*. To obtain this value it is necessary to choose a point on the equator as the point of reference. The meridian chosen to be the *prime meridian* passes through the observatory in Greenwich, England, set up to aid British navigators at a time when England was the world's principal maritime nation. The intersection of this meridian with the equator is the initial point from which longitude is measured. This system is shown in Fig. 2-3.

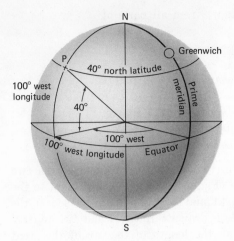

FIG. 2-3   The system of terrestrial latitude and longitude.

Longitude is therefore defined as the angular distance between the meridian of Greenwich and any other meridian measured east or west to 180°. The continental United States lies in the western hemisphere from about 65°W to 125°W longitude and from 25°N to 50°N latitude.

THE HORIZON SYSTEM

2.4 THE CELESTIAL SPHERE

The ancient astronomers thought of the stars as set in a crystal sphere. Today the stars are not thought of as fixed on a sphere. Instead a hypothetical surface, the *celestial sphere,* is set up onto which all the stars can be projected.

Since it is now known that the stars are at various distances from the earth, and since the galaxies are still more distant, the celestial sphere must be large enough to include all the bodies in the universe. Its radius is therefore considered to be infinite in length. The center may be anywhere we wish, but since the earth is so small compared with the size of an infinite sphere it may just as well be the point where the observer is located. In other words, each person has his own celestial sphere, but it is essentially the same as that of a person nearby. The celestial sphere is, therefore, a sphere of infinite size with its center at the eye of the observer and on which all objects in space can be projected.

In a simple system of celestial coordinates, only direction is considered. Distance is not included, since it is not needed at first. Thus the stars are located only by their directions with respect to each other. We may say that two stars are so many degrees apart and talk about their angular distances from each other or from some initial point or points—a bright star, for example.

Two initial points may be determined by the direction of gravity. If a heavy weight is hung on a nearly weightless string, called a plumb line, and hypothetically extended to infinity in both directions, it will intersect the celestial sphere in two points. The point overhead is called the *zenith* and the point underneath is the *nadir*.

In Fig. 2-4, S_1 and S_2 are two stars seen by an observer at the center, O. The stars appear to be projected onto the celestial sphere at S'_1 and S'_2. The angular distance between S_1 and S_2 is therefore the angle S_1OS_2, which is equal to $S'_1OS'_2$. Similarly, if G_1 is a galaxy, it will appear to be projected onto the celestial sphere at G'_1; and the angle between S_2 and G_1 is S_2OG_1 or S'_2OG_1. It is obvious that the distances to the stars or to the galaxy are not considered.

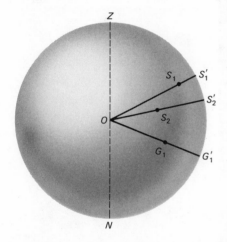

FIG. 2-4 The celestial sphere, showing angular measurements, S_1 and S_2 are stars; G_1 is a galaxy.

The zenith Z, the point at the top of the sphere, is located as the intersection of a plumb line and the celestial sphere in the upward direction. Similarly, the nadir N is located by the downward direction of the plumb line.

2.5 CELESTIAL REFERENCE CIRCLES

The zenith and nadir are poles of a great circle, the *celestial horizon*. If great circles are drawn through the zenith and nadir, they will therefore be perpendicular to the celestial horizon; they are called *vertical* (perpendicular) *circles*. The zenith and nadir are 90° from all points of the celestial horizon.

Two starting points on the celestial sphere have now been located by the direction of a plumb line. Thus each point on the earth has its own zenith and nadir, and therefore its own horizon. This is because the direction to the center of the earth is different for all points on the earth's surface. Because of the irregularities of the landscape, the terrestrial horizon does not always coincide with the celestial horizon, which is a perfect circle. The sea horizon is more than

90° from the zenith because of the "dip of the horizon" due to the curvature of the earth and the height of the observer above the water.

If a camera is loaded with high-speed film, pointed toward the North Star, and the shutter is left open for an hour or so (not so long that the film will be fogged), it will be found that the stars trace arcs of circles on the film, as shown in Fig. 2-5. These arcs are parts of small circles, which increase in size away from Polaris, the North Star. They are all centered at the *north celestial pole*.

FIG. 2-5 Star trails around the north celestial pole. The bright overexposed object near the center is Polaris. *(Yerkes Observatory photograph)*

If a great circle is drawn from the zenith through the north celestial pole and continued completely around the sphere, it will intersect the horizon in two points, the north point and the south point. This great circle, called the *celestial meridian*, is one of the fundamental circles of astronomy and navigation. It is evident that each point on the earth has its own celestial meridian concentric with the terrestrial meridian that passes through the terrestrial poles and the place of observation.

2.6 ALTITUDE AND AZIMUTH

Any celestial body can be located by means of two coordinates, based on the horizon and meridian as references. One of these coordinates, called *altitude*, is the arc of a vertical circle from the celestial horizon to the object, as shown in Fig. 2-6. It is also the smallest angle at the eye of the observer between the

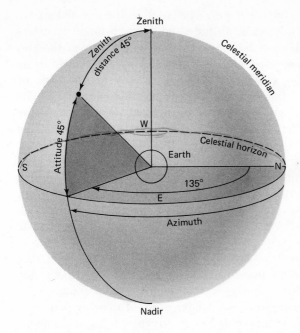

FIG. 2-6 Diagram showing altitude 45° and azimuth 135° of a star.

celestial horizon and the object. It can be measured by the surveyor's transit or the marine sextant.

Altitude is marked "plus" if the object is above the horizon and "minus" if it is below the horizon. Usually negative altitudes mean that the body is not visible, but the air bends the light from an object just below the horizon in such a way that it is visible even though its true altitude is negative by a small angle. The rising and setting sun and moon appear to be on the horizon, but are actually a little below it. Thus, it is possible to compute the altitude of a celestial body to see if it is below the horizon and therefore invisible at a time when an observation is planned.

It is sometimes more convenient to use the angular distance of a star from the zenith instead of its altitude. This is called the *zenith distance* and each is the complement of the other:

$$z = 90° - h \qquad \text{and} \qquad h = 90° - z \qquad (2\text{-}1)$$

where z, the zenith distance, and h, the altitude, are the standard symbols.

Altitude is only one coordinate. There are an infinite number of points on the celestial sphere with the same altitude, all lying on a small circle parallel to the horizon. To locate a particular object a second coordinate is necessary. The angle between the north point and the intersection of the vertical circle through the body with the horizon is called its *azimuth*. Azimuth is measured as an arc of the celestial horizon from the north point toward the east to the intersection of the vertical circle with the horizon. This angle may have any value from 0° to 360°. For example, north is 0°, east 90°, south 180°, and west 270°. These are the four cardinal points. Southeast is 135°, northwest 315°, and so on.

As seen from a given place on the earth, only one point on the celestial sphere has a given altitude and azimuth. That is, altitude and azimuth uniquely determine a point. If the north point can be located as described above, the two coordinates of a star can be measured easily with a surveyor's transit. It is equipped with a horizontal plate, which can be set accurately in the plane of the horizon. This plate serves as the horizon, the altitude reference plane. The zero of the horizontal plate is aligned with the north point, which locates the celestial meridian (hereafter referred to as the meridian) and serves as the azimuth reference point.

However, the horizon system has certain disadvantages. At a given instant both coordinates are different from those for the same object at other observing stations. For example, the sun may be overhead (altitude 90°) in the tropics, but at the same instant it will be several degrees south of the zenith at a station farther north, such as altitude 80°, azimuth 170°. Also, the azimuth and altitude of an object change continuously.

But the advantages of this system outweigh its defects. The very fact that altitudes and azimuths change continuously makes it possible to determine the terrestrial latitude and longitude of an observer, provided certain fundamental data are available. This is a basic problem of navigation. Tables of planetary positions for every day of the year are published in advance, using the equator system to be described in the next section. The problems of navigation are not within the scope of this text, but there are many books on navigation available to the interested student.

THE EQUATOR SYSTEM

2.7 EQUATOR AND POLES

There is another system, the *equator system*, which is used on star charts and for the setting of telescopes and in which the coordinates of a celestial object are the same for an observer anywhere in the world. Hence it is important that the student of astronomy become familiar with it.

The fundamental points in the equator system are the two celestial poles. They are directly above the north and south terrestrial poles, and are therefore in line with the earth's axis of rotation (Fig. 2-5). For an observer at the terrestrial pole, the celestial pole is at his zenith, and the Pole Star is overhead. If a camera is pointed toward the zenith and the film exposed for 24 hours during the long polar night, each star describes a small circle around the celestial pole. The great circle 90° from the pole, called the *celestial equator,* is the fundamental circle of the equator system and, in this case, coincides with the horizon. The celestial equator is the projection of the terrestrial equator onto the celestial sphere. The apparent motion of the stars, which is due to the rotation of the earth, is called *diurnal* (or daily) motion.

For an observer on the terrestrial equator, the celestial poles are on the horizon and the celestial equator is an east-west vertical circle. All the star paths are perpendicular to the horizon and are bisected by it. Therefore the sun, moon, and all the stars rise and set at right angles to the horizon. Every star is above the horizon for 12 hours out of each 24-hour day and can be seen from the terrestrial equator at some time during the year.

From the north or south pole only half of the stars can be seen, the other half being permanently below the horizon. At intermediate latitudes, the diurnal circles all make angles with the horizon depending on the latitude of the place of observation. The angles decrease from 90° for an observer at the equator to 0° at the poles. They are equal to 90° minus the latitude. Figure 2-7 shows the diurnal circles as seen from the north pole (parallel sphere), the equator (right sphere), and an intermediate latitude (oblique sphere).

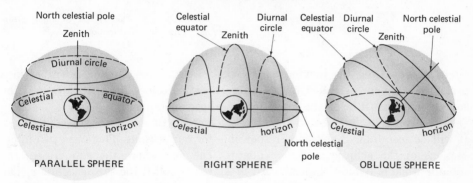

FIG. 2-7 Position of diurnal circles as seen from the terrestrial pole (parallel sphere), equator (right sphere), and intermediate latitude (oblique sphere).

2.8 RIGHT ASCENSION AND DECLINATION

An infinite number of great circles can be drawn through the poles. From property 5 (Section 2.2), they will all be perpendicular to the celestial equator. These great circles are called *hour circles*, because they are related to time (see Fig. 2-8).

Only one hour circle can be drawn through a given point, such as the sun, the moon, a planet, or a star. That is, only one hour circle can be drawn through a given celestial body. The angular distance of a body from the celestial equator, measured along its hour circle, is called *declination*. Declination is defined with respect to the equator and poles as altitude is defined with respect to the horizon and the zenith. Similarly, there are an infinite number of points with the same declination. Therefore it is necessary to select a reference point on the celestial equator in order to define another coordinate, as was done with azimuth in the horizon system. The point selected is based on the motion of the sun.

Centuries of observation have shown that the sun seems to describe a great

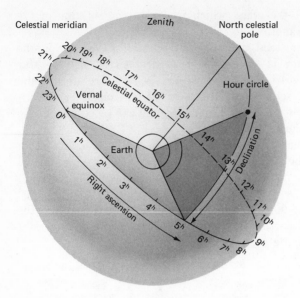

FIG. 2-8 The system of right ascension and declination.

circle as seen from the moving earth. This circle, which the sun describes in a period of one year, is called the *ecliptic*. The word comes from the fact that the sun and moon must be on this circle for an eclipse to occur.

The ecliptic is inclined to the celestial equator at an angle of about 23½°. It will be seen later that this is responsible for our seasons. There are two intersections of the ecliptic and the equator. The one where the sun crosses the equator when it is moving northward in March is called the *vernal equinox*. The other, where the sun crosses the equator moving southward in September, is the *autumnal equinox*. Vernal means spring, and autumnal means fall. Equinox means that the day and night are of equal length. Figure 2-9 shows the ecliptic and its two intersections with the celestial equator.

The vernal equinox is the starting point for the coordinate called *right ascension*. Its right ascension is 0° and, since it is on the equator, its declination is also 0°. Declination is measured in degrees from the equator, plus to the north and minus to the south up to 90° at the celestial poles. Right ascension is measured eastward, beginning at the vernal equinox and up to 360°. But since right ascension is related to time, it is customary to express it in hours, up to 24 hours (written 24^h).

The point of highest declination on the ecliptic is +23½° and the lowest is −23½°. The former point is called the *summer solstice* and the latter is the *winter solstice*. Solstice means "sun stands still." Summer begins when the sun reaches the summer solstice (about June 21), and winter begins when the sun reaches the winter solstice (about December 21). Spring begins when the sun is at the vernal equinox (around March 21), and autumn begins (about September 23) when the autumnal equinox is reached.

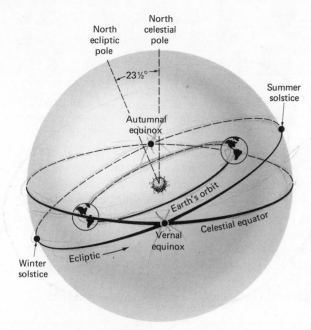

FIG. 2-9 The ecliptic and the celestial equator, showing their intersections at the two equinoxes. The two circles are inclined to each other by an angle of 23½°.

The right ascension and declination of the four principal points of the ecliptic are given in Table 2-1 (see also Fig. 2-10).

TABLE 2-1

name	right ascension	declination
Vernal equinox	$0^h = 0°$	$0°$
Summer solstice	$6^h = 90°$	$+23\frac{1}{2}°$
Autumnal equinox	$12^h = 180°$	$0°$
Winter solstice	$18^h = 270°$	$-23\frac{1}{2}°$

Since $24^h = 360°$, it is easily seen that:

$$1 \text{ hr of time} = 15° \text{ of arc}$$
$$1 \text{ min} = 15'$$
$$1 \text{ sec} = 15''$$

The statement that the sun rises in the east and sets in the west is not strictly true. At all latitudes the celestial equator cuts the celestial horizon at exactly the east and west points. Therefore, when the sun is on the equator at the beginning of spring and autumn, when its declination is 0°, it rises at the east point and sets at the west point. When the sun is north of the celestial equator, as it is from the beginning of spring to the beginning of autumn, it rises to the north of the east point, describes its diurnal circle, a small circle parallel to the equator but

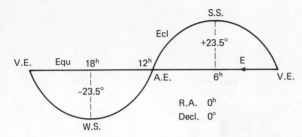

FIG. 2-10 The equator drawn as a straight line; the ecliptic drawn as a curve. Equinoxes and solstices shown with their right ascensions and declinations. East is to the left.

north of it, and sets to the north of west. From autumn to spring, the sun is south of the equator, rises to the south of east and sets to the south of west. The exact directions of sunrise and sunset depend on both the sun's declination and the latitude of the observing station. Their calculation is a problem in spherical trigonometry.

The reason the days are longer in summer than in winter is that more than half of the sun's diurnal circle is above the horizon from March to September. During the other half of the year the larger part of the diurnal circle is below the horizon.

In the next chapter, use is made of these coordinates in discussing star maps.

2.9 SIDEREAL TIME

As the earth rotates from west to east, the celestial sphere appears to rotate from east to west and successively numbered hour circles come into coincidence with the celestial meridian. Thus time and right ascension are related.

If a clock is set to read 0^h when the hour circle through the vernal equinox, the zero hour circle, coincides with the meridian, and if it is regulated to read 0^h or 24^h whenever the vernal equinox is on the meridian, the clock will show the hours successively as the corresponding hour circles cross the meridian. Such a system of timekeeping, called *sidereal time*, can be defined as the right ascension of the meridian.

The angle between the meridian and an hour circle, measured toward the west, is called the *local hour angle* and is equal to the time that has elapsed since any point on the hour circle was on the meridian. The hour angle of the meridian is of course always zero. Hour angles measured toward the east are negative.

Sidereal time may also be defined as the angle between the meridian and the hour circle through the vernal equinox; that is, sidereal time is the hour angle of the vernal equinox. It is also the time since the vernal equinox crossed the meridian. All three of these definitions of sidereal time are identical in meaning.

The hour angle of a star is equal to the difference between the sidereal time at any instant and the right ascension of the star; that is,

$$\text{H.A.} = t = \text{S.T.} - \text{R.A.} \tag{2-2}$$

The hour angle of a star can then be computed, if its right ascension and the sidereal time are known. Figure 2-11 shows the relation between the right ascension of a star and its hour angle.

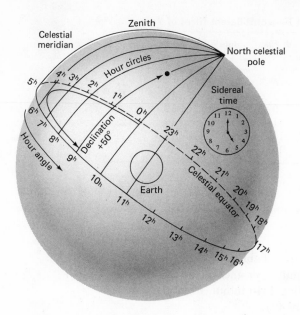

FIG. 2-11 The position of a star whose right ascension is 9^h and declination is $+50°$, for sidereal time 5^h.

EXAMPLE: Compute the hour angle of Vega at sidereal time $15^h 27^m$. The right ascension of Vega is $18^h 35^m$.

SOLUTION: Substituting in equation (2-2),

$$t = 15^h 27^m - 18^h 35^m = -3^h 08^m$$

Vega is therefore east of the meridian. The hour angle may be converted to degrees by the relations given on page 25. $3^h = 45°$; $08^m = 120' = 2°$. Adding, $-3^h 08^m = -47°$.

The sun always moves eastward along the ecliptic and its right ascension, which continually increases, is known for any given time. Sidereal time therefore changes with respect to *solar time,* which is defined as the hour angle of the sun. The two kinds of time do not coincide except on March 21, when the sun's right ascension is 0^h.

Apparent noon is defined as the time when the sun is on the local meridian. The sidereal time at apparent noon is equal to the right ascension of the sun, as can be seen by substituting in equation (2-2). Because the length of the year is not exactly 365 days, the sun is not at the vernal equinox at the same time by our clocks every year. Table 2-2 gives the approximate sidereal time at noon and 9 P.M. for different times of year.

2.10 COMPUTING SIDEREAL TIME

The computation of sidereal time is important for the proper use of star maps. By means of the following procedure, an accuracy of 5 minutes or less can be obtained. This is sufficient for most purposes.

TABLE 2-2   Sidereal Time at Different Times of Year

date	noon	9 P.M.	date	noon	9 P.M.
	(standard meridian)			(standard meridian)	
Jan. 21	20^h	5^h	July 21	8^h	17^h
Feb. 21	22	7	Aug. 21	10	19
Mar. 21	0	9	Sep. 21	12	21
Apr. 21	2	11	Oct. 21	14	23
May 21	4	13	Nov. 21	16	1
June 21	6	15	Dec. 21	18	3

EXAMPLE: Suppose it is required to compute the sidereal time for San Diego at 10:30 P.M. Pacific Standard Time on February 1.

SOLUTION: It can be seen from Table 2-2 that the sidereal time increases at a rate of 2 hours per month, or 24 hours = 1440 minutes in one year. Dividing 1440 minutes by 365¼ days, the gain is 3^m 56^s per day (roughly 4^m).

In the example, February 1 is 11 days after January 21. At the rate of 4^m per day, in 11 days the sidereal time gains 11 × 4^m, or 44^m. Adding 44^m to the tabular value for January 21, the sidereal time at noon on February 1 is 20^h 44^m. At 9 P.M., it is 5^h 44^m. At 10:30, 1^h 30^m later, the sidereal time is 5^h 44^m + 1^h 30^m = 7^h 14^m. This would be accurate to within a few minutes if San Diego were on a standard meridian (see Section 6.7). For Pacific Standard Time, the standard meridian is the 120th, 8^h behind the meridian of Greenwich, or 3^h behind the 75th meridian of Eastern Standard Time.

San Diego is 3° or 12^m of time east of the 120th meridian and the sun and the vernal equinox cross its meridian (the 117th) 12^m before they do the 120th. Hence 12 minutes must be added to 7^h 14^m, and the sidereal time at 10:30 P.M. PST on February 1 for San Diego is 7^h 26^m.

Summary for procedure in computing sidereal time:

1. Look up the tabular value for the date just before the required date. Use either noon or 9 P.M. as desired.
2. Add 4 minutes per day for the number of days following the tabular date.
3. Add the interval of time that has elapsed between noon or 9 P.M. and the time of observation.
4. Correct for the number of minutes from the nearest standard meridian, plus if east, minus if west. For daylight saving time, subtract one additional hour.

If a more accurate computation is required, for example to the nearest second, a standard almanac, such as the *American Ephemeris and Nautical Almanac*, published by the U.S. Nautical Almanac Office, should be consulted. The almanac outlines the necessary procedure.

2.11 MERIDIAN ALTITUDES

It may be desirable to compute the altitude of the sun or other body when it is on the meridian. This requires that the declination of the object be known; also the latitude of the place where the observation is to be made. It is easy to show that the latitude of the place of observation is equal to the altitude of the celestial pole. This is done as follows:

It is necessary first to define the required terms. In Fig. 2-12, the latitude, L = angle ECO or arc EO is the angle between the terrestrial equator and O, the place of observation. It is determined from a map or a survey. Neglecting the fact that the earth is not exactly a sphere, the zenith is on a line from the center C through O projected to the celestial sphere. Even considering the true shape of the earth, this statement is correct if C is on the plane of the terrestrial equator and the astronomical latitude (to be defined later) is used.

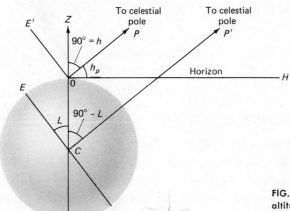

FIG. 2-12 Illustration proving that the altitude of the celestial pole, h, is equal to the latitude of the observer, L.

The horizon is perpendicular to the direction of the zenith as defined by the plumb line. Hence angle $ZOH = 90°$. Also, the polar axis is perpendicular to the equator and angle $ECP' = 90°$. The altitude of the pole, h, is the angle between the horizon and the direction to the pole = angle HOP. Since the celestial sphere is infinitely far away, the lines through C and O, which are parallel, appear to meet on the celestial sphere at the pole.

We must now prove that $h = L$. This is a simple problem in plane geometry.

1. The angles ZOP and ZCP' are equal, because OC is a transversal (straight line) cutting two parallel lines. Therefore, corresponding angles, ZOP and ZCP', are equal.

2. $h = 90° - ZOP$, since $ZOP + HOP = ZOH = 90°$.
3. $L = 90° - ZCP'$, since $ECZ + ZCP' = ECP' = 90°$.
4. $ZOP = ZCP'$, since they are corresponding angles.
5. $L = h$, since they are complements of equal angles.

This relation, the altitude of the pole equals the latitude, $h = L$, is very convenient and is used in many different problems of astronomy, navigation, engineering, and so on. It is used as the basis of a formula to calculate the noon altitude of the sun or the meridian altitude of any other astronomical body.

In Fig. 2-13, let SZN be the meridian from the south point S to the north point N. It passes through the zenith Z and the pole P. The altitude of the pole, $NOP = h$; the latitude, $ECO = L$. It has just been proved that $L = h$.

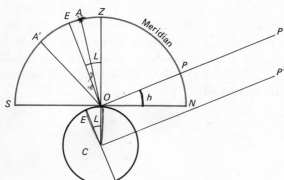

FIG. 2-13   Meridian altitudes of stars A and A'.

Let the center of the sun or any other object be at A. From Fig. 2-13, its declination $d =$ angle EOA. Hence, its altitude h above the southern horizon is

$$h = \text{angle } SOA = SOE + EOA \tag{2-3}$$

and

$$h = 90° - L + d \tag{2-4}$$

If the object is above the celestial equator, its declination is plus; if below, d is minus. The equation (2-4) can be used to find the altitude of any object above the southern horizon when it is on the meridian south of the zenith. In this derivation we have assumed the observer at O is in the northern hemisphere; therefore the pole is above the northern horizon.

EXAMPLE: Compute the meridian altitude of the bright star Sirius as seen from San Diego.
SOLUTION: From a table of bright stars (Chapter 16), the declination of Sirius is $-16° 39'$. The latitude of San Diego is $32°45'$N. Substituting in equation (2-4), the meridian altitude of Sirius is:

$$h = 90° - 32° 45' - 16° 39' = 40° 36'$$

QUESTIONS AND PROBLEMS

Group A

1. Observe a group of stars during a 4-hour period and determine (a) in which direction they move and (b) approximately what fraction of the sky they travel across during this time.

2. Name reference points and coordinates that would locate points (a) on a football field, (b) on a checkerboard, (c) on a road between two cities, (d) in a room. *Answer:* (a) References: goal lines, sidelines, and hashmarks. Coordinates: distances measured from a goal line, a sideline, or a hashmark.

3. Must any great circle that passes through the earth's north pole pass also through the south pole? Justify your answer by listing the appropriate property of a great circle.

4. On the earth's surface: (a) If the circumference is 25,000 miles, what is the distance between latitudes 30°N and 60°N? (b) Is the distance between two points at latitude 30°N the same as that between two points at 60°N? Explain. *Answer:* (a) 25,000 miles divided by 12 = 2100 miles, approximately.

5. Describe in your own words, as you would to a neighbor, where your celestial meridian is.

Group B

6. Compute the sidereal time at (a) 10:30 A.M. on May 30 and (b) 11:30 P.M. on December 10. Assume standard meridian for each and daylight time for (a), standard time for (b). *Answer:* (a) $2^h 06^m$.

7. Does each point on the earth's surface have a unique celestial horizon? Explain.

8. What is the azimuth of your instructor in this course as measured from your usual set in the classroom? Sketch his position, your position, and other relevant information.

9. The celestial equator system of coordinates is similar to the terrestrial system. What celestial terms correspond to the following terrestrial terms: (a) equator; (b) north pole; (c) latitude; (d) longitude; (e) prime meridian?

10. The right ascension of Sirius is $6^h 43^m$. At sidereal time $5^h 13^m$, what is its hour angle expressed in time and in degrees? Is Sirius east or west of the meridian? *Answer:* $-1^h 30^m = 22.5°$ east.

11. Assuming the sky to be dark and stars visible between 6 P.M. and 6 A.M. standard time, what would be the minimum and maximum right ascensions of stars at the observer's zenith between those hours on (a) March 21, (b) Sept. 21, and (c) Nov. 15? *Answer:* (a) 6^h and 18^m.

Group C

12. Draw a circle representing the earth and a concentric larger circle representing the celestial sphere. Show that no stars within 45° of the south celestial pole would be visible from latitude 45°N.

13. Compute the altitudes of Sirius and Vega (a) when on the meridian of an observer at 45°N latitude, and (b) for your latitude. Their declinations are −16° 39′ and +38° 44′, respectively.

14. What are the altitude, azimuth, declination, and hour angle of the sun on the following dates and at local times for an observer at 45° latitude: (a) noon, June 21; (b) 6 P.M., March 21; (c) noon, Dec. 21; (d) same times at your latitude? *Answer:* (a) h = 68½°; Az = 180°; decl. = 23½°; t = 0ʰ = 0°.

15. (a) How much of the meridian is above the horizon? (b) Is this answer the same for all great circles on the celestial sphere? (c) How much of the meridian is between the pole and the zenith at your latitude?

16. Compute the sidereal times for the following: (a) 8 P.M. PDT on July 5 in Portland, Ore., longitude 123°W; (b) 7:30 A.M. EST on Dec. 31 in Pittsburgh, Pa., longitude 80°W. *Answer:* (a) 13ʰ 44ᵐ.

*17. What fraction of the celestial sphere can be observed during the course of one year (a) latitude 60° and (b) at your latitude: *Answer:* (a) Set up a celestial sphere for latitude 60° and estimate the fraction below the horizon.

*18. Equation (2-4) reads: h = 90° − L + d for computing the meridian altitude of a star. (a) Show that this equation does not hold for stars north of the zenith. (b) Derive the correct formula for stars between the zenith and the pole. (c) Derive the corresponding formula for a star below the pole. Use a diagram.

19. Show that the sides of angles *HOP* and *OCE* in Fig. 2-12 are mutually perpendicular.

20. Do you think celestial coordinates are important in astronomy? Give reasons for your answer.

3 constellations and star maps

3.1 THE CONSTELLATIONS

The first astronomers carved in stone the forms of several well-known and easily identified configurations, among them Ursa Major (Big Dipper), Ursa Minor (Little Dipper), and the Pleiades (Seven Sisters). The stars are represented by small depressions in the stone. Several dozen of these rock carvings have been found in France. They are prehistoric, dating from near the end of the Stone Age.

It is not definitely known how the constellations were named. Some of their shapes resemble animals, such as a serpent or a scorpion; but others do not bear any resemblance to their names, such as the constellation Aquarius, shown in Fig. 3-1.

It is certain that the first constellations were named by people living north of the equator, since there are two bears, a lion, several dogs, some sea creatures, and others; but there is no elephant or hippopotamus. There is, however, a centaur. The giraffe was added later. There is also a gap around the south celestial pole not visible in the Mediterranean regions, where there were no named constellations in the list published by Claudius Ptolemy (90?–168?) in A.D. 150. Southern constellations were named after the beginning of the 17th century. We now recognize 88 constellations in all. They are listed in Table 3-1.

The old star maps, published in the 1800s, had drawings of the supposed creatures and mythological figures in the neighborhood of the constellations that carried their names. Some modern maps have lines drawn between certain stars in a group to help outline the figure it is supposed to represent, as in Fig. 3-2.

A chart that shows the positions and brightness of the stars is called a *star map*. When locating stars by means of these maps, it is helpful to think of the

33

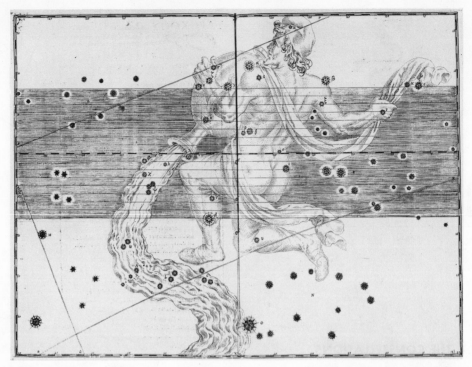

FIG. 3-1 The constellation Aquarius, the "Water Bearer." *(Courtesy of the American Museum of Natural History)*

stars as being mounted on the inside of an imaginary, transparent shell surrounding the earth. This is the celestial sphere.

The north celestial pole is marked approximately by the Pole Star, Polaris (see Fig. 3-3), which is within one degree of the true pole and is therefore almost

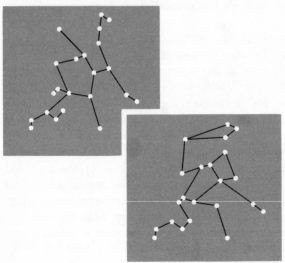

FIG. 3-2 The customary way of representing the constellation Hercules (upper) and a modern rearrangement of lines to help recognize the figure (lower).

directly above the north terrestrial pole. The north and south celestial poles are in line with the earth's axis of rotation. The celestial equator, midway between the two poles, is a projection of the earth's equator onto the celestial sphere. It is drawn as a straight line on Maps 1 to 4. (See front and back covers.)

The sun's apparent path among the stars, the ecliptic, is also drawn on Maps 1 to 4. It crosses the celestial equator at the vernal and autumnal equinoxes. The summer and winter solstices are marked on the maps by arrows. The right ascensions and declinations of these four points are listed in Table 2-1.

Ptolemy published one of the first star maps. He used a system of coordinates based on the ecliptic, not on the equator as is customary today. He also noted in his catalog the star positions in the constellations, such as the shoulder of the hunter or the knee of the giant. Ptolemy and Hipparchus (160?–125? B.C.) also used numbers to indicate the brightness of the stars. The brightest stars were

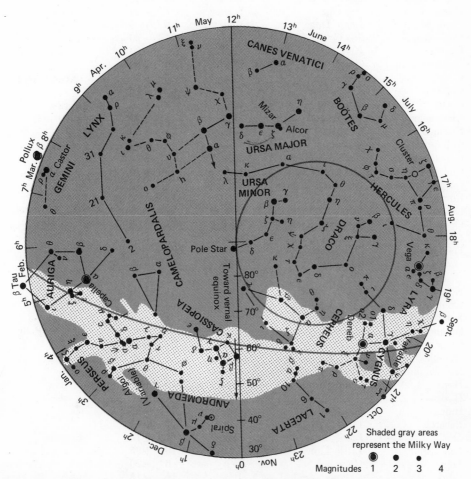

FIG. 3-3 North circumpolar stars with precessional circle of north celestial pole and curve showing the galactic equator. Time is 8 P.M. standard time or 9 P.M. daylight time in months shown.

TABLE 3-1   A List of Constellations

name in Latin[a]	name in English	abbr.	approx. position	bright star(s) (if named)
Andromeda	Princess	And	1^h +40°	(Galaxy)
Antlia	Air Pump	Ant	10 −35	
Apus	Bird of Paradise	Aps	16 −75	
Aquarius	Water Bearer	Aqr	23 −15	
Aquila	Eagle	Aql	20 + 5	Altair
Ara	Altar	Ara	17 −55	
Aries	Ram	Ari	3 +20	
Auriga	Charioteer	Aur	6 +40	Capella
Bootes	Herdsman	Boo	15 +30	Arcturus
Coelum	Graving Tool	Coe	5 −40	
Camelopardalis	Giraffe	Cam	6 +70	
Cancer	Crab	Cnc	9 +20	(Praesepe)
Canes Venatici	Hunting Dogs	CVn	13 +40	
Canis Major	Big Dog	CMa	7 −20	Sirius
Canis Minor	Little Dog	CMi	8 + 5	Procyon
Capricornus	Sea Goat	Cap	21 −20	
Carina	Keel of Ship	Car	9 −60	Canopus
Cassiopeia	Queen	Cas	1 +60	
Centaurus	Centaur	Cen	13 −50	(Nearest star)
Cepheus	King	Cep	22 +70	
Cetus	Whale	Cet	2 −10	Mira
Chamaeleon	Chameleon	Cha	11 −80	
Circinus	Compass	Cir	15 −60	
Columba	Dove	Col	6 −35	
Coma Berenices	Bernice's Hair	Com	13 +20	
Corona Australis	Southern Crown	CrA	19 −40	
Corona Borealis	Northern Crown	CrB	16 +30	
Corvus	Crow	Crv	12 −20	
Crater	Cup	Crt	11 −15	
Crux	Southern Cross	Cru	12 −60	
Cygnus	Swan, N Cross	Cyg	21 +40	Deneb
Delphinus	Dolphin	Del	21 +10	
Dorado	Swordfish	Dor	5 −65	
Draco	Dragon	Dra	17 +65	Thuban
Equuleus	Little Horse	Equ	21 +10	
Eridanus	Po River	Eri	3 −20	Achernar
Fornax	Furnace	For	3 −30	
Gemini	Twins	Gem	7 +20	Castor, Pollux
Grus	Crane	Gru	22 −45	
Hercules	Hercules	Her	17 +30	(Cluster)
Horologium	Clock	Hor	3 −60	
Hydra	Sea Serpent	Hya	10 −20	
Hydrus	Water Snake	Hyi	2 −75	
Indus	Indian	Ind	21 −55	
Lacerta	Lizard	Lac	22 +45	
Leo	Lion	Leo	11 +15	Regulus
Leo Minor	Little Lion	LMi	10 +35	
Lepus	Hare	Lep	6 −20	
Libra	Balance (Scales)	Lib	15 −15	

TABLE 3-1 (Continued)

name in Latin[a]	name in English	abbr.	approx. position		bright star(s) (if named)
Lupus	Wolf	Lup	15	−45	
Lynx	Lynx (Bobcat)	Lyn	8	+45	
Lyra	Lyre (Harp)	Lyr	19	+40	Vega
Mensa	Table Mountain	Men	5	−80	
Microscopium	Microscope	Mic	21	−35	
Monoceros	Unicorn	Mon	7	− 5	
Musca	Fly	Mus	12	−70	
Norma	Carpenter's Level	Nor	16	−50	
Octans	Octant	Oct	22	−85	(South pole)
Ophiuchus	Holder of Serpent	Oph	17	0	
Orion	Hunter	Ori	5	+ 5	Betelgeuse, Rigel
Pavo	Peacock	Pav	20	−65	
Pegasus	Winged Horse	Peg	22	+20	
Perseus	Perseus (Hero)	Per	3	+45	Algol
Phoenix	Legendary Bird	Phe	1	−50	
Pictor	Easel	Pic	6	−55	
Pisces	Fishes	Psc	1	+15	
Piscis Austrinus	Southern Fish	PsA	22	−30	Fomalhaut
Puppis	Stern of Ship	Pup	8	−40	
Pyxis	Compass of Ship	Pyx	9	−30	
Reticulum	Net	Ret	4	−60	
Sagitta	Arrow	Sge	20	+10	
Sagittarius	Archer	Sgr	19	−25	
Scorpius	Scorpion	Sco	17	−40	Antares
Sculptor	Sculptor's Tools	Scl	0	−30	
Scutum	Shield	Sct	19	−10	
Serpens	Serpent	Ser	17	0	
Sextans	Sextant	Sex	10	0	
Taurus	Bull	Tau	4	+15	Aldebaran
Telescopium	Telescope	Tel	19	−50	
Triangulum	Triangle	Tri	2	+30	
Triangulum Aus.	Southern Triangle	TrA	16	−65	
Tucana	Toucan	Tuc	0	−65	
Ursa Major	Big Bear	UMa	11	+50	Mizar
Ursa Minor	Little Bear	UMi	15	+70	Polaris
Vela	Sail of Ship	Vel	9	−50	
Virgo	Virgin	Vir	13	0	Spica
Volans	Flying Fish	Vol	8	−70	
Vulpecula	Fox	Vul	20	+25	

[a] Carina, Puppis, Pyxis, and Vela were originally a single constellation, Argo Navis, the ship of the Argonauts, commanded by Jason.

called *first magnitude,* the next brightest were *second magnitude,* and so on, down to *sixth magnitude* for the faintest stars visible in the Egyptian sky.

On the star maps of today the same system of magnitudes is used, although the system has now been put on a mathematical basis. On the star maps in this book the brightness, down to fourth magnitude only, is indicated by the size of the filled circles used to locate the stars. The scale is shown at the bottom of each map. Star positions are indicated by a grid of lines with scales for right ascension at the top and bottom and for declination at the sides. Lines connecting the stars in their respective constellations are used to help in the identification. Most of the stars in each constellation are designated by Greek letters, with usually the brightest star having the letter α, the next brightest β, and so on. One exception to this is that in the Big Dipper the stars are lettered beginning with α for the Pointer star in the Bowl nearer the pole and running along the configuration to the end of the Handle. The Big Dipper is a portion of the constellation Ursa Major (Great Bear). A group of stars inside a constellation, such as the Little Dipper, the Pleiades, and others, is called an *asterism*.

Sometimes several stars are given the same Greek letter, such as μ_1 and μ_2 in the Scorpion and π_1 to π_6 in Orion. Sometimes Latin letters or numbers are used, such as G in the Tail of the Scorpion and numbers 41 and 43 in Eridanus. All of these and others will be found on star maps. Letters beginning with R are used to designate variable stars—those whose light varies either periodically or irregularly (see Chapter 17). In the maps in this book variable stars also carry the word variable, and, if known, occasionally the period of variation is given.

The ancient astronomers paid particular attention to 12 constellations along the ecliptic. A zone about 8° on each side of the ecliptic is called the *zodiac* and the 12 constellations are called the signs of the zodiac. They are shown in order in Fig. 3-4. Because of the motion of the earth around the sun, the sun appears to be in Pisces at the beginning of spring and then moves through the zodiac at the rate of one constellation per month, in the order Pisces, Aries, Taurus, and so forth.

In addition to Greek letters, the brightest stars have names. Some of these are listed in Table 3-1.

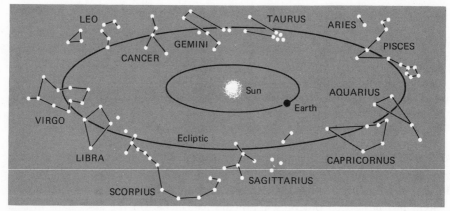

FIG. 3-4   The 12 signs of the zodiac.

3.2 THE MILKY WAY

There is a region in the sky where the stars are more numerous than in any other part. Because of its whitish appearance, this has long been called the *Milky Way*. In 1609 Galileo found that it contains great numbers of faint stars invisible without a telescope. In the Milky Way the number of stars increases greatly towards the south, where they are so thick these regions have been called *star clouds*. The sun is a star in the Milky Way system, now called the Galaxy, from a Latin word meaning milk. The region where the stars are most numerous and where there are great clouds of gas and dust is now known to be the center of this great system, containing some 100 billion (10^{11}) stars. It is the location of an axis about which the sun moves in a period of about 220 million years.

It is also easily apparent that the stars have a tendency to group together. Such groups are called *star clusters*. Some, like the Pleiades, which were known in the Stone Age, are loosely formed *open clusters*. Others, like the Great Star Cluster in Hercules, contain thousands of stars, most of them invisible without a telescope. They are closely packed in a spherical shape and are called *globular clusters*. The brightest cluster of this type is in the constellation Centaurus and is marked with the Greek letter ω, since it appears to the unaided eye as a faint star of about the fourth magnitude (see Map 3).

The Milky Way is shown on the maps in this book as an irregular, shaded area, part of which appears on all the maps. The Milky Way follows a great circle around the sky, the *galactic equator,* and is shown as a curved line in Fig. 3-3. See also Table 3-2.

TABLE 3-2   Right Ascensions and Declinations of the Galactic Equator

right ascension	declination	right ascension	declination
0ʰ	+62.1°	12ʰ	−62.1°
1	+62.6	13	−62.6
2	+61.4	14	−61.4
3	+58.3	15	−58.3
4	+52.4	16	−52.4
5	+41.6	17	−41.6
6	+22.2	18	−22.2
7	− 5.4	19	+ 5.4
8	−30.5	20	+30.5
9	−46.3	21	+46.3
10	−55.0	22	+55.0
11	−59.8	23	+59.8
		24	+62.1

3.3 NORTH CIRCUMPOLAR STAR MAP

For the beginner who is just learning the constellations, the first map to use is Fig. 3-3, which shows the North Circumpolar Stars. This region of the sky is always above the horizon for observers north of latitude 60°. The zone of circumpolar stars decreases in size to zero for observers at the equator. For any observer, it has a radius equal to his latitude. For example, from San Diego in southern California the circumpolar circle has a radius of 32¾°, from Chicago 40°, and so forth.

To use the map in Fig. 3-3, face north and hold the map so the center is in line with Polaris, the Pole Star. Polaris is most easily located by using the two Pointer Stars in the Big Dipper. They are marked α and β in Ursa Major and on the map are connected by a line tipped with an arrow pointing toward the pole. Polaris is about 30° from α and α is about 5° from β. The length of the Big Dipper is about 30°. These angular distances should give the beginner a good idea of the scale to be used in estimating angles in the sky. α and β Ursae Majoris, and Polaris, α Ursae Minoris, are all three about equally bright second-magnitude stars.

The next step depends on the time of year and the time of night the observations are being made. Perhaps the best way is to turn the map until you see the Big Dipper in its proper place with respect to the Pole Star. The map in Fig. 3-3 is designed for 8 P.M. standard time or 9 P.M. daylight saving time (DST). So if the time is 9 P.M. in May, the map is to be held with the top straight up. If the month is November and the time 8 P.M. standard time, the map is to be turned upside down. Other months are marked, and the map must be held with the given month up. The constellations will be in the positions shown for 8 P.M. standard time.

If the time is not 8 P.M. standard time, the map must be turned to compensate for the earth's rotation, which is 15° per hour. The sky appears to rotate around the pole in a counterclockwise direction; that is, from east to west above the pole for observers in the northern hemisphere. As an aid to the adjustment of sky maps for rotation of the sky during the night, the map in Fig. 3-3 is marked in hours at 15° intervals. The hours are shown from 0^h to 24^h around the rim.

The line marked 0^h from the celestial pole at the center of Fig. 3-3 toward the bottom of the page is a portion of a great circle to the vernal equinox, which is not on the map. If this line were extended on around the celestial sphere, it would complete a circle passing through the north and south celestial poles. This is true of all other lines drawn from the pole through the other hours marked on the rim. These imaginary circles are the hour circles shown in Fig. 2-11, where they were discussed in more detail.

Astronomers use 24-hour clocks that keep time in accordance with the passage of these hour circles overhead. The hour indicated on the clock always corresponds to the hour circle coinciding with the meridian at that time; in other words, sidereal time (star time).

For a north-south scale, 10° of declination are marked on Fig. 3-3 on the zero-hour circle. The declination of the Pole Star is approximately 90°, and it would be near the zenith of an observer at the north pole, where the latitude is

90°. Likewise, a star whose declination is 40° would pass through the zenith of an observer at latitude 40°, and similarly for any given latitude.

Sidereal time was defined in Section 2-9. When the hours marked on the map coincide with the meridian, the sidereal time will be the same as the right ascension of the hour circle. For example, at 9 P.M. DST in May, the 12^h circle will pass from the north through the pole to and beyond the zenith. The meridian then runs through the Pole Star and between the stars γ and δ in the Dipper (Ursa Major), which will be in the north above the pole, and also through the east corner of Cassiopeia below the pole.

In the old maps the Dipper is shown as the back of the bear, with the Handle as the bear's tail. The stars are about second and third magnitudes. With today's meaning of magnitude, the magnitudes are between 1.68 for ϵ, the brightest, and 3.44 for δ the faintest star. Translated into light, ϵ is five times brighter than δ. It is an interesting experiment to see whether an observer can distinguish between the stars of the Dipper by brightness and can arrange them in order, not of position, but of magnitude. There are also slight differences in color. (See Chapter 16 for a more complete discussion.)

In addition to the seven bright stars, there is a fainter star close to ζ (Mizar), the middle star in the Handle. Mizar's faint companion is called Alcor. They are sometimes called the Horse and Rider and are supposed to have been used by the Indians and other peoples as a test for vision. Mizar itself is composed of two stars very close together, but visible separately in a telescope. The brighter component of Mizar is also double, as is proved by observations with the spectrograph. So Mizar is at least a triple star, with Alcor a companion star close by. It has been discovered recently that these two stars form a seven-star system.

Another important constellation shown in Fig. 3-3 is Ursa Minor, the Little Bear, which contains the asterism called the Little Dipper. This asterism contains the same number of stars as the Big Dipper, but the shape is slightly different and the stars are not as bright. The most important star, Polaris, is the last star in the Handle of the Little Dipper, and is near the pole.

Across the pole from Ursa Major is the W-shaped constellation Cassiopeia. It lies in the Milky Way about 30° from the celestial pole. A fainter constellation, Draco the Dragon, winds around the Little Dipper with its tail between the two Dippers. The star α on the map was the pole star at the time of the building of the pyramids in Egypt. It is between Mizar and the two end stars in the bowl of the Little Dipper. Its name is Thuban and it was very close to the pole in 2800 B.C. A passage in the Great Pyramid of Cheops was pointed toward Thuban. This fact helps to date its construction.

Cassiopeia was represented as the Queen sitting in a chair and the constellation is sometimes called Cassiopeia's Chair. A nearby group, Cepheus, was the King. Their daughter, Andromeda, is also nearby. Andromeda does not have any very bright stars; but inside its boundaries can be found the Great Galaxy, a system of stars similar to, but larger than, the one of which the sun is a part. It is marked "Spiral" on the map in Fig. 3-3 and on Map 1 on the front end papers.

It is very faint, but visible to the unaided eye on a dark sky and appears as a fuzzy patch of light with binoculars and small telescopes.

3.4 THE STARS OF AUTUMN

Suppose a study of the constellations overhead and in the south is to be made in early October and that the time is 9 P.M. DST. Turn the map in Fig. 3-3 until the hour circle marked 21^h is held up toward the zenith. That is, the sidereal time is 21^h. This can be checked from the table for computing sidereal time in Chapter 2 (Table 2-2). Now Map 1 on the front end papers is to be used. Face south and hold the map so the 21-hour circle is on the meridian and the top of the page is toward the zenith. The declination of the zenith is equal to the latitude of the observer, as noted in Chapter 2.

At 9 P.M. DST on October 7, the bright star Deneb, at the top of Cygnus (the Northern Cross), will be nearly overhead. The Cross may be identified by its shape, with Deneb the most northerly star of four in the upright and three stars forming the crossarm. It will be tipped slightly with the north part to the left (east) and the bottom to the right (west). It lies in the Milky Way where the stars are closely packed together except for a dark area, the Coalsack. The Coalsack is at the northern end of a long section of the Milky Way, which seems to split near Deneb and continues to below the southern horizon, where its two parts seem to join again near the Southern Cross in the southern hemisphere. In a telescope, the star β Cygni (Albireo) at the foot of the Northern Cross is a double star with two components of slightly different colors.

To the west of the center of Cygnus is the bright star Vega in the constellation Lyra. Vega is the brightest star in the summer sky and is blue or white. Lyra (the Lyre or Harp) is composed of an equilateral triangle, with Vega as the brightest star. The eastern corner is a well-known double-double star, ϵ. Some people with especially keen vision can see it with the naked eye as two stars. With binoculars the two stars are easily seen. In a larger telescope, each star can be seen as two distinct stars only a second or two apart.

The third star in the triangle of Lyra is ζ, a third-magnitude star. It also forms one corner of a parallelogram, with β, γ, and δ as the other three stars. An object of interest in this configuration is the Ring Nebula, invisible except with telescopes larger than 3 inches in diameter. A nebula is a mass of diffuse gas in space. This one is a luminous planetary nebula illuminated by a star at its center (see Fig. 3-5). A planetary nebula is not like a planet, except that it has a disklike shape displaying some detail, as the figure shows. It is believed that the material in this object was ejected from the central star by a violent explosion, which was first visible in the year 1054. The ring is actually a sphere, which appears as a ring from a distance of 1800 light-years. Its color is greenish when seen visually and the central star is not visible except in photographs, because of its low luminosity and blue color. (In addition to a strong green line in its spectrum, there is also a blue line not visible to the human eye.)

The Milky Way runs in the same direction as the upright of the Cross—that

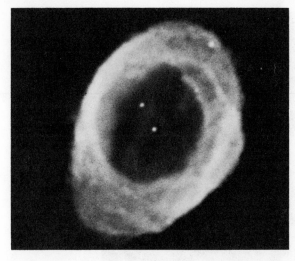

FIG. 3-5 The Ring Nebula, a planetary nebula in Lyra. (Photograph from the Hale Observatories)

is, southwesterly in October. Its apparent division into two parts is not real, but the effect is produced by the absorption of light from the background stars by enormous clouds of dust and gas in space.

Lying in and along the Milky Way are several other constellations. The one with the first-magnitude star, Altair, is Aquila (the Eagle). This constellation is easily identified by the bright star with a fainter star on each side. Near Aquila in the Milky Way is Sagitta (the Arrow), a faint group of four stars. Forming the third angle of a triangle with Aquila and Sagitta is Delphinus (the Dolphin, sometimes called Job's Coffin), which lies alongside the Milky Way. This little group looks something like a kite with a short tail. The Arrow is aimed toward the east side of Cygnus. Still farther south is an inconspicuous constellation, Scutum (the Shield), and below that, in and along the Milky Way is Sagittarius (the Archer). This constellation is in the most thickly populated area of the sky and is in the direction of the center of our galaxy, which is hidden by thick clouds of dust and gas. The center of the galaxy cannot be seen with optical instruments but is being probed by radio telescopes.

To the west of Sagittarius is Scorpius (the Scorpion), a long, winding constellation somewhat resembling the scorpion for which it is named. The first-magnitude star in Scorpius is Antares, which means Rival of Mars. It is definitely reddish in color, about the same as Mars, but is a little fainter when Mars is seen alongside at opposition. Antares is about 500 times the diameter of the sun; its redness is caused by its temperature, which is lower than that of the sun. Its average density is comparable to a high vacuum in the physics laboratory of about 1 mm of pressure or about 0.001 atmosphere.

East of Antares, the Milky Way stars are so thick they resemble a luminous cloud. This is called the Great Star Cloud in Sagittarius (see Fig. 3-6).

West of Lyra are some faint constellations that belong to the stars of summer (see Map 3). Hercules is named for the famous hero of mythology and includes four stars in the Keystone, which forms the central part of Hercules' body. On

FIG. 3-6   The Great Star Cloud in Sagittarius. *(Yerkes Observatory photograph)*

the western side of the Keystone is the Great Cluster, which is marked on the map. This is a globular cluster containing so many stars that it cannot be resolved at the center with even the biggest telescopes. It may be barely visible to the unaided eye in a dark sky and should be visible in a pair of binoculars. It probably contains more than 100,000 stars, most of the visible ones being brighter and hotter than the sun, but seen at a distance of over 30,000 light-years (Fig. 3-7). There are other clusters of this type near the southern part of the Milky Way, in Scorpius and Sagittarius, but none are found in the Milky Way itself. They surround the stars of the galaxy, but are seen mostly in the Sagittarius region because of the eccentric position of the sun.

East of overhead in October are some other faint constellations. To name a

FIG. 3-7 The globular cluster M13 in Hercules. *(Photograph from the Hale Observatories)*

few: Pegasus (the Winged Horse) has a central square 15° on a side, of which one star is in Andromeda (see Map 1). Pisces (the Fish) is represented by two fish tied together by their tails in the old star maps. Below Pisces is Aquarius (the Water Bearer), and in the southeast is the first-magnitude star Fomalhaut in Piscis Austrinus (the Southern Fish). Between Aquarius and Sagittarius is Capricornus (the Goat), also one of the signs of the zodiac.

3.5 WINTER AND SPRING SKIES

Maps 1 and 4, on the end papers, show the stars of winter and spring. Suppose the time is early February and the time is 8 P.M. The sidereal time on February 7 is $5^h 00^m$, and the brightest stars of the winter sky are near the meridian. A conspicuous feature of the sky at this time is the Winter Triangle. The three vertices of the triangle are in three different constellations. Sirius, the Dog Star, in Canis Major (the Big Dog), is the brightest star in the sky.

Procyon, in Canis Minor (the Little Dog), is only one sixth as bright as Sirius and only slightly less blue (see Map 4). The third star is the decidedly reddish star, Betelgeuse, about the color of Antares and of about the same brightness. Its light varies by a small amount, not detectable by visual observations. Betelgeuse is a very large star, but its diameter is not easy to measure. It may be nearly 1000 times larger than the sun and varies in size.

The Winter Triangle is used by navigators for their star fixes because the stars are very bright and are therefore easily identified.

Probably the most beautiful and certainly the best-known constellation in the winter sky is Orion (the Great Hunter). It is located near the winter branch of the Milky Way, which is not very bright in this part of the sky, but can be seen as patches of light much fainter than the part visible in the fall. Orion has many hot, blue stars, which form an "association." The stars in this association are now thought to be young stars that are using up their internal energy at a relatively fast rate and in a few million years will have lost their ability to shine. The long-exposure photographs show a great deal of bright gas and dust in this region of space. (See Chapter 1, Fig. 1-1.)

The central part of Orion is composed of three second-magnitude stars in line representing the belt of the hunter. Extending south from the belt are three fainter stars called the Sword. The central star in the Sword is θ on the star maps. It is actually composed of many parts; about six are listed, but there are many more faint stars very close to what appears to the eye to be one star. Here is the Great Nebula, which is luminous because it is close to and illuminated by these hot, bright stars. It is easily visible with binoculars, and a telescope reveals considerable detail, as shown in Fig. 3-8.

The summer solstice is located at right ascension 6^h, declination $+23.5°$, about 15° north of Betelgeuse.

Just north and a little west of Orion is Taurus (the Bull). Its brightest star is Aldebaran, a slightly reddish giant star, whose color has been described as orange. It is a little hotter than Betelgeuse, about 4000°K, and has a diameter about 25 times that of the sun, or 22.5 million miles. Aldebaran is found at one tip of a V-shaped cluster, the Hyades, which contains about 200 stars. It is not actually a member of the cluster, but is at a different distance and is not moving at the same speed.

Taurus also contains an even better-known cluster, the Pleiades or the Seven Sisters. There are six bright stars easily visible to the unaided eye, but two or three more may be visible in the dark sky at a mountain station. It is possible that the seventh bright star has faded to below naked-eye visibility. There are 250 or more members of this cluster visible in large telescopes. Twenty-five of them are perhaps visible with a good pair of binoculars. The cluster is embedded in glowing gas that can be detected by long-exposure photography.

North of Taurus is Auriga (the Charioteer), with the first-magnitude star Capella. This star is sometimes called the sailor's star, since it is visible in late summer in northern latitudes and was supposed to warn sailors of stormy weather to come. Capella has the same temperature as the sun, but it has a diameter about ten times larger and is 100 times brighter. The constellation Auriga consists of a pentagon of naked-eye stars. Of particular interest are two of the three stars in a triangle near Capella. The star at the vertex nearest Capella is ϵ Aurigae, a long-period eclipsing star. It is composed of one supergiant, hot star and an even larger star not visible to the eye. It is so cool that it radiates only infrared light and is therefore detectable only because it eclipses its companion every 27 years. The infrared component is probably at least 3000 times the diameter of the sun and is one of the largest stars known.

The fainter of the other two stars in the triangle is ζ, which is composed of two stars that eclipse each other every $2\frac{2}{3}$ years. The fainter star is about 40 times larger than the brighter one. The eclipses of this pair last about one month, while the eclipses of ϵ last two years!

Directly north of Procyon are the celestial twins, Castor and Pollux, in

FIG. 3-8   The Great Nebula in Orion. *(Yerkes Observatory photograph)*

Gemini. Castor is a well-known double star and has a ninth-magnitude companion, all three visible in a telescope of moderate size. Castor is actually composed of six stars.

3.6 STARS OF SPRING AND SUMMER

Following Gemini comes Leo (the Lion). It consists of a group of stars called the Sickle, with the bright star Regulus at the end of the handle. This group forms the forepart of the lion—his head, mane, and front legs. Then come the back, hind legs, and finally the tail, marked by a naked-eye star Denebola (β on Map 4). This constellation probably looks more like the animal it is supposed to represent than any other. When it is visible in the evening sky, spring is not far away.

Following the Lion are other fainter constellations seen in the evenings of spring and early summer. The autumnal equinox is about 15° south of Denebola. It is in Virgo (the Maiden), which contains one bright star, Spica (see Map 3). South of the autumnal equinox are Crater (the Cup), and Corvus (the Crow). North of it the Little Lion (Leo Minor) is not on the map because its stars are faint. North of Virgo are Coma Berenices (Bernice's Hair), and Canes Venatici (the Hunting Dogs). Their stars are also quite faint.

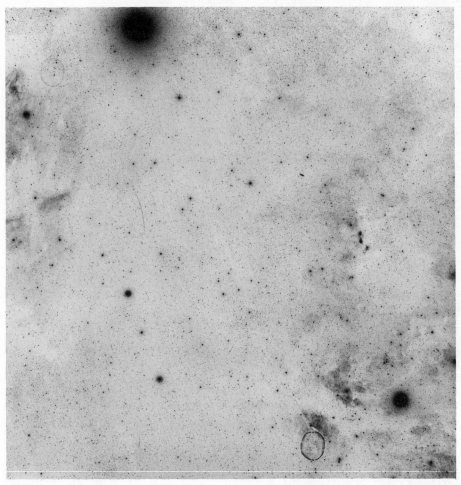

FIG. 3-9   Two photographs of the same region of the Milky Way, showing the Great Rift in Cygnus: left, in blue light; right (page 49), in red light. In these negative prints the stars are seen as dark spots against a lighter background. Notice the difference in the number and brightness of stars and nebulae in the two colors. The bright blue star at the upper left is Deneb. (*Hale Observatories; copyright National Geographic Society–Palomar Observatory Sky Survey*)

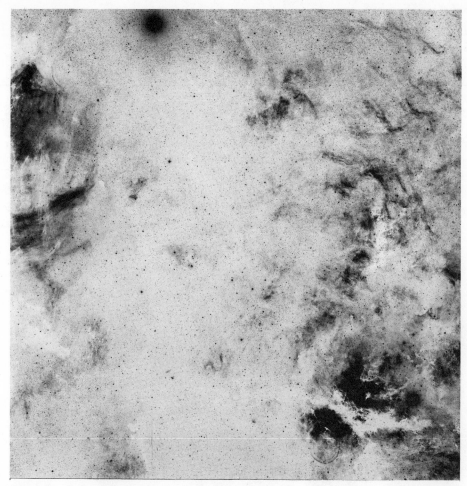

FIG. 3-9 (Continued)

One more first-magnitude star worth mentioning is Arcturus, the bright star of Bootes (the Herdsman), sometimes called the Driver of the Bears, which are to the northwest. Arcturus was the star selected to open the World's Fair in Chicago in 1933. Its light left the star 40 years earlier at the time of the first Chicago fair; reaching the earth in 1933, it was focused onto a photoelectric cell that operated relays and turned on the lights of the fair every evening when the sky was clear.

Following Bootes (pronounced with three syllables, Bo-o-tes) is the semicircle of stars called Corona (the Northern Crown). It is easily identified, since it has a second-magnitude star on one side. South of Corona is Serpens, the long, winding constellation that runs down to the east of Antares in Scorpius. Thene constellations are followed by the stars of summer that were described at the beginning of Section 3.4.

3.7 OTHER MAPS

The student who has a telescope and wishes to have more detailed star maps will find them in the monthly magazine *Sky and Telescope* or the bimonthly *Review of Popular Astronomy*. Both magazines print star maps in each issue with instructions for their use and also positions of the sun, moon, and planets, which are constantly changing.

An even more detailed set of maps is found in *Norton's Sky Atlas*, which also lists objects of interest such as clusters, nebulae, and galaxies, double stars, and variable stars. A more recent set, still more detailed, is the *Skelnate-Pleso Atlas*, published in Czechoslovakia and available at moderate cost. The newest issue of this set of maps has been done in color to indicate the color-temperature of the stars. An older and still more complete set of maps with magnitudes down to below naked-eye visibility is the *Bonner Durchmusterung*, also with a catalog of positions. This set was published in Germany almost a century ago, but has

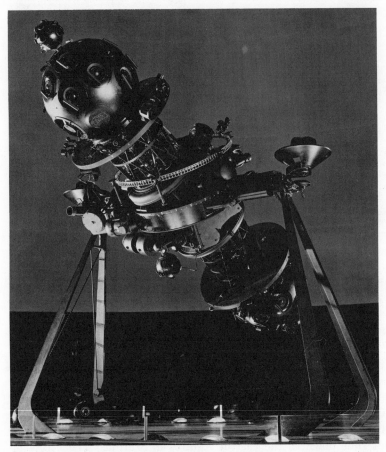

FIG. 3-10   The Zeiss projector in the Hayden Planetarium, New York City. (Courtesy of the American Museum of Natural History)

been reissued. The maps and catalog give positions of stars for 1855 or 1875, and considerable computation is necessary to bring the star positions up to date.

The most complete of all is the *Palomar Sky Atlas*. This is a set of photographic prints of negatives taken with the Mount Palomar 48-inch Schmidt camera. It shows all stars down to about eighteenth magnitude taken in red and blue light. Each print has its companion in the other color (Fig. 3-9). The *Sky Atlas* is particularly useful for observations with large telescopes.

3.8 THE PLANETARIUM

The planetarium is an instrument with which star images can be projected onto the inside surface of a spherical dome. It is a very useful device for instruction in constellation study, in teaching right ascension and declination, and the motions in the solar system. It is also possible to simulate a shower of meteors, and to project a comet onto the sky among the stars. There are several large planetariums in the United States (Fig. 3-10), and others of smaller size are being installed in schools, colleges, and universities. In many planetariums it is possible to show the appearance of the sky at any date in the past, present, or future.

QUESTIONS AND PROBLEMS

Group A

1. Observe the moon against the background of stars at intervals of a few hours. Sketch the position of the moon with respect to nearby stars and also to landmarks on the horizon. Repeat the following evening at the same time. Indicate the new positions on your sketch. Describe the motion of the moon with respect to the stars and with the horizon.

2. Locate some bright stars and constellations. Estimate the altitude and azimuth of each.

3. Star globes show east and west as on a globe of the world. On star maps, such as in the end papers, east and west are reversed. Why?

4. If you wanted to describe the size of an object in the sky to a friend, would it be better to compare it to a familiar object, such as a golf ball, or in terms of the full moon? Explain.

5. Look up the derivation of the word *zodiac* in the dictionary. Why do you suppose this name was chosen?

6. It is often helpful to estimate angular distances in the sky by comparing them with the known angle between familiar stars. The angle between the pointer stars in the Big Dipper is 5°. Estimate the following angles: (a) the angle between Polaris and α Ursae Majoris; (b) the length of Orion's belt; (c) the length of Cygnus. *Answer:* (a) 25°.

7. Which of the following stars are circumpolar for your latitude? (a) Deneb; (b) γ Cassiopeia; (c) γ Ursae Majoris?

8. Assuming the earth is at the position shown in Fig. 3-3, which

constellation would be on your meridian at (a) midnight, (b) 6 P.M., and (c) 6 A.M.? (*Hint:* The sun is on your meridian at noon.) *Answer:* (a) Capricornus.

9. Draw a diagram like Fig. 3-3 showing the zodiac and the earth's orbit around the sun. Indicate the position of the earth on March 21, June 21, Sept. 21, and Dec. 21.

10. If you observe a constellation rise at 6 P.M. on Nov. 1, where should you look for it at 6 P.M. on Jan. 1?

11. Assuming the map in Fig. 3-4 is properly oriented for 8 P.M. standard time, how many degrees should it be rotated, and in which direction, for the following times on the same night? (a) 10:40 P.M. standard time; (b) 6:28 P.M. daylight time.

Group B

12. Show that the radius of the zone of circumpolar stars is equal to the latitude of the observer. Use a diagram.

13. Show that the declination of the zenith is equal to the latitude of the observer. Draw a diagram.

14. On what day of the year is the vernal equinox on the meridian at 10 P.M. standard time for your longitude?

15. In which month is a part of the Milky Way at the zenith of an observer at 40°N latitude, at 9 P.M. standard time?

16. Draw a large diagram showing the orbits of Mercury, Venus, Earth, Mars, Jupiter, and Saturn as a series of concentric circles with the sun in the center. Also draw one still larger circle for the zodiac as in question 9, but label the right ascensions from a star map. Place the vernal equinox halfway between Pisces and Aquarius at the top of the chart. Then indicate the position of each planet in its orbit based on the following data (as seen from the earth): The date is March 21; Mercury is in Aries; Venus is in Aquarius; Mars is in Leo; Jupiter is between Aries and Taurus; and Saturn is in Sagittarius. Fill in the table:

	aspect	phase	rises	sets	crosses meridian
Mercury	eastern elongation	half	8 A.M.	8 P.M.	2 P.M.
Venus					
Mars					
Jupiter					
Saturn					

4 theories of planetary motion

4.1 APPARENT MOTIONS

Everybody knows that the sun and the moon appear to rise in the east and set in the west. It may not be so well known, however, that the stars do likewise. The explanation is simple, if it is assumed that the earth rotates on an axis and that the motions we see of the heavenly bodies across the sky are a result of the earth's motion. This apparent movement would be the same if the earth were motionless and the stars were fixed on a sphere that rotated around the earth. Since the earth seemed stationary, the ancient astronomers assumed it was the stars that moved.

The ancient astronomers were well aware of the phenomena of the rising and setting of the stars and had studied their motions on the celestial sphere for centuries. They recognized groups of stars that kept their shapes. These were the *constellations*, of which 48 had been named before A.D. 150. Even today, the stars in the constellations are called fixed stars. The ancient astronomers also noticed that there are some bodies that move among the constellations. They were the *planets:* the sun, the moon, Mercury, Venus, Mars, Jupiter, and Saturn.

Viewed from the earth, the sun and moon are seen to move among the stars from day to day, always in an easterly direction. This motion is called *direct motion.* The other five of the ancient astronomers' planets also move in direct motion most of the time, but occasionally they stop moving eastward and move westward for a short distance and over a period of a few weeks. This is called *retrograde motion.* If the apparent paths are plotted on a star map, they make loops, which may be open or closed. Figure 4-1 shows part of the loop made by Mars in 1965. These loops were difficult to explain, and it was not until the

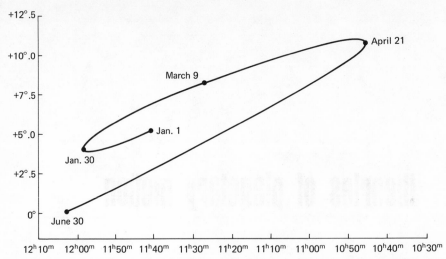

FIG. 4-1   Part of the loop on the celestial sphere traced by the planet Mars in 1964 and 1965. January 30 and April 21 were stationary points; March 9 was opposition. Coordinates are right ascension and declination.

time of Copernicus (1473–1543) and Kepler (1571–1630) that they were fully understood.

4.2 EARLY PLANETARY SYSTEMS

In the Greek astronomy, the earth was considered to be the largest body in the universe and fixed in space. There were various ways to explain how the earth was supported, most of which were quite fantastic. To explain the motions of the planets, each of the seven was thought to be fixed on a separate sphere that rotated around the earth once a day. This system explained the daily motions but did not quite explain the motions of the planets among the stars. There was an eighth sphere, to which the stars were fixed, that also rotated around the earth in one day. These spheres were transparent, so the remainder of the sky could be seen through them.

One very complicated system was proposed by a Greek astronomer, Eudoxus (409–356 B.C.). His system used 27 spheres to explain both the daily motions and the direct and retrograde motions of the planets. This system was later modified to 34 spheres by Callippus (about 350 B.C.).

The most famous system has been called the *Ptolemaic system,* because it was published by Claudius Ptolemy in his book *Megale Syntaxis tes Astronomias,* or *Almagest* for short. (*Almagest* means "the greatest book.") It was published about A.D. 140. The theory was based on the work of two men who lived in Asia Minor: Apollonius in the third century B.C. and Hipparchus, about 150 B.C.

In the Ptolemaic system, as in an earlier one, the earth was spherical, but stationary. Each planet had its own circle, a *deferent,* whose center was the earth. The smallest deferent was that of the moon; then came increasingly larger ones for Mercury, Venus, the sun, Mars, Jupiter, and Saturn in that order (see Fig. 4-2). The seven planets circled the earth in one day, the moon and sun eastward on their deferents, the moon in one month, and the sun in one year, in addition to their daily motions.

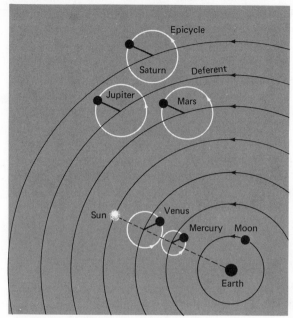

FIG. 4-2 The Ptolemaic system of the universe, later called the solar system.

In order to explain the motions of the other five planets, the starlike planets, each was assumed to move, not in its deferent, but in a smaller circle, called an *epicycle*, whose center moved eastward along the deferent. The sun and moon did not need epicycles, since their motions were always direct. Since Mercury is never more than 28° from the sun and Venus is never more than 46°, the centers of their epicycles stayed in line with the sun.

To explain the direct motion, all five planets when farthest from the sun moved on their epicycles in the same direction as that of the epicycle on the deferent—that is, in direct motion. But when nearest the earth, the planets were moving in the opposite direction. The backward motion on the epicycle exceeded the forward motion along the deferent, resulting in retrograde motion. During the retrograde motion, the planets were on the side of the epicycle nearest the earth. This accounted for their being brightest at the time. Mercury's fastest retrograde motion coincided with the time when it was closest to the earth in the center of the retrograde loop. This position occurred for Mercury once in about 116 days. Venus described a similar, but larger loop, once every 586 days (1½ years).

For Mars, Jupiter, and Saturn the centers of the epicycles did not stay in line with the sun and the planets came to the middle of their loops every 780, 399, and 378 days, respectively. At those times each planet was on the side of the deferent opposite the sun. Mars is conspicuously brighter at those times. We now call this position *opposition*. The loop of Mars is very long, sometimes as long as 19°, and may take almost three months to complete. Saturn's loop is 6.5° long and takes almost five months.

The biggest difficulty with the Ptolemaic system was that the loops made by the planets are not always the same. Sometimes they are closed, like the one made by Mars in 1964 and 1965. Sometimes they are open. Also they are not always the same size. In order to explain these irregularities, smaller epicycles were added to the rims of the larger ones, until, according to some historians, there were 55 in all. This system was so cumbersome that King Alphonso X of Spain remarked that if he had been consulted at the creation of the universe, he would have designed it on a simpler plan. The fault was not with the creation, but with the interpretation. The Greeks thought of the planets as perfect bodies and the circle as the perfect geometrical figure; therefore the planets could move only in circles. Their system was not improved upon for more than 1500 years.

Few distances were known, although Aristarchus, a Greek scholar of the third century B.C., calculated the distance to the moon at 240,000 miles and the sun at 4.8 million miles. Kepler increased this latter distance to about 14 million miles.

4.3 THE COPERNICAN SYSTEM

The first change in the theories of the universe came in the 16th century. Nicholas Copernicus, a Polish mathematician and astronomer, discovered that Aristarchus had suggested a universe with the sun at the center—a *heliocentric* instead of a *geocentric* universe. Copernicus adopted this system, which simplified the problem of explaining the motions of the planets by having them describe orbits around the sun (see Fig. 4-3). Although the system adopted by Copernicus is not accepted today, our present heliocentric system is called the Copernican system. It was changed and improved almost 100 years later by Johannes Kepler.

Aristarchus' plan had other advantages besides its simplicity. It also explained why the planets vary in brightness during the year. If the earth is assumed to be in motion, its distance from the other planets changes continuously, causing them to change in brightness.

Copernicus also believed that the planets move around the sun in circles. To explain some of the irregularities, such as the sizes and shapes of the loops, he placed the sun slightly out of center. This could also explain why Mars shows a difference in brightness at oppositions from year to year. But Copernicus also used epicycles. Some writers think he reduced the number, but recent historical research seems to indicate that the number may have been increased.

Although the Copernican system was widely adopted, Tycho Brahe rejected it. Thomas Digges (?–1595), an English astronomer and mathematician, had suggested that the stars are not all at the same distance on the eighth sphere, but are scattered throughout space. Probably he thought the brighter stars are bright because they are near. Accepting this concept, Tycho tested the theory by

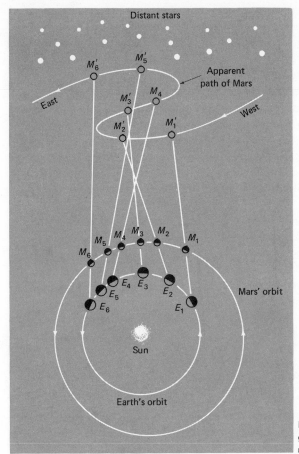

FIG. 4-3 Explanation of the retrograde motion of Mars by the Copernican system.

looking for *stellar parallax,* an apparent movement of the nearer stars with respect to the distant background stars as the earth moves in its orbit. Tycho could not detect any parallax, because the stars are too far away to show the effect with his instruments. It was not until about 1830 that such motions could be observed, and then only with large telescopes.

It remained for the invention of the telescope to lead to radical changes in astronomy, and for Kepler's work to really explain the apparent complexities of the solar system. Two discoveries by Galileo became of great importance for the heliocentric theory. First, he showed that Jupiter has a family of four moons circling it as the planets were assumed to circle the sun. Then he found that Venus shows all the phases that the moon exhibits (see Fig. 4-5). This was in contradiction to the Ptolemaic system, as an examination of that system easily shows. That is, since Venus and Mercury were always between the earth and the sun in that system, they should show only new and crescent phases. But Galileo pointed out that Venus also shows gibbous and full phases and therefore must at times be more distant than the sun. In fact, Venus shows a gibbous phase, greater than quarter, for 445 days in its phase period of 584 days. Galileo announced to the world his discovery of the phases of Venus by an anagram in Latin, which, when the letters were rearranged, said: The Mother of Loves imitates the form of Cynthia. This means: Venus imitates the moon.

With the adoption of a heliocentric system, the sun was no longer considered to be a planet and the earth became the third planet from the sun and the moon was called a satellite of the earth. Not until the time of Newton was the explanation of planetary orbits under the law of gravitation possible.

In the heliocentric theory the retrograde motions can be explained easily. Because of the stronger attraction by the sun, the planet nearest the sun moves most rapidly in its orbit. Thus Mercury moves faster than Venus, which moves faster than the earth. All the others move more slowly, the speed decreasing with increasing distance.

Figure 4-3 shows the reason for the apparent motions of Mars as seen from the earth. When the earth overtakes Mars, as it does during the oppositions, Mars seems to move backwards among the stars. The open or closed loops depend on the planes in which the planets move, and the shape of the loops depend on where the earth and the planet happen to be at the time of opposition. The variation in the size of the loops and the apparent change of brightness of the planet from one opposition to another was explained by Copernicus by means of eccentric circular orbits and small epicycles.

4.4 ELONGATIONS AND PHASES

Elongation is the angle between the centers of two astronomical bodies as seen from the center of the earth. In Fig. 4-4, E is the earth, S the sun, and P and P' are planets. Elongation is the angle SEP (or SEP'), whose vertex is at the center

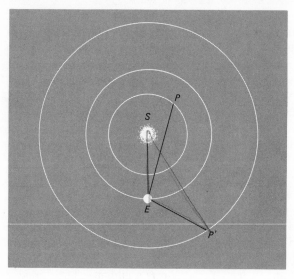

FIG. 4-4 Elongation—angle SEP or SEP'—of a planet at P or P'.

of the earth. Since the earth is so small compared to the distances to the sun and the planets, elongation may be measured from any station on the earth's surface without appreciable error. A planet's elongation is its angular distance from the center of the sun. When the elongation is minimum, the planet is said to be in *conjunction*. Since Mercury and Venus are closer to the sun than the earth is, they may come to *inferior conjunction* when they are between the earth and the sun, or to *superior conjunction* when they are beyond the sun (Fig. 4-5).

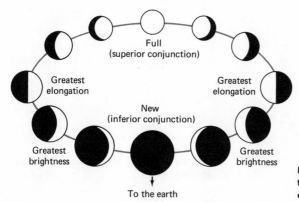

FIG. 4-5 Relation between elongations and phases of the inferior planets, Mercury or Venus.

It can be seen from Fig. 4-5 that at superior conjunction the disk of an inferior planet facing the earth is fully illuminated. At inferior conjunction the dark (unilluminated) disk is presented to the earth and the phase is new. Other phases, similar to those of the moon, are shown in the figure at the elongations illustrated. At both superior and inferior conjunction the planets are invisible, because they are too close to the sun.

At the greatest elongations of Mercury and Venus the phases are like the quarter-moon, because we see only half of the illuminated face. Between inferior conjunction and greatest elongation the phases are crescent, and between greatest elongation and superior conjunction the phases are gibbous. They show the same phases in reverse after superior conjunction. Thus the phases of these two planets are like the phases of the moon.

All the other planets are outside the earth's orbit and come to conjunction only on the far side of the sun (Fig. 4-6). After conjunction their elongations increase. An elongation of exactly 90° is called *quadrature*, and one of 180° is called *opposition*. Their phases change very little from full, except Mars which shows a slightly gibbous phase when near quadrature. When two bodies are in conjunction or opposition as seen from the earth, they are said to be in *syzygy* (siz'-i-je).

The definitions of conjunction and opposition should be changed slightly. For example, when Venus is in inferior conjunction, its elongation may be as large as 7°, because the inclination of its orbit puts it above or below the line

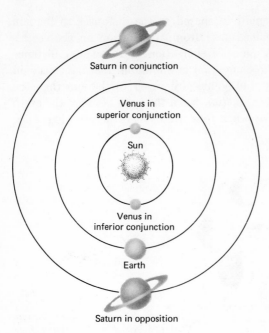

FIG. 4-6.   Planets in conjunctions and op-positions.

joining the earth and the sun. The planet is said to be in conjunction when its elongation is a minimum. The same may be said of opposition, where the elongation may be not quite 180°. Elongation is the shortest angular distance, measured on the arc of a great circle, between two bodies (Fig. 4-7).

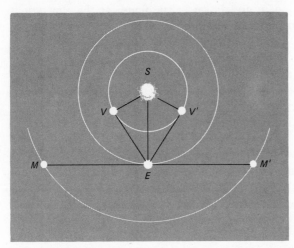

FIG. 4-7   Venus at greatest elongation, V or V'. Mars in quadrature, M or M'.

The *synodic period* of a planet is the length of time between two successive conjunctions of the same kind, or between two successive oppositions (Fig. 4-8). It is possible to determine the times of opposition or conjunction by observation. At conjunctions, Mercury and Venus are not visible, but by interpolations carried

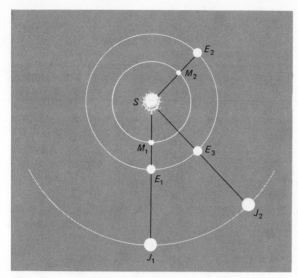

FIG. 4-8 In one sidereal period, 88 days, Mercury advances in its orbit from M_1 back to M_1. In one synodic period, 116 days, the earth goes from E_1 to E_2; Mercury makes one complete orbit and advances to M_2. Jupiter moves once around its orbit in 5.2 years; in one synodic period it moves from J_1 to J_2 in 1.1 years.

out over many years (or even centuries) these times have been computed, and the synodic periods are known with great accuracy. Because of the elliptical orbits, the times of conjunction or opposition do not occur with exact regularity. The average (mean) synodic periods are given in Table 4-1.

TABLE 4-1   Sidereal and Synodic Periods

planet	sidereal period	synodic period	planet	sidereal period	synodic period
Mercury	88.0^d	116^d	Jupiter	11.9^y	399^d
Venus	224.7	584	Saturn	28.6	378
Earth	365.2	—	Uranus	84.0	370
Mars	687.0	780	Neptune	163.9	367
Ceres	4.6^y	467	Pluto	247.3	367

The synodic periods result from the differences between the sidereal periods of the planets and that of the earth. The more distant the planet is from the sun, the slower it travels and the longer is its sidereal period (the time required for a planet to complete one orbit around the sun). The inner planets gain a lap on the earth between similar conjunctions, whereas the outer planets lose a lap between oppositions. The relation between sidereal and synodic period can be developed by determining the fraction of a lap gained or lost each day, as follows:

Let Sy stand for the synodic period of a planet, Si the sidereal period, and E the sidereal period of the earth (all expressed in the same units, days for the nearby planets or years for the distant planets). Then $1/Si$ is the fraction of its orbit that a planet travels in one day and $1/E$ is the fraction of its orbit the earth travels in one day. $1/Sy$ is the fraction of a lap the planet gains or loses on the earth each day.

For the inner planets:

$$\frac{1}{Sy} = \frac{1}{Si} - \frac{1}{E} \tag{4-1}$$

For the outer planets:

$$\frac{1}{Sy} = \frac{1}{E} - \frac{1}{Si} \tag{4-2}$$

EXAMPLE: The period of Venus is approximately **225** days. Compute its synodic period.

SOLUTION: Substituting in Equation (4-1) above:

$$\frac{1}{Sy} = \frac{1}{225} - \frac{1}{365} = 0.00444 - 0.00274 = 0.00170$$

Solving for Sy,

$$Sy = \frac{1}{0.00170} = 588 \text{ days}$$

A more accurate value is obtained by substituting the exact values of the periods of Venus and the earth.

4.5 KEPLER'S THREE LAWS

In 1600 Kepler went to Prague, then the residence of the emperor of the Holy Roman Empire, to become a student of Tycho Brahe. Tycho had been forced to leave his observatory on the island of Hven and had gone to Prague as official astronomer. There he spent the rest of his life refining his tables of planetary motion. After Tycho's death in 1601, Kepler had access to the storehouse of Tycho's observations of the planets, in particular those of Mars.

Kepler derived three laws that revolutionized the theories of the solar system. His first two laws were published in his book *Commentaries on the Motions of Mars* in 1609, eight years after Tycho's death. The third law followed in 1619 in a book called *The Harmony of the Worlds*. In the three laws Kepler abandoned the Greek idea of circles and turned to ellipses for an explanation of the apparent irregularities found in the speeds with which the planets move across the sky, and particularly in the size of their retrograde motions. (See Fig. 4-9.)

The first law is as follows:

1. *The orbits of the planets are ellipses with the sun at one focus of each ellipse.*

The ellipse is a closed curve in a plane in which the sum of the distances from any point on the curve to two internal points is always the same. The two internal points are called *foci* (singular, *focus*). An easy way to draw an ellipse is shown in Fig. 4-10. A string is looped around two tacks, one at each focus of

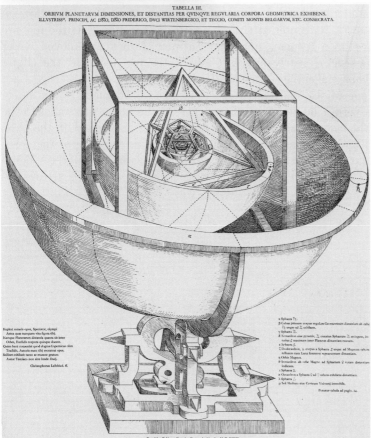

FIG. 4-9 Kepler and his model of the universe, which used five regular solids tangent to five spheres. The radii of the spheres were proportional to the distances of the planets from the sun. This model did not fit the facts. (*Yerkes Observatory photograph; N.Y. Public Library, Science and Technology Division, from Kepler's original works*)

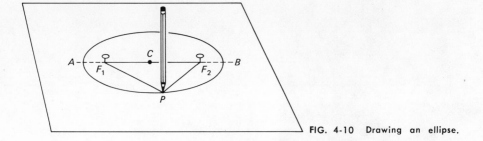

FIG. 4-10 Drawing an ellipse.

the ellipse, leaving some slack, so a pencil can be moved around the tacks while keeping the string taut. This keeps the sum of the distances from the pencil to the foci equal. The shape of the ellipse depends on the separation of the foci and the length of the string.

The *major axis* of the ellipse is a straight line from A to B passing through the two foci F_1 and F_2. Its center at C is the center of the ellipse. If the foci are moved apart, the ellipse becomes long and narrow; if they are moved closer together, it becomes more nearly circular; and if the foci coincide, the ellipse becomes a circle. Thus, the circle is a special form of ellipse.

The ellipse is one of the conic sections and is a curve produced by the intersection of a plane with a hollow cone (Fig. 4-11). If the plane is parallel to the base of the cone, the curve is a circle. If the plane makes an angle with the base, the figure is an ellipse until the plane becomes parallel to one edge of the cone; in that case the figure is a parabola. Finally, if the angle is still greater, the figure is a hyperbola. All of these types of conic section are important in astronomy.

Circle

Hyperbola

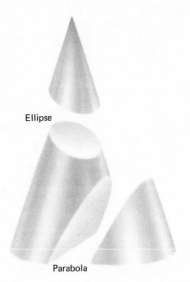

Ellipse

Parabola

FIG. 4-11 The four types of conic sections.

From Kepler's first law, since the ellipse is a plane figure, it follows that the sun and the orbit of each planet lie in one plane. However, the six planets known in Kepler's time and all nine known today do not all move in the same plane; rather these planes are all slightly inclined to each other and thus to the plane of the earth's orbit around the sun, the ecliptic. The sun, of course, is in each of the nine planes.

Figure 4-12 shows the elliptical orbit of a planet with the sun at one focus. The other focus is empty. Since the orbit is an ellipse, the distance from the sun to the planet changes continually. The size of the orbit is measured by the *mean distance*, the average of the greatest and smallest distances.

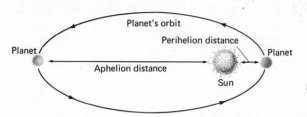

FIG. 4-12 The elliptical shape of a planet's orbit with the sun at one focus.

The point in each orbit where the planet is closest to the sun is called the *perihelion*. The point of greatest distance is the *aphelion*. The stem of these words is from the Greek word *helios*, meaning sun. The prefixes *peri-* and *ap-* mean near to and away from, respectively. The mean distances, eccentricities, and perihelion distances for the planets are all different (see Table 13-1).

The earth's perihelion distance is about 91.5 million miles and its aphelion distance about 94.5 million miles. The mean distance, the average of these two distances, is just short of 93 million miles. This distance is called the *astronomical unit* (abbreviated *a.u.*) and is now accepted to be 92,976,000 miles or 149,598,000 km. It is extremely difficult to determine exactly but is known with an accuracy of about 300 miles (500 km).

The shape of an ellipse is given by its *eccentricity*, a number that expresses the amount of flattening. It is determined by dividing the distance between the two foci by the length of the major axis. For the earth, the eccentricity e is

$$e = \frac{3,111,000 \text{ miles}}{185,952,000 \text{ miles}} = 0.0167$$

The size and shape of a planet's ellipse are determined, therefore, from a knowledge of the distance from the sun to the planet at various points in the orbit.

The eccentricity can also be obtained indirectly without knowledge of the actual distance. For example, since the distance from the earth to the sun varies, the apparent size of the sun also varies; that is, the apparent diameter is inversely proportional to the distance. If the apparent angular diameter of the sun at perihelion is 32.49′ and at aphelion is 31.42′, the average angular diameter is 31.955′. The eccentricity of the earth's orbit is the difference between the largest and smallest angles divided by their sums. That is, for the earth's orbit,

$$e = \frac{1.07'}{63.91'} = 0.0167$$

as before.

Kepler's second law states:

2. *The line joining each planet to the sun (called the* radius vector) *sweeps over equal areas of space in equal intervals of time.*

Kepler's second law cleared up some unexplained variations in the planetary motions and made it unnecessary to use the cumbersome system of epicycles. It can be seen from Fig. 4-13 that, since the areas are equal for a given time interval, the speed of the planet must vary; that is, a planet moves fastest in its orbit at perihelion and slowest at aphelion. This can be demonstrated by watching the motion of the sun along the ecliptic. It appears to describe a longer arc in one day in January than it does in July. Perihelion occurs in early January and aphelion in early July.

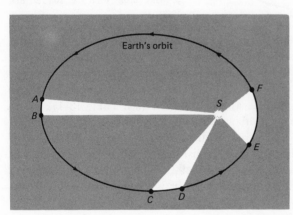

FIG. 4-13 Kepler's second law of areas explains why a planet's orbital velocity varies. The three shaded sections are equal in area and represent periods of 15 days each.

Kepler's third law, sometimes called the harmonic law, states:

3. *The squares of the sidereal periods of any two planets are proportional to the cubes of their mean distances from the sun.*

The third law may be stated in mathematical form as:

$$\frac{P_1{}^2}{P_2{}^2} = \frac{A_1{}^3}{A_2{}^3} \tag{4-3}$$

where P_1 and P_2 are the sidereal periods of two planets expressed in days or years, but both in the same units. A_1 and A_2 are their mean distances in miles or astronomical units. If P is expressed in years and A in astronomical units, the denominators are both equal to 1 if the second planet is the earth. The equation may be written

$$P^2 = A^3 \tag{4-4}$$

EXAMPLE.   Compute the sidereal period of Mars, if its mean distance is 1.52 a.u.

SOLUTION: Substituting in equation (4-4),

$$P^2 = (1.52)^3 = 3.51$$
$$P = \sqrt{3.51} = 1.88 \text{ years} = 686 \text{ days}$$

It should be noted that as A increases, P also increases.

In order to deduce his three laws, Kepler had to determine the distances of the plants from the sun and to show that the orbits are not circles, but ellipses. His method was as follows:

He first had to calculate the sidereal periods by the formula relating the sidereal periods and the synodic periods. The mean synodic periods were known from observations made over many centuries. Next, Kepler used the observations made by Tycho Brahe and picked out pairs of observations made at intervals of the sidereal periods. Consider Mars as the best example:

The sidereal period of Mars is 687 days. Suppose, in Fig. 4-14, Mars is at the point M_1 when it is observed at the beginning and at the end of this period of time. The earth is at E_1 and E_2 and has made nearly two revolutions around the sun. So the angle E_1SE_2 is 43° short of 720°. Mars is observed in the two directions marked on the figure and is seen against the background of the stars. From Tycho's observations, the angle between the sun and M_1 was determined. The fact that Mars is not in the plane of the earth's orbit can be neglected, since it is only about 2° out of the plane.

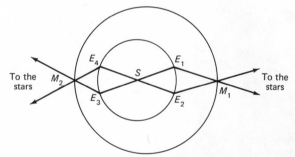

FIG. 4-14 Kepler's method of showing that the orbits of the planets are ellipses and of computing their distances from the sun in terms of the earth's distance (the astronomical unit).

Kepler did not know at first that the earth's orbit is not a circle, but assumed it to be circular. He assumed the distance from the sun to the earth as the unit of distance, which was unknown at that time. Then it was a matter of geometry and trigonometry to calculate the lengths of the sides of the triangles and find the ratio of Mars' and the earth's distances from the sun in terms of the earth's distance.

Notice that in the figure the distances of Mars from the sun are different on the two sides of the figure. On the right, the planet is at aphelion; at the left, it is at perihelion. By pairing the observations made by Tycho all around the orbit, Kepler was able to show that the distances are different and that the assumption of the elliptical shape of the orbit would fit his calculations of the distances.

This method was used for all the planets and his third law gave the relationship between the sidereal periods and their mean distances from the sun.

4.6 NEWTON'S LAWS

The work of such scientists as Tycho, Galileo, Kepler, and others paved the way for Sir Isaac Newton (Fig. 4-15), who was called by the French geometrician and astronomer, Lagrange, "the greatest genius that ever existed." Newton was born on December 25, 1642 (old-style Julian calendar), or January 4, 1643 (present Gregorian calendar). He made great contributions to the field of astronomy, optics, mechanics, and mathematics before he was 24 years old. However, he did not publish his work until 20 years later, when his famous book, usually called the *Principia* (*Philosophiae Naturalis Principia Mathematica*) was printed at the expense of the astronomer Edmund Halley (1656–1742) in 1687.

FIG. 4-15   Isaac Newton. *(Yerkes Observatory photograph)*

In the *Principia*, Newton stated three laws of motion, which are the basis of mechanics:

1. *A body remains at rest or continues to move with constant velocity in a straight line unless it is acted on by an unbalanced outside force.* This fundamental property of all matter is called *inertia*.

2. *The change of speed or direction of a body is directly proportional to the external force producing the change, inversely proportional to the mass of the body, and takes place in the direction of the external force.* This law may be written: $a = f/m$, or its equivalent $f = ma$, where f is the external force, m is the mass of the body, and a is the acceleration produced by the force.

3. *For every action there exists an opposite and equal reaction.* That is, actions never exist alone, but are always accompanied by other actions.

There are two important consequences of these laws:

1. It is obvious that if a body at rest is pushed by equal forces from opposite directions, it will remain at rest. Also, balanced forces do not change the uniform velocity of a body. But if the forces are unbalanced, the body will change its speed or direction, or both.

   For example, a car travels at constant velocity if the retarding forces are balanced by the driving force of the wheels. Increasing the driving force so it exceeds the retarding forces increases the velocity. To change the direction of the car's motion, another force must be applied to the car. This force is exerted by the road on the front wheels when they are turned to follow a curve in the road. If the road is icy, it will not exert the force needed and the car continues in a straight line.

2. Since the planets continually change their directions of motion, some unbalanced force must be acting on them to counteract their inertial tendency to move in a straight line.

   Newton was able to show that the force changing their speeds and directions is directed toward the sun. This law is completely incompatible with Ptolemy's epicycles, because it requires a force directed toward the center of each of the loops. The Greeks thought that the planet's motions were an intrinsic part of their nature. Figure 4-16 shows how a force changes the speed or direction of a moving body.

Thus, Newton's first law of motion defines inertia. Velocity is the rate of change of position and includes both speed and direction. Change of velocity is called *acceleration*.

The second law relates the mass of a body and the rate at which its velocity changes; that is, acceleration. Force is defined by this law as the product of the mass and the acceleration. All three—force, velocity, and acceleration—may be

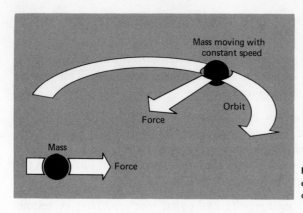

FIG. 4-16   Diagram showing a change of speed by a force (bottom) and a change of direction (top).

represented by vectors. A vector is a quantity that has size (magnitude) and direction. It can be represented by a straight line with an arrow indicating direction; the length of the line is proportional to the magnitude of the vector.

The third law, that actions exist in pairs, can be illustrated simply. For example, if the earth pulls on a given small mass, the mass pulls on the earth with an equal force, but in the opposite direction. Because the mass of the earth is so much greater, the force on the small mass accelerates it by 32 ft/sec/sec. The acceleration of the earth by the pull of the small mass is infinitesimally small and the earth is not moved by a measurable amount.

The effect of the pull between the earth and the moon is to swing the moon around a point between the two bodies in an orbit that is nearly a quarter of a million miles in radius. But the opposite pull of the moon moves the earth in an orbit only 2900 miles in radius. That is, the earth keeps opposite the moon in a small orbit around their center of mass.

Kepler's first and second laws were discovered by the use of geometry, based entirely on observation, and at the time had no theoretical foundation. The unifying principle basic to these laws had not yet been discovered. Newton realized from his three laws that the planets are pulled in orbits around the sun by a force he called *gravitation*. He stated the *law of gravitation* as follows:

Every particle in the universe attracts every other particle with a force that is directly proportional to the product of their masses and inversely proportional to the square of the distance between them.

Expressed as a formula:

$$F = G \frac{m_1 m_2}{d^2} \tag{4-5}$$

where F is the gravitational force, m_1 and m_2 are the masses of the two bodies, and d is the distance between their centers of mass. G is a universal constant, a factor of proportionality between the two sides of the equation to balance the units. If the force is measured in newtons, the mass in kilograms, and the distance in meters, $G = 6.67 \times 10^{-11}$ n m²/kg².

It has been shown in various ways that this law holds for all bodies in the solar system and probably in the universe, although this has not been definitely proved. The constant G is therefore called the universal constant of gravitation.

4.7 WEIGHT AND MASS

Weight is a measure of the attraction between a celestial body and a given object. It has been found that on the surface of the earth, the acceleration of a falling body changes slightly with location. Therefore, the weight of a given object varies over the surface of the earth. Also, from Newton's law of gravitation, if the distance of an object from the center of mass of the earth changes, the attraction by the earth changes, and so does the weight. Similarly, the weight of a body on the moon or on a planet is different from its weight on the earth.

Mass is a measure of inertia, the resistance to a change in motion. This property of an object never changes, regardless of the location. For example, a spaceship far removed from any gravitational source, such as the earth, would not have any weight. Yet the force required to impart a given acceleration would be exactly the same as on the earth, because the mass is still the same. Inside the spaceship, objects would be weightless while the ship was in uniform motion (no acceleration) even though the ship might be traveling at a high velocity with respect to the earth. There would be neither up nor down. If the ship were accelerated forward by its rocket engine, the rear of the passenger compartment would overtake the occupants and press against them, exerting the necessary force to accelerate them at the same rate as the ship. The occupants would have the impression that they had fallen toward the rear of the ship. There would seem to be a force, just like gravity, holding them against the rear compartment wall. The sensation would be the same as that of a person inside a ship that was standing on its stern on the earth. The rear compartment wall would seem to be the floor. The "weight" would be proportional to the rocket's acceleration.

We can apply Newton's law of gravitation to the calculation of the weight of an object on the earth by rewriting it in the following form:

$$W = G \frac{mE}{r^2} \tag{4-6}$$

where m is the mass of the object, E the mass of the earth, and r the radius of the earth.

Notice that the weight is proportional to the mass of the attracting body, in this case the earth. Since the masses of the planets differ, the weight of an object depends on the planet on which it is located. The weight is also proportional to the mass of the object and thus heavier objects have more inertia than lighter ones. This agrees with common experience.

In modern practice the units used in the metric system are the meter, the kilogram, and the second (mks), instead of the older centimeter, gram, and second (cgs) units. In mks units, the unit of force is the newton, a force that produces an acceleration of 1 m/sec/sec on a mass of 1 kg.

Kepler's first law places the sun at one focus of each of the orbits of the planets. Kepler knew this as an observational fact, but he did not know why it should be so. Newton's universal law gave the reason. Newton showed that the force is inversely proportional to the square of the distance. Thus the force increases as the bodies approach each other. Furthermore, Newton showed mathematically that the attracting body cannot be at the center, but must be located at the focus of the ellipse, thus proving Kepler's first law and also that the second law is correct.

Kepler's third law is only approximately true, but it gave a satisfactory result with the known data of the time. It remained for Newton and the law of gravitation to refine it. There are several ways of stating Newton's modification of this law. One way is:

$$\frac{P_1{}^2}{P_2{}^2} = \frac{A_1{}^3(M + m_2)}{A_2{}^3(M + m_1)} \tag{4-7}$$

where the symbols are the same as for Kepler's statement, except for M, the mass of the sun, and m_1 and m_2, the masses of the two planets. Since the mass of the largest planet, Jupiter, is only 0.001 that of the mass of the sun, it can be seen easily that no serious error is made by neglecting the masses m_1 and m_2, as Kepler did.

Another way is to let the second plant be the earth, as before. Then

$$P^2(M + m) = A^3 M \tag{4-8}$$

where P is the sidereal period of a planet expressed in sidereal years, A is its distance from the sun in astronomical units, and $(M + m)$ is the combined mass of the sun and the planet. The earth's mass has been neglected, since it is only about three millionths the mass of the sun.

4.8 ENERGY

Newton found that force and mass are important concepts for analyzing motion. Two other closely related concepts, energy and work, were developed later and are also needed for theories in mechanics.

Work is done whenever a force moves through a distance in the direction in which the force acts. Force acts on stationary objects as the force of gravity does on a parked car, but no work is done if there is no displacement. Even if the car is moving, no work is done by gravity if the highway is level, because the downward direction of the force is perpendicular to the motion. However, if the car is going downhill, part of the motion is downward in the direction of gravity and work is done. The car accelerates, increasing its energy of motion, which is called *kinetic energy*. Kinetic energy is proportional to the mass and the square of the velocity of the moving body. That is,

$$\text{K.E.} = \tfrac{1}{2}mv^2 \tag{4-9}$$

Energy comes in many forms, all closely related to work. Whenever work is done, it is converted into some form of energy; on the other hand, energy is the ability to do work. This relationship is illustrated by a bullet that is fired straight up and then returns to earth. To simplify the discussion, assume that there is no atmosphere to produce friction. When the gun is fired, the *chemical energy* of the powder is converted to kinetic energy, which supplies the work to move the bullet against the force of gravity. As its altitude increases, its velocity decreases and hence its kinetic energy decreases, until both are finally reduced to zero when the bullet attains its maximum height. During its fall back to earth, the force of gravity acts to increase the velocity and therefore the kinetic energy. The work done is exactly the same as it was during the ascent, since the bullet regains its initial kinetic energy. The greater the altitude reached, the more kinetic energy it is potentially able to acquire by falling back to earth. It has *potential energy* due to its position above the earth. Potential energy is the capability of a mass to do work because of its position.

The foregoing illustrates one of the great laws of science, the *law of conservation of energy:* energy may be transformed from one kind to another, but it cannot be created or destroyed; the total energy is constant. In the case of the bullet, the balance between potential and kinetic energy changes with altitude, but the total amount does not change. In the actual case of a bullet traveling through the earth's atmosphere, the kinetic energy that is lost to atmospheric friction is converted to an equal amount of *heat energy*. Another form of energy that is of particular importance in astronomy is *radiant energy*, such as light and radio waves.

Planets have potential energy by virtue of their positions in the gravitational force field of the sun. The greater the distance of a planet from the sun, the greater is its potential energy. There is no friction in space, so the sum of the kinetic and potential energies of a planet always remains constant, although the balance between them is continually changing because of their elliptical orbits. The potential energy of a planet is greatest when it is farthest from the sun, at aphelion; consequently its kinetic energy is at a minimum at that time. When the planet is closest to the sun, at perihelion, its potential energy is least and its kinetic energy is at a maximum. This is in agreement with Kepler's second law, since the velocity is smallest at aphelion and largest at perihelion.

QUESTIONS AND PROBLEMS

Group A

1. The explanation by ancient astronomers of retrograde motion in terms of epicycles indicated their ignorance of what fundamental property of matter? Explain.

2. Show the various phases to be expected of Venus according to the Ptolemaic system by means of a diagram like Fig. 4-2. Draw Venus at several positions in its epicycle, shading the side not illuminated by the sun.

3. The retrograde loops made by the planets are not always the same. (a) How was this explained by the Ptolemaic system? (b) How by the heliocentric system?

4. Draw a diagram of Venus and the earth at several positions

in their orbits, and the corresponding lines of sight from earth to Venus, as in Fig. 4-3 for the earth and Mars. Indicate the positions of Venus where it appears to be in direct motion and its positions where it appears to be in retrograde motion.

5. What planets can be at (a) inferior conjunction; (b) superior conjunction; (c) quadrature; (d) opposition? *Answer:* (a) Venus and Mercury.

6. Why cannot Mercury and Venus ever be in quadrature?

7. Why do the synodic periods of planets decrease from Mars to Pluto?

8. Galileo is said to have found a very slight difference in the times of fall of light and heavy weights from the Leaning Tower of Pisa. What caused this difference? Should not a heavy weight fall faster than a lighter one, since it is accelerated by a greater force? Explain. Why should a feather and a hammer dropped on the moon fall in equal times?

9. Illustrate each of Newton's three laws of motion with an example of your own.

10. If a hunter were deposited on the middle of a frozen lake covered with frictionless ice, how could he get to shore? He has a loaded shotgun and two dead ducks. State the laws of Newton that apply.

11. Rockets used to launch satellites consist of several stages that drop off in succession during launch. Why does the final stage have a higher velocity than a single-stage rocket of the same total weight and fuel?

12. Is the gravitational attraction between two bodies affected by the presence of matter between them? Illustrate by an example.

13. Assume you are standing on a platform scales in an elevator. Indicate in each of the following cases whether the scale reading would read (i) less than, (ii) equal to, or (iii) greater than your true weight. (a) The elevator moves upward at constant velocity. (b) The elevator moves downward at constant velocity. (c) The elevator accelerates upward. (d) The elevator accelerates downward.

14. An astronaut in a space vehicle orbiting the earth feels weightless, and loose objects float around inside the cabin. Does this mean that the earth's gravity does not exert a force on the vehicle and its contents? Explain.

15. The ancient Greeks had many fanciful explanations of how the earth was supported. How do you explain this?

Group B

16. The velocity of a planet is inversely proportional to its distance from the sun at perihelion and aphelion. Show that Mercury's perihelion velocity is about 1.5 times its aphelion velocity.

17. The perihelion distance of Mercury from the sun is 28.6 million miles and its aphelion distance is 43.4 million miles. What is the eccentricity of its orbit? *Answer:* e = 0.206.

18. Compare the forces (a) between Jupiter and the sun, and (b) between Jupiter and Saturn, with the force between the earth and the sun.

19. Compute the weight of a 200-pound man in each of the following situations: (a) in space 4000 miles above the earth's surface; (b) on a planet 10 times the mass of the earth, but of equal radius; (c) on a planet of equal mass but radius twice that of the earth.

*20. (a) Compute the sidereal period of Mercury from its mean distance (Table 13-1) by using Kepler's third law. (b) Compute its synodic period.

*21. Verify equation (4-8) from the tabular data for Jupiter and Saturn.

22. The thrust of a space vehicle's rockets increases its speed by 100 ft/sec/sec. What will the acceleration be if (a) the thrust is doubled, (b) thrust remains the same but the mass is doubled, and (c) both thrust and mass are doubled? *Answer:* (a) doubled.

23. How much more kinetic energy does a car have when traveling at 60 mph than at (a) 20 mph and (b) at 15 mph? *Answer:* (a) 9 times.

24. What is the ultimate source of the heat energy in an electric toaster? Explain.

25. A crater was formed by the impact of a meteorite. What happened to the original kinetic energy of the meteorite?

5 planet earth

5.1 THE SHAPE AND SIZE OF THE EARTH

It is important to think of the earth as a planet early in a course in astronomy, since it is frequently used as a standard of comparison for size, mass, structure, and other things. How does it compare in size with the other planets? Is it the only planet on which life is possible or are there others in the solar system, or in orbit around other stars, capable of sustaining life?

First of all, what are the proofs that the earth is round? By 2500 B.C. the Egyptians had explored southward up the Nile River beyond what is now Aswan, near the first cataract of the river. By 2000 B.C. the Phoenicians had sailed out of the Straits of Gibraltar and northward to the British Isles. In 300 B.C. Pytheas, a Greek sea captain, had explored northward from Gibraltar. He had noticed that the Pole Star rose higher in the sky as he sailed north. This would be the case if the earth were round.

Other explorers sailing south along the west coast of Africa noticed that the Pole Star dropped lower in the north and that formerly unknown constellations rose above the southern horizon. Since all believed the earth to be flat, none of the explorers could explain these movements of the stars.

A similar change in the altitude of the sun with change of latitude led Bion, a Greek philosopher in the fourth century B.C., to predict that the sun should be visible for 24 hours a day in the summer, if an explorer could go far enough north. This was the first prediction of the midnight sun.

It is common knowledge that the lower part of a ship disappears below the horizon before the upper part does. Also, the shadow of the earth, which falls on the moon during a lunar eclipse, is always curved and is the shape expected when the shadow of one sphere falls on another. And, what is most convincing of all, photographs of the earth taken from space vehicles show it as a sphere (Fig. 5-1).

FIG. 5-1 The "gibbous earth," as photographed from the American space-ship, Apollo 10, from the distance of the moon. The west coast of North America can be seen. The remaining land masses are obscured by clouds. (NASA)

In order to answer questions about other planets, we must measure the size of our own earth, study its composition, and investigate as accurately as possible its internal construction in order to compare it with other planets within reach of our telescopes. Not only the size, but the exact shape should be determined.

Is the earth a perfect sphere or is it only approximately a sphere?

About 200 B.C. Eratosthenes, a Greek astronomer living in Egypt, assumed that the earth is a sphere and proceeded to determine its circumference. It was noticed that on the day of the year when the sun was highest at noon, its rays fell directly down a vertical well at Syene in Egypt. On the same day in Alexandria, about 500 miles (800 km) north, a vertical post cast a shadow that made an angle of 7.2° with the vertical. Since 7.2° is ¹⁄₅₀ of 360°, the number of degrees in a circle, the distance between Syene and Alexandria must be ¹⁄₅₀ of the circumference of the earth (Fig. 5-2). The unit of measurement of distance used in Egypt was the stadium, about 0.1 mile. The distance between the two places in Egypt was 5000 stadia, which would make the circumference 250,000 stadia or about 25,000 miles—a figure very close to that measured today.

From the circumference, the diameter of the earth can be calculated by the familiar formula of plane geometry, $D = C/\pi$. The diameter of the earth from Eratosthenes' figures is hence 25,000 miles divided by $\pi = 7960$ miles (12,700 km).

In about A.D. 30 Strabo, a Greek geographer, deduced that the earth is round and that its circumference is about 18,000 miles. This figure, some 7000 miles too

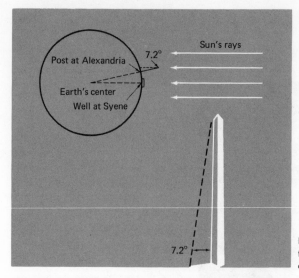

FIG. 5-2 The geometry of Eratos-thenes' determination of the diameter of the earth.

small, was the accepted figure in 1492, and was the distance used by Columbus when he made his attempt to sail from Spain to India. It is possible that if Columbus had known the real size of the earth, he might have hesitated to make such a long journey across many thousands of miles of unknown ocean. Perhaps the course of history owes a great debt of thanks to the error of a scientist!

It was shown in Chapter 2 that the altitude of the celestial pole is equal to the latitude of the observer. This provides us with a way of calculating the circumference and diameter of the earth from the measure of a north-south arc of a meridian. The measurements are made by a process called triangulation, used by surveyors. The results are as follows:

If the earth were a perfect sphere, the length of a degree of latitude should be constant. That is, a change of altitude of the celestial pole by 1° should occur at equal distances of about 69 miles on the surface of the earth. This is shown in the left-hand drawing of Fig. 5-3 for a change of 20° of latitude.

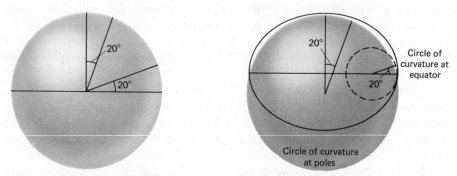

FIG. 5-3 Distances on an arc of a circle are equal. Distances on an arc of an ellipse are greater at the poles than at the equator for a change of latitude of 20°.

However, if the meridians on the earth are arcs of ellipses, the length of 1° at the poles would be greater than at the equator. This is because the radius of curvature at the poles is greater than that at the equator, as shown in the right-hand drawing.

The result of measurement of the miles per degree of latitude is shown in Table 5-1. These figures have been obtained by measuring the distance traveled north or south for a change of 1° in altitude of the pole. From these measures and others for intermediate latitudes a slice of the earth through the poles would show that the curvature varies from place to place and that the shape of a north-south cut is an ellipse.

TABLE 5-1

at latitude	miles per degree of latitude
0° (equator)	68.7
20	68.8
40	69.0
60	69.2
90 (poles)	69.4

East and west measures, however, show that the earth's equator is a circle (neglecting such irregularities as mountains and valleys). In other words, the equator at sea level is a circle and the earth is an oblate spheroid, a figure obtained by rotating an ellipse about its minor axis.

Also by computation, the diameter of the earth through the poles, the polar axis about which the earth rotates, is 7899.98 miles and the equatorial diameter is 7926.68 miles. The difference of 26.70 miles divided by the equatorial diameter is called the oblateness of the earth. That is, $26.70/7926.68 = 1/297$.

The oblateness of the earth is caused by rotation, as it is for all the planets. Each part of the earth travels in a circle around the polar axis. To maintain this motion a centripetal (center-seeking) force must be exerted on each part to divert it continuously from its natural tendency to travel in a straight line. In accordance with Newton's third law of motion, there is an opposing inertial reaction away from the axis of rotation. A similar inertial reaction would be observed by a person standing on a turntable; the faster the rotation, the greater the inertial reaction away from the center. Since this acts as a real force to the observer in circular motion, it is often called a centrifugal (center-fleeing) force. To indicate that it is the inertial reaction to the centripetal force, it will be referred to as the *centrifugal reaction force*.

The directions of the gravitational force and the centrifugal reaction force on parts of the earth's crust are shown in Fig. 5-4. Any force may be resolved into components that, when added together as vectors, are equivalent to the force they replace. In the case of the rotation of the earth, the centrifugal reaction

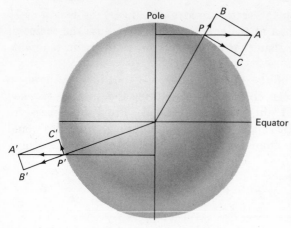

FIG. 5-4 The centrifugal reaction force, PA, and its two components: PB opposes the force of gravity and diminishes the weight of an object; PC is in the direction of the equator and produces the earth's oblateness.

force, *PA*, may be broken down into a component directed away from the center of the earth, the other at right angles to it; that is, along a tangent to the earth's surface.

In Fig. 5-4 the component *PB* acts opposite to the direction of gravity and decreases the weight of a body at point *P*. The other component, *PC*, is in the horizontal plane and is directed toward the earth's equator in both hemispheres. It is this component that produced the bulge at the earth's equator at a time when the earth was somewhat plastic and the particles of the crust were free to move. Table 5-2 shows the magnitude of the outward force of rotation and its two components at four latitudes.

TABLE 5-2

latitude	outward force on 100 lb	vertical component	horizontal component
0°	0.37 lb = 6 oz	6 oz	0 oz
30	0.30 = 5	4.3	2.5
60	0.18 = 3	1.5	2.6
90	0. = 0	0	0.

The American satellite Vanguard in 1958 showed orbital variations at the poles of the earth. These variations were interpreted to mean that the earth bulges a little at the north pole and is a little flatter at the south pole; the earth is not a regular spheroid. The amount is only 50 feet up or down. That is, the earth is slightly pear-shaped. It is thought this deformation is caused by the greater weight of the antarctic continent, which pushes toward the center of the earth, and by the lack of a continent at the north pole.

From the dimensions of the earth, we may compute its volume. But, since density is best expressed in grams per cubic centimeter (g/cm³), the figures for the dimensions given in miles must be converted to the metric system; 1 mile = 1.6093 km.

Since the earth is an oblate spheroid, its average diameter can be used as approximately the diameter of a sphere of equal volume. To find the average, the polar diameter is taken once and the equatorial diameter twice. We then have

$$\text{Average diameter} = \frac{7899.98 + 2 \times 7926.68}{3}$$
$$= 7917.78 \text{ miles}$$

Multiplying by 1.6093 km/mile, we find

$$\text{Average diameter} = 1.2743 \times 10^4 \text{ km}$$
$$= 1.2743 \times 10^9 \text{ cm}$$

since 1 km = 1000 m = 100,000 cm = 10^5 cm. Therefore the volume of the earth, from the formula of solid geometry, is as follows:

$$\text{Volume} = \tfrac{1}{6}\pi d^3$$
$$= \tfrac{1}{6} \times 3.14159 \times (1.2743 \times 10^9 \text{ cm})^3$$
$$= 1.0834 \times 10^{27} \text{ cm}^3$$

5.2 LATITUDE

The fact that the earth is shaped like an oblate spheroid, with minor variations, requires that latitude be redefined. If the earth were a sphere, latitude could be defined as the angle between the plane of the equator and a line from the center of the earth to the zenith. This would equal the number of degrees in the arc of a meridian between the equator and the place of observation.

Astronomical latitude is defined as the angle between the plane of the equator and the direction of gravity. The plumb line does not point toward the center of the earth, but is perpendicular to a level surface tangent to the earth's surface at the point P in Fig. 5-5. This does not change the proof that the altitude of the celestial pole is equal to the latitude, if the new definition of astronomical latitude is used. This kind of latitude is used because it is determined by measures of the altitude of the pole above a horizontal surface. *Geocentric latitude* is defined, as for a sphere, as the angle between the plane of the equator and a line from the center of the spheroid to the observer.

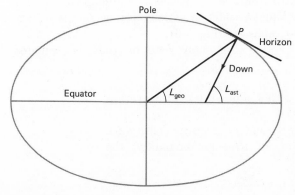

FIG. 5-5 Astronomical latitude (L_{ast}) differs from geocentric latitude (L_{geo}) because of the oblateness (greatly exaggerated for clarity). Arrow shows direction of gravity.

Geographical latitude is astronomical latitude corrected for station errors. Since gravity is used for astronomical latitude, it is possible that the plumb line may be disturbed and its direction changed by an irregularity in the density of the earth's crust nearby. The difference between the actual direction of gravity and its direction if the earth were a smooth spheroid is called a *station error*. These errors are small, at most a few seconds of arc.

5.3 THE EARTH'S INTERIOR

The structure of the earth's interior can be studied by earthquake waves; this branch of science is called seismology. An earthquake center sends out waves in all directions. One type of wave, the primary wave (or P-wave), is compressional (longitudinal); that is, the vibrations are parallel to the direction in which the wave travels. This type of wave can be generated by alternately pushing and releasing the end of a long spring. Each push compresses the end of the spring where the force is applied and the compression then travels on to the other end. This is followed by the expansion that occurs when the spring is released. If this is continued, an alternating series of expansions and compressions travels down the spring, creating a sequence of P-waves.

Another type of wave, the secondary wave (or S-wave), is transverse; its vibrations are perpendicular to the direction in which the wave travels. S-waves can be generated in a spring (or a rope) by shaking the end back and forth, perpendicularly to its length. This transverse motion is transmitted from particle to particle onto the other end of the spring. As each particle is pulled at right angles to the spring, it pulls the next one along with it. Since the transmission of this pull from particle to particle can occur only when there are bonds between them, these waves are transmitted only in solids. The compressional motion of the P-waves, on the other hand, is transmitted by pushing and does not depend on a bond; so P-waves are transmitted in gases and liquids as well as in solids.

The speed of each wave varies with depth. Both types are refracted (bent) as they pass from one medium to another of different density. P-waves travel faster than S-waves and so the P-waves are the first to arrive at a seismic station (hence the names primary and secondary). These facts permit the seismologist to study the structure of the earth's interior by means of the elapsed time for each wave to travel from the earthquake center to various stations on the surface where they are recorded by seismographs (see Fig. 5-6).

The internal structure of the earth, after many years of study, is estimated to be as follows:

1. The crust is a shallow layer a few miles thick with average density 2.7 g/cm^3.
2. The mantle is 1800 miles thick and of average density 5.0 g/cm^3. It meets the crust in an irregular boundary, the Mohorovičić discontinuity.

3. The outer core is liquid, about 1300 miles thick, density 11.0 g/cm³.

4. The inner core is solid, radius 800 miles, and density 14.0 g/cm³.

The crust is not uniform in density or thickness. Its composition is studied by means of samples of rock from different depths. The rocks on the surface vary in density from 2.5 g/cm³ for the lightest sedimentary rocks to 3.4 g/cm³ for the heaviest. The crust is thinnest under the oceans, which cover more than 70 percent of the surface. Because of the dissolved minerals ocean water contains, its density is slightly greater than that of pure water, which is 1 g/cm³. The thickness of the crust under the oceans may be as small as 3 to 5 miles; under the mountains it may be 30 miles.

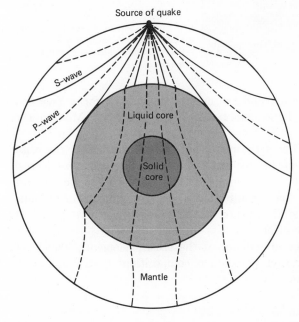

FIG. 5-6   Paths of earthquake waves from the quake center to seismological stations on the surface. All waves are bent, owing to changes in density along their paths. The sudden change in density at the boundaries of the earth's layers causes the P-waves to be sharply refracted. The S-waves do not penetrate the liquid core.

According to the theory of isostasy, the crust of the earth is sufficiently plastic to adjust itself to pressure variations. That is, the thickness of the crust is governed by the laws of equilibrium.

Mountains consist of granitic rocks of low density that "float" on the underlying basalt and mantle. The basalt layer under the mountains is thinner than it is under the oceans. So the material underlying the mountains (called the *root*) is somewhat indented into the mantle, and the crust in the mountain areas is thicker than it is under the oceans, where there is a layer of sediment, followed by a layer of basalt 3 to 5 miles thick.

Think of the earth as consisting of vertical columns. Each part of each column is subject to pressure from all sides, which must always be in balance. This results in an adjustment such that the weight of all columns is the same.

The root of a mountain, which projects into the mantle, is less dense than the material of the mantle it displaces, thereby compensating for the extra weight of the mountain. In all cases the dividing line between the crust and the mantle is irregular, as shown in Fig. 5-7.

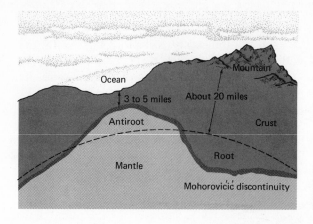

FIG. 5-7 Seismic waves have shown that the earth's crust is deeper under the mountains than it is beneath the oceans.

The boundary between the crust and the mantle was named for a Yugoslav scientist, Andrja Mohorovičić. It is called the Moho, short for Mohorovičić discontinuity.

Studies of the crust are far from complete, but they indicate that the crust is composed primarily of the elements oxygen and silicon. The continental crust is of granitic composition rich in aluminum. The floors of the oceans are basaltic, contain less aluminum than granite, but a larger amount of magnesium and iron. Reports of lunar material indicate that the lunar sea where Apollo 11 landed may have a similar composition.

Figure 5-8 shows the relative abundance of the crustal elements. This is of course an estimate, since the exact percentage composition is still unknown.

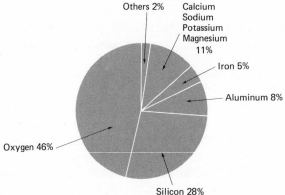

FIG. 5-8 The estimated abundance of elements in the earth's crust.

Oxygen, the most abundant element, is in combination with hydrogen in water and is also one of the elements in other compounds that make up the rocks. So it is not surprising that the crust is thought to contain 46 percent oxygen.

Silicon is abundant in sand and indeed in most rocks; its abundance is estimated as 28 percent. This is followed by aluminum and iron and smaller amounts of all the other elements. Because of erosion, all elements are being washed from the continents into the sea. It is likely that in the near future a project will be organized to recover valuable minerals from the oceans.

5.4 THE MASS AND DENSITY OF THE EARTH

The first determination of the mass of the earth was made in 1774 by Nevil Maskelyne (1732–1811), one of the British astronomers royal. His method was to study the attraction of a mountain, Schehallion, in Scotland, by suspending a plumb line in various positions and at different distances around the mountain. As Fig. 5-9 shows, the plumb bob was attracted toward the mountain. The deviation from the vertical was small, but measurable. By estimating the mass of the mountain and its distance from the various positions of the plumb bob, Maskelyne computed the mass of the earth. However, because of the difficulty of estimating the mass of the mountain and the location of its center of mass, the result was not accurate.

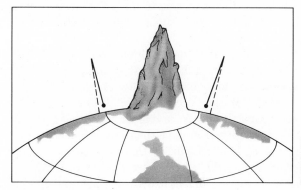

FIG. 5-9 Maskelyne's method of measuring the mass of the earth. The plumb bob is deflected from the vertical by the attraction of the mountain.

Two other, more accurate, methods are now available. The older was used by Henry Cavendish (1731–1810) in England in 1798 and the other by P. von Jolly in Germany in 1881.

In the Cavendish experiment a pair of small masses is suspended from a very fine wire, as shown in Fig. 5-10. The deflection of the small masses toward a pair of heavy ones is measured by a torsion (twist) of the wire. The mass of the earth can be deduced from the known masses and the amount of torsion.

The accepted mass of the earth is 5.977×10^{27} grams or 5.977×10^{21} metric tons of 1 million grams each. The uncertainty is 4 in the third decimal place.

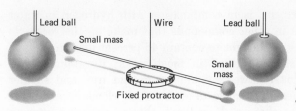

FIG. 5-10 The Cavendish torsion balance method of determining the mass of the earth.

From the mass and the volume given in Section 5.1, the density of the earth is

$$\text{Density} = \frac{5.977 \times 10^{27} \text{ g}}{1.0834 \times 10^{27} \text{ cm}^3} = 5.52 \text{ g/cm}^3 \tag{5-1}$$

This is the average density, since it is based on the volume and mass of the entire earth.

Compare this density with the density of some common substances: water, 1 g/cm³; rock, about 3 g/cm³; copper, 8.9 g/cm³; lead, 11.3 g/cm³; mercury, 13.6 g/cm³. It is obvious that, since the crust has a density between that of water and rock, there must be material inside the earth that has a density greater than average to compensate for the lighter crust.

The high density at the center of the earth may be assumed to be the result of pressure from above. Since the crust is composed of seawater and rock, its average density is assumed to be about 2.7 g/cm³. The density at the top of the mantle is about 3.5 g/cm³ at the Moho and increases to 5.5 g/cm³ at its inner boundary some 1800 miles below the crust. It is thought to be composed of silicates with large amounts of iron and magnesium.

The most probable composition of the outer core is liquid iron; that of the inner core is still unknown. The densities are 11.0 and 14.0 g/cm³, respectively. The solid body of the earth responds to tidal forces as the oceans do, but the amount is much less due to its rigidity. At the times of the highest spring tides, the surface rises about 9 inches. From this it is estimated that the earth is almost perfectly elastic, although its rigidity is greater than that of steel.

Pressures increase from an estimated 12,000 lb/in² (6 tons/in²) at the bottom of the Pacific Ocean to 50 tons/in² at a depth of 20 miles in the crust. At the base of the mantle, the pressure is estimated to be 10,000 tons/in², and at the center of the earth, 25,000 tons/in². Estimates of the central temperature vary so much that it can be said the temperature at the center is still unknown, although one estimate is that it is less than 10,000°K.

The temperature of the crust increases downward at the rate of about 1.8°F for every 100 feet. Heat flows from the core through the solid mantle and crust, but since the conductivity is very low, cooling of the core must be at a very low rate. Calculations show that only about 20 percent of the heat in the crust comes from the interior. The remainder comes from radioactive elements, such as uranium and thorium, that spontaneously disintegrate through a chain of various other radioactive elements, such as radium and radon gas, finally becoming stable lead. During this process some of the mass is converted to heat.

The age of the earth can be estimated in several ways. There is the rate of deposit of materials in the oceans from erosion of the continents. Probably a more accurate estimate is from the decay of radioactive elements. Although investigations by different methods give a variety of results, the probable age of the earth is estimated at approximately 4.6×10^9 years.

5.5 GASES

Before we consider the earth's atmosphere, it will be helpful to discuss some of the properties of gases. They are by far the most common form of matter in the universe: The larger planets have extensive atmospheres; stars are composed entirely of gas; and there is a considerable quantity of gas in space. Since most gases are invisible, their structure must be inferred from their properties. The *kinetic theory of gases*, developed during the 19th century, has been very successful in accounting for the behavior of gases.

According to this theory, gases are composed of molecules that are in rapid motion. On the average, they are widely separated compared to their very small size. Except at high pressure, gas molecules occupy less than 0.001 of the volume of the space around them. Their motion is completely random, so they are continually colliding with each other or any other obstacles in their paths, such as the walls of a container in which a gas is confined. A cubic centimeter of air at sea level contains approximately 2.7×10^{19} molecules. A typical molecule travels an average of 0.3 mile (0.5 km) and has 6 billion (6×10^9) collisions each second. These incessant collisions of the molecules of a gas with any surface immersed in it produce an average pressure on the surface.

Heating the gas increases the average kinetic energy of its molecules. This energy may be in the form of internal vibrations, rotational motion, or motion of the molecules from one point to another, called *translational* motion. It is the translational kinetic motion that is responsible for the pressure exerted by a gas. When the temperature is increased, the number of molecules striking a surface per second and the impact of each collision also increase, resulting in an increase in pressure. Similarly, lowering the temperature decreases the pressure.

Scientists have found it convenient to use a temperature scale called the *absolute* or *Kelvin* scale. Its zero point, called *absolute zero,* is equal to $-273.16°C$ or $-459.69°F$. The Kelvin scale has the same degree unit as the Celsius scale. To convert from the Kelvin scale to the Celsius scale add $273°$. For example, water freezes at $0°C = 273°K$ and boils at $100°C = 373°K$. (See appendix B.)

The pressure of a gas also depends on its volume. Compressing it decreases the average distance between the molecules, and the number of impacts per second on a surface immersed in it increases correspondingly. The gases near the center of a star are under extreme pressure and are therefore greatly compressed. The star would collapse if it were not for the counteracting high pressure exerted by these gases due to their extremely high temperature and great density.

The continual motion of the molecules of a gas tends to make it disperse. In the atmosphere of a planet or star, this tendency is opposed by the gravitational attraction. To escape, a molecule must acquire a critical velocity, called the *velocity of escape,* to overcome this attraction. The velocity of escape is determined by the diameter and mass of the particular planet or star but does not depend on the mass of the body escaping. It is the same for a molecule or a space vehicle. The escape velocity at the surface of the earth is just under 7 miles (11.3 km) per second (25,000 mph).

We have so far considered only the average velocity of the molecules of a gas. Because of the random nature of the collisions, they acquire a wide range of velocities. At any given instant some are traveling much faster than the average and some are traveling much slower. Also, in a mixture of gases, the average velocity of the lighter molecules is greater than that of the heavier molecules. At atmospheric temperatures only hydrogen and helium have a significant proportion of molecules that reach the escape velocity. The earth is therefore losing some of its hydrogen and helium.

5.6 THE ATMOSPHERE

The "ocean of air," the atmosphere, around the earth can be studied by a variety of methods. The oldest method was that of taking samples of air at all levels, from sea level up to the highest mountain that could be climbed. The next method was by balloons, both captive and free. Free balloons are sent up regularly from weather stations to determine the direction and velocity of winds. Manned balloons have gone up to about 15 miles. Free balloons can be equipped with radio apparatus to help in locating them when they come down. They have been sent up to about 19 miles with equipment for taking samples of air and for making records of radiation, temperature, and other conditions at various levels.

Information about the extent of the atmosphere comes from meteor trails to about 100 miles and by triangulation of northern lights (auroras) to about 600 miles. Most recently, satellite launchings have given more data, including the fact that the atmosphere extends farther from the earth than formerly thought. It mingles with the solar atmosphere. However, the density is so low at 1000 miles that it is difficult to say where the atmosphere ends.

To facilitate study, scientists have classified the atmosphere into layers, and there are many systems of classification. Only the most important layers will be discussed here. For details the student is referred to advanced texts on meteorology.

The lowest layer is the *troposphere* (see Fig. 5-11). This layer is most important to life on the earth, since it contains all the elements necessary to support life and because nearly all the weather is determined by it. It extends to 5 miles above sea level at the poles and 10 miles at the equator. The rotation of the earth may be the cause of this difference, since rotation causes flattening of the air mass as well as the central mass of the earth. It may be that the seasons with their different temperatures also affect the height of the troposphere.

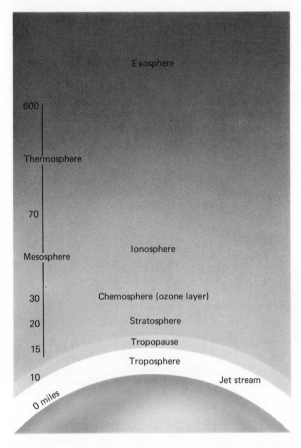

FIG. 5-11 The important layers of the earth's atmosphere. The heights are not to scale.

The pressure at the bottom of the troposphere averages 14.7 lb/in², and the density of air at sea level is about $\frac{1}{800}$ that of water. The density varies with pressure, because air is compressible.

The composition of the troposphere is: nitrogen 78 percent, and oxygen 21 percent; water vapor, carbon dioxide, argon, neon, and so forth make up the remaining 1 percent. Smog is not considered part of the true atmosphere, because it is composed of solid particles. Moist air contains up to 2 percent of water vapor by weight.

In the upper part of the troposphere are streams of high-speed particles. Westerly winds, called *jet streams,* have velocities that at times reach 300 mph. Discovered accidentally during World War II, they are both useful and bothersome to long-distance air travel. Westbound planes should fly at levels that avoid these contrary winds.

It has been discovered that jet streams are associated with the air masses in the lower atmosphere. When cold air masses move south, the jet streams move south also and mark the boundary between masses of cold and warmer air. They have been called the heat exchangers for the world's weather. If the air did not circulate between the equator and the poles, one would get hotter and the other colder. The jet streams equalize them. It is believed that cold air from the north pole moves south high in the troposphere and the currents of the troposphere and stratosphere mix (see Fig. 5-12). High-speed westerly winds blow along gaps in the upper troposphere, producing the polar-front jet. There is also a southern jet stream, which is not as high or as fast as the polar jet. The polar jets blow over densely populated regions, where they are accessible to commercial aviation during most of the year. But in summer they move north with the polar front and cannot be used by aviators.

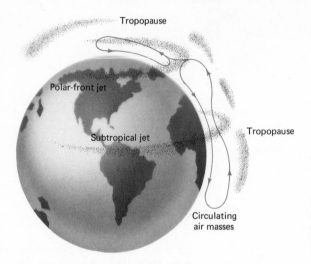

FIG. 5-12 The jet streams, believed to be the "heat exchangers" of the atmosphere. They occur where air masses break through the tropopause to mix with the stratosphere.

Jet streams may also determine the location and severity of typhoons and hurricanes. Study of these streams, therefore, may be of great help in predicting these destructive storms, and also help decide whether the earth is getting colder or warmer.

The average temperature at the bottom of the troposphere is about 56°F, although temperatures well over 100°F and down to −125°F have been recorded. At the top, the temperature has dropped to about −85°F. The upper boundary of the troposphere is marked by the *tropopause,* a thin layer of somewhat higher temperature. This makes it possible for a pilot to determine when he has reached that level by reading an outside thermometer.

The layer just above the troposphere is the *stratosphere,* which extends upward to about 20 miles. Here the temperature is fairly constant at about −65°F, except in the ozone layer, where it is near +32°F. Its low density makes it difficult for clouds and storms to develop. Therefore the stratosphere has little effect on the weather except at its lower levels, where air currents mix with those of the troposphere to form the jet streams.

Above the stratosphere is the *ionosphere,* which extends upward from about 20 miles to over 500 miles. This layer gets its name from the many ions present at that altitude. An ion is a remnant of an atom or molecule after one or more electrons have been removed, or to which one or more electrons have been added. The former is therefore positively charged and the latter has an excess negative charge. The energy required to ionize the nitrogen and oxygen atoms is supplied by the more energetic portion of the solar radiation. The energy of the photons is inversely proportional to their wavelengths, and only the shorter wavelengths (gamma rays, X rays, and short ultraviolet rays) have enough energy to remove electrons. Many free electrons are present in the ionosphere, the electron density ranging from some 10,000 electrons per cubic centimeter at the bottom to about 1 million in the middle.

The splitting of oxygen molecules into atoms requires less energy and absorbs the long ultraviolet wavelengths. The free oxygen atoms unite with the diatomic (two-atom) oxygen molecules to form ozone, a molecule of three oxygen atoms. This results in an ozone layer (also called the *chemosphere*) beginning in the stratosphere at about 15 miles altitude and extending into the ionosphere. This layer is very beneficial to animal life, because it converts the ultraviolet light from the sun into harmless heat (infrared) rays. It therefore helps to prevent sunburn. (See Fig. 5-13.)

THE IONOSPHERE

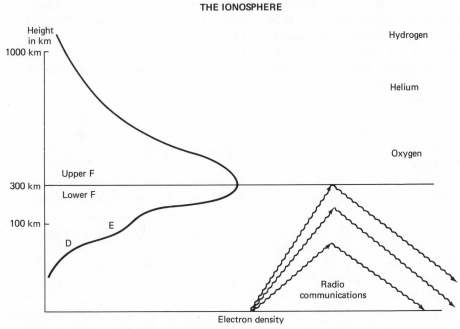

FIG. 5-13 Layers of the ionosphere. (NASA)

The lowest layer of the ionosphere is also called the *mesosphere,* meaning middle sphere. Here the temperature is about 32°F at the bottom, rises to perhaps 170°F, and drops to about −85°F at the top.

The pressure decreases very rapidly above the stratosphere. Half of the mass of the atmosphere is below an altitude of 3.5 miles. At 60 miles it is only 1/400,000 the density at sea level. The number of particles per cubic centimeter here is 7×10^{13}.

The Kennelly-Heaviside layer is another useful layer of the ionosphere. Radio waves are reflected from this layer. It is further divided into levels named D, E, F_1, and F_2. They reflect radio wavelengths longer than about 17 meters and make long-range radio communication possible. Radio waves can travel around the earth by successive reflections alternately between the ocean or the ground and the ionized layers. Wavelengths shorter than 17 meters, however, penetrate the ionosphere and make possible the reception of short waves from space.

The ionosphere is not stationary, but moves up and down because of the solar influence. The D-layer at about 50 miles reflects the radio broadcast band, the E-layer at 75 miles turns back the short waves, and the F-layers, which vary from about 120 miles by day and 200 miles by night, also turn back short waves. Wind velocities, studied by meteor trails and aurora curtains, are estimated at about 1000 mph at 400 miles. The temperature is above 2000°F. The hot layer at the top of the ionosphere is sometimes called the *thermosphere*.

In addition to the light of the moon and bright stars, the night sky glows with the faint light of unresolved stars. Also the atoms and molecules store up energy during the day and radiate it at night. More spectacular are the auroras—luminous bands and streamers of light that sporadically appear over the northern and southern parts of the earth. There is also a permanent aurora, which radiates in the infrared region of the spectrum. Auroral streamers do not exist below 60 miles, but curtains have been triangulated up to about 600 miles. They are therefore confined to the ionosphere.

The topmost layer of the atmosphere is called the *exosphere*, or, freely translated, the escape sphere. The lightest atoms, especially hydrogen and helium, have velocities greater than the velocity of escape. Helium is escaping from the exosphere at about the rate it is being released from natural gas wells in the crust.

The satellites launched since 1957 have given much information about the ionosphere and exosphere and also led to the discovery of the Van Allen radiation belts. These belts are almost certainly produced by radiation from the sun and vary considerably in height and intensity with the sunspot cycle. They were discovered in 1958 during the IGY, when the sun was extremely active. Later they decreased to about one fifth of their 1958 intensity during the following solar minimum. There is a region near the poles where there are no radiation belts.

In 1969 it was found that one of the radiation belts extends downward above the Atlantic Ocean. This is called the South Atlantic Anomaly because it interferes with electronic equipment carried by the orbiting astronomical observatory, OAO-2.

The radiation belts consist mostly of a mixture of electrons and protons, called *plasma*. Van Allen believes that the belts are caused by streams of plasma particles from the sun, called the *solar wind*. As the plasma reacts with the earth's magnetic field (Fig. 5-14), a shower of charged particles is produced. The *magnetosphere* may be considered a layer of the earth's atmosphere above the exosphere.

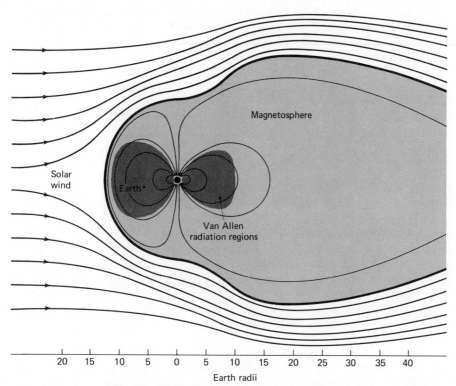

FIG. 5-14   The earth's magnetosphere. Based on a figure from Modern Physics by J. Williams, F. Trinklein, H. C. Metcalfe, and R. Lefler. Copyright 1972 by Holt, Rinehart and Winston. Reproduced with permission of Holt, Rinehart and Winston.

At the end of 1961 a new picture of the structure of the atmosphere was announced by Robert Jastrow, director of the theoretical division of the National Aeronautics and Space Administration (NASA). It was based on a new discovery of a thick layer of helium gas surrounding the ionosphere and extending to 1500 miles. The discovery of this helium zone came originally from France. Data from Explorer VIII confirmed it.

The distribution of the various gases in Jastrow's proposed structure of the atmosphere is as follows:

nitrogen and oxygen to 72 miles;
oxygen to 600 miles;
helium layer 600 to 1500 miles;
hydrogen to 6000 miles;
interplanetary gas and particles of solar origin beyond 6000 miles.

QUESTIONS AND PROBLEMS

Group A

1. At your latitude, what is the altitude of (a) Polaris and (b) the celestial equator where it crosses your meridian?

2. (a) Give two reasons why a person's weight at the north pole would differ from his weight at the equator. (b) If a man weighs 150 pounds at the equator, how much would he weigh at the north pole?

3. When a rock is whirled on a string, it pulls outward on the string, away from the center of rotation. This outward force would break the string if whirled fast enough. (a) Explain in terms of Newton's laws. (b) Describe the motion of the rock, if the string breaks.

4. Judging from Fig. 5-5, where on the earth is the difference between geocentric and astronomical latitude (a) the greatest and (b) the least? *Answer:* (a) 45°.

5. Suppose the earth were shaped like a football, with the poles at the two ends. At what locations on the surface would there be (a) the smallest number and (b) the greatest number of miles per degree of astronomical latitude?

6. About what percent of the earth's crust is composed of metals?

7. (a) What evidence indicates that the earth's core is denser than its crust? (b) Why is density alone not a sure way of identifying the elements in the crust?

8. What are two ways of increasing the pressure of a gas?

9. Compare the advantages and disadvantages of digging a hole down to the mantle on land or in the ocean.

10. How fast would a hydrogen molecule have to travel to have the same kinetic energy as an oxygen molecule traveling at 1 mile/sec? (An oxygen molecule weighs 16 times as much as a hydrogen molecule.) *Answer:* 4 miles/sec.

11. What are two possible sources of error in estimating the age of the earth from the rate of deposit of erosion materials?

12. Why are no ions produced in the earth's atmosphere below 20 miles?

13. (a) How does the height of the troposphere vary with location on the earth? (b) The mean atmospheric pressure at sea level is about the same over the entire surface of the earth. Is this consistent with the variation in the height of the troposphere?

Group B

14. Using the diameters of the earth through the poles and the equator, calculate the eccentricity of a section through the poles. Draw a figure. *Answer:* e = 0.08.

15. If the temperature of the earth increased uniformly with depth from the surface to the center, compute the central temperature. Assume $T = 0°C$ at the surface. Is your answer reasonable?

6 time and position

6.1 LATITUDE AND LONGITUDE

The determination of time is one of the duties of astronomy. Navigation, whose major function is the determination of position, depends on astronomy for accurate time. The Royal Greenwich Observatory was established in England, near London, for the express purpose of determining time and the accurate positions of stars, the sun, the moon, and the planets for use in navigation.

The position of a place on the earth requires the calculation of its latitude and longitude. If both are known, the location is uniquely determined.

Astronomical latitude was defined in Chapter 5 as the angle between the plane of the terrestrial equator and the direction of gravity. The celestial meridian is that great circle through the celestial poles and the zenith of a given place. The terrestrial meridian is one half of a great circle concentric with the celestial meridian, but on the earth itself. Since all celestial meridians converge at the celestial poles, so also the terrestrial meridians pass through the terrestrial poles. The meridian through Greenwich, called the prime meridian, was selected as the fundamental meridian from which longitude is measured. Thus *longitude* is defined as the angle between a given meridian and the prime meridian.

For example, since the longitude of San Diego is 117°, the angle between its meridian and the prime meridian is 117°. This angle may be measured either as a spherical angle at the pole or as an arc of the equator. Longitudes east or west of Greenwich are measured up to 180°. The longitude of Rome is 12° 5′E. New York is 74° 0′W. There is no point on the earth except New York at longitude 74° 0′W and latitude 40° 25′N. Figure 2-3 shows a point at latitude 40°N, longitude 100°W.

It has been shown in Chapter 2 that the altitude of the celestial pole is equal to the astronomical latitude of any place of observation. This suggests a simple way of determining latitude. However, there is no bright star exactly at the pole, so latitude cannot be determined directly. But it is possible to determine latitude from two altitudes of Polaris when the star is observed on opposite sides of the pole. The average of these two altitudes is the altitude of the pole and hence equal to the latitude. For these observations Polaris, or any other star in the neighborhood must be observed at intervals of 12 hours, once on each side of the pole.

Table 6-1 gives the declination and angular distance of Polaris from the pole for various dates. The reason for the change of declination will be discussed later in this chapter.

TABLE 6-1   Declinations of Polaris

date	declination	polar distance	date	declination	polar distance
1900	88° 46′	1° 14′	1960	89° 05′	0° 55′
1910	88 50	1 10	1970	89 08	0 52
1920	88 53	1 07	1980	89 11	0 49
1930	88 56	1 04	1990	89 14	0 46
1940	88 59	1 01	2000	89 17	0 43
1950	89 02	0 58	2010	89 20	0 40

6.2 EFFECT OF REFRACTION

There is one correction that must be made to the observed altitude. The air above the earth bends (refracts) the light from the star in such a way that the star appears to be higher in the sky than it would if the earth had no atmosphere. The amount of refraction, and therefore the amount of correction necessary, depends on the altitude. The change of refraction with altitude is shown in Table 6-2.

TABLE 6-2   Refraction at Various Altitudes

altitude (degrees)	refraction (min of arc)	altitude (degrees)	refraction (min of arc)
0°	34′ 50″	50°	0′ 48.2″
5	9 45	60	33.2
10	5 16	70	20.9
20	2 37.0	80	10.2
30	1 39.5	90	0.0
40	1 08.6		

Refraction increases approximately as the zenith distance. It depends on temperature and barometric pressure. For a more complete table of refraction, an advanced text on navigation or the tables in Bowditch's *American Practical Navigator* should be consulted. The corrections should be applied separately for all observations.

6.3 MERIDIAN ALTITUDES

Equation (2-4) gives the relation between altitude h and declination d of a star and the astronomical latitude L of the place of observation for stars on the meridian south of the zenith. This equation can be restated to find the latitude by a single observation, as follows:

$$L = 90° + d - h \tag{6-1}$$

The altitude h is measured. This gives the apparent altitude, which must be corrected for refraction (see Section 6.2) before the computation of the latitude is made. The declination d is obtained from a table of star positions for the year the observation is made.

As can be shown easily by a figure like Fig. 2-11, for a star between the zenith and the pole the equation is

$$L = h - (90° - d) = h + d - 90° \tag{6-2}$$

For a star between the pole and the horizon,

$$L = h + (90° - d) = 90° + h - d \tag{6-3}$$

All three equations are for the northern hemisphere.

Thus the easiest and most accurate way of determining astronomical latitude is by observing the altitudes of stars as they cross the meridian. Geographical latitude requires a slight correction for station errors. Of course the declinations must be known from the *American Ephemeris and Nautical Almanac* or some other table of star positions. The astronomical transit or meridian circle (see Chapter 9) will show a difference of latitude between two stations only about 10 feet apart.

SAMPLE PROBLEM: On November 26, 1965, at 6 P.M. the altitude of Polaris was measured as 33° 40.0′ and on November 27, 1965, at 6 A.M. its altitude was observed to be 31° 52.9′. In both cases the star was on the meridian. What was the latitude of the place of observation and what was the declination of Polaris?

SOLUTION: (Look at Fig. 6-1). Interpolating from Table 6-2, for altitude 33.7° the refraction is 1′ 28.1″ or 1.5′. For altitude 31.9° it is 1′ 33.7″ or 1.6′. The corrected altitudes are 33° 38.5′ and 31° 51.3′. The mean is 32° 44.9′, which is the latitude.

The polar distances of the star are the differences of their altitudes from the mean or 0° 53.6′. This is easily shown from a figure. The declination equals

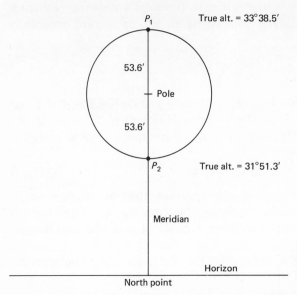

FIG. 6-1 Determination of latitude by observations of Polaris. Altitudes have been corrected for refraction.

90° minus the polar distance, or 89° 6.4′. This can be checked from Table 6-1, which gives 89° 6.5′ for the average between 1960 and 1970.

SAMPLE PROBLEM: Sirius was observed on the meridian of the observatory in the preceding problem on November 26, 1965. The position of Sirius was: R.A. 6^h 41^m; Decl. $-16°$ 36′. What was the latitude of the observatory if the star's measured altitude was 40° 40′?

SOLUTION: Correct the measured altitude for refraction from Table 6-2. Since the declination and altitude are given only to the nearest minute, use 1′ from the table for the refraction. Hence the corrected altitude was 40° 39′.

Substituting in equation (6-1), $L = 90° + d - h$,

$$L = 90° + (-16° \ 36′) - 40° \ 39′$$
$$= 90° - 16° \ 36′ - 40° \ 39′ = 90° - 57° \ 15′ = 32° \ 45′$$

as in the preceding problem.

6.4 TIMEKEEPERS

The determination of the other coordinate, longitude, is based on its relation to time. *Time* is the measure of the instant when an event occurs, or it is the interval between two events. For example, it is apparent noon when the sun is highest in the sky, or it is 8 A.M. 4 hours before noon. Today, time can be obtained by observation to a small fraction of a second and it may be indicated by clocks that have an accuracy of a millisecond (0.001 second) or less.

Time has been kept by astronomers for thousands of years. At first its determination and the time services were in the hands of the soldiers or of the priests, but later the responsibility was given to astronomers in their observatories because they had the instruments necessary for its accurate determination.

Centuries ago the day was divided into hours and the night into watches, the latter because it was necessary to keep watch for security. The Hebrews had three equal night watches, the Greeks five, and the Romans four. Timekeeping in the daytime was fairly simple by observing the sun. The Egyptians developed a system based on the direction and motion of shadows. A stake called a *gnomon* was set vertically in the ground. When the shadow cast by the stake was shortest, it was noon. The hours could be marked off on a horizontal area or a plate, which was graduated into equal divisions. The sundial, the gnomon, and the 12 divisions of the day came into Greece from Babylon.

The shadow at noon points to the north in the northern latitudes and to the south in southern latitudes. The Egyptians eventually set up ornate obelisks and sundials for timekeeping. Figure 9-1 shows an Egyptian obelisk, now in New York City. Naturally this method was fine for clear days, but it would not work on cloudy days or at night. So sandglasses and, later, water clocks, or clepsydras, were built that could be used at all times. Sandglasses had to be turned over at the end of an hour and hence were called hourglasses. The water clock had to be kept filled, but an automatic reservoir to maintain constant level was later invented. Also, candles that burned at a given rate were marked into equal intervals to show the hours.

The Egyptians also used a series of stars at night as time indicators. In the Great Pyramid there was a "grand gallery," which was open to the sky before the pyramid was capped and finished. The hours were determined by observing the passage of certain stars across the opening of the gallery, but an intricate table of correlating data was necessary because of the changing position of the stars with respect to the sun during the course of a year.

The first mechanical clocks were run by weights. But there was no satisfactory way of regulating the rate at which these mechanisms ran, so time by these clocks was not very accurate. In 1360 there were only about 20 mechanical clocks in Europe, most of them in churches and monasteries. Timekeeping was in the hands of the priests in Egypt and in Europe for the purpose of announcing the times of the religious offices and services. Clocks were not used in public buildings until about 1450.

6.5 THE DAY AND THE YEAR

As the earth rotates, all the visible stars appear to cross the celestial meridian. This crossing is called a *transit*. An instrument set up so stars can be seen only when they are on the meridian is also called a transit. A meridian circle is a transit with a slightly larger and more accurate circle for measuring altitudes. If a clock is set to run perfectly and to read zero when the vernal equinox is at the center of the transit (that is, when it is on the meridian), the length of time between successive transits of the vernal equinox is called one *sidereal day*. Time by the clock is called *sidereal time*. Since the meridian is local, each place has its own sidereal time. In this system the vernal equinox is the timekeeper and the earth is the timepiece.

For civil affairs it is customary to use the sun as the timekeeper. The interval of time between successive transits of the sun is called a *solar day*. When the sun is on the meridian, it is said to be *apparent noon*. If time is referred to a given meridian, the sun transits the local meridian at local apparent noon. The hour angle of the apparent (real) sun is *local apparent solar time,* and is abbreviated L.A.T. or L.A.S.T. Each meridian therefore also has its own local solar time.

Both solar and sidereal days are divided into hours, minutes, and seconds, but the lengths of the two kinds of days are unequal.

Since the earth revolves around the sun, the difference between the times of transit of the vernal equinox and the sun across each meridian change continually. If both are on the meridian on March 21 at noon, on March 22 the sun will have moved about 1° to the east and will transit the meridian after the vernal equinox does. Each day the sun moves farther to the east and the vernal equinox crosses the meridian earlier by solar time. In one year it gains an entire lap of 360° or 24^h over the sun. In other words, there are about 365 solar days and 366 sidereal days in a year. The gain of a reference star on the vernal equinox over the sun is shown in Fig. 6-2.

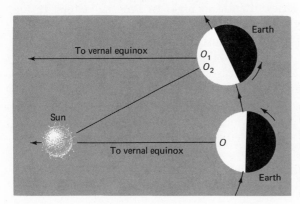

FIG. 6-2 For an observer at O, the sun and the vernal equinox are in conjunction. One sidereal day later, the observer is at O_1, because the earth has made a complete rotation. But the sun is not overhead until the observer arrives at O_2. The interval of time from O through O_1 to O_2 is one solar day.

Stated in another way, the earth requires about 4 minutes longer to complete one rotation with respect to the sun than it does with respect to the equinox. More accurately, a sidereal day of 24 sidereal hours is equal to only 23^h 56^m 4.091^s of solar time. Or, the sidereal day is about 236 seconds shorter than the average solar day. One sidereal second = 1/86,400 sidereal day = 0.9972 average solar second.

According to Kepler's second law, the earth's rate of speed around the sun is variable. Also the sun does not appear to move along the celestial equator. When the ecliptic and equator are shown on a flat surface as in Fig. 6-3 with

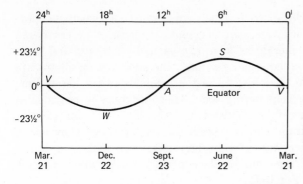

FIG. 6-3 The ecliptic and celestial equator projected on a flat surface. V = vernal equinox; A = autumnal equinox; W = winter solstice; S = summer solstice.

the equator as a straight line, the ecliptic is a curve. This figure conforms to the directions on the star maps on the end papers in this text. Since the equator and ecliptic are great circles, their circumferences on the celestial sphere are equal, although they do not appear so when projected onto a flat surface. Even if the sun moved along the ecliptic at a uniform rate, its apparent motion relative (parallel) to the equator would not be uniform.

For these reasons, the solar days are not all equal in length. Table 6-3 shows the length of the apparent solar day at four times of year. The days are longer in both January and July because in each of these months the sun is near one of the solstices, where its apparent speed relative to the equator is greater, thus increasing the distance between O_1 and O_2 shown in the figure. The other effect, the earth's variable speed, causes the days to be longer in January than in July, because the earth is at perihelion in January.

TABLE 6-3   Variation in Length of Apparent Solar Day

date	length of day		
January 1	24^h	00^m	29^s
April 1	23	59	42
July 1	24	00	12
October 1	23	59	41

6.6 MEAN SOLAR TIME

For the same reasons the lengths of the seconds of apparent solar time are also variable. It would be difficult to build a clock to keep this kind of time. So it has been the custom for many years to use a fictitious sun, the *mean sun*, by which our clocks can be regulated. The mean sun is an imaginary sun, an average sun, that is assumed to move eastward along the equator at a uniform rate such that it makes a complete circle around the sky in exactly the time it takes the apparent sun to move around the ecliptic. The hour angle of the mean sun is called *local mean time*. It is the interval of time since the mean sun was on the local meridian.

Because each meridian on the earth has its own mean time, when the railroads became important in the American economy it became necessary to standardize the time. Standard meridians were chosen and each city and town in a zone kept the same time, called *standard time*. The standard meridians are spaced 15° apart, beginning at Greenwich, and a standard zone keeps time exactly 1 hour from that in the neighboring zone. The four zones in continental United States are the 75th meridian for Eastern time, the 90th for Central time, the 105th for Mountain time, and the 120th for Pacific time. They are 5, 6, 7, and 8 hours from Greenwich, respectively. However, some individual states or parts of states adopt zone times out of their normal zones, which became somewhat irregular, as shown in Fig. 6-4.

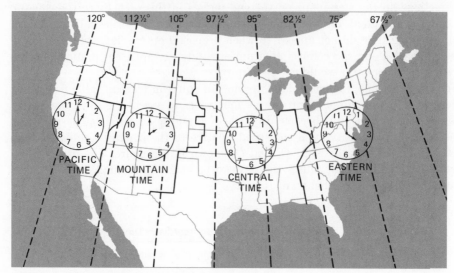

FIG. 6-4   Time zones in the continental United States.

When daylight saving time is used, now specified by law from the last Sunday in April to the last Sunday in October, each zone using daylight time adopts the standard time of the zone adjacent to the east. That is, Central time adopts Eastern Standard Time, Mountain time becomes Central Standard Time, and so on for all time zones.

It is also apparent that if a person sailed or flew around the earth, his date on completion of the journey would not agree with that of the place he had sailed from. For example, if a plane flew westward from New York on Monday on a round-the-world trip, and if it took 24 hours, the pilot would need to set his clock back 1 hour for every hour of the trip to agree with the time zones over which he was flying. When he got back to New York, his clock would show no elapsed time but it would be Tuesday on landing. To avoid this, it is agreed to omit one day when traveling westward—that is, to turn the calendar ahead one day. An imaginary line at approximately longitude 180° is called the *International Date Line*. When traveling eastward across this line, one day must be repeated. The line is somewhat irregular to avoid as many land masses as possible, as shown in Fig. 6-5.

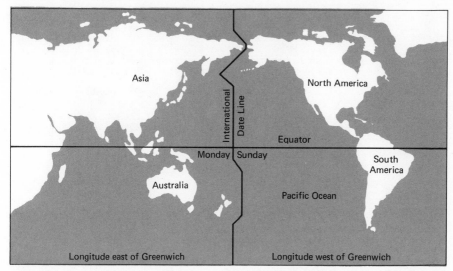

FIG. 6-5 The International Date Line, which is near longitude 180°.

Since the apparent sun moves irregularly along the ecliptic, it does not come to the meridian at equal intervals of time. The mean sun does, which was the reason for inventing this imaginary sun. The difference in time between transits of the mean sun and the apparent sun—that is, between their hour circles—is called the *equation of time*, which is defined as apparent time minus mean time.

Figure 6-6 is a graph of the equation of time during the year. The plus sign means that the apparent sun is ahead of the mean sun and crosses the meridian before the mean sun. Table 6-4 also gives the equation of time for the beginning of each month. Notice from Fig. 6-6 that it is zero four times during the year, about the middle of April and June, in early September and late December. At those times the two suns are on the meridian at the same time, although their declinations are different.

On many sundial plates and terrestrial globes, a figure shaped roughly like a figure 8 is shown. It is called an analemma and shows the sun's declination and the equation of time throughout the year. Some sundials incorporate the analemma in their construction (Fig. 6-7).

From the graph and the table, it can be seen that in December the apparent sun is ahead of the mean sun until late in the month. The equation of time changes from $+11^m$ on December 1 to -3^m on January 1. Since the earth is traveling faster than average, the apparent sun gains on the mean sun, moving eastward by 14 minutes with respect to the mean sun during the month. The shortest day, when the sun is above the horizon for the shortest time, comes about December 21, at the time of the winter solstice. Most people think that the sun sets earliest and rises latest at that time. This is not the case. The sun sets earliest about December 5 and rises latest about January 5. The reason

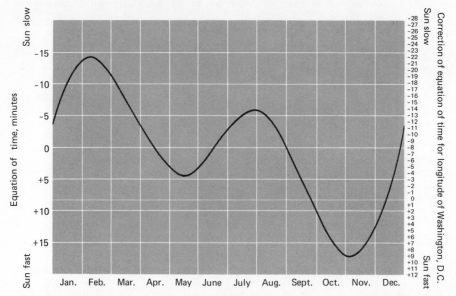

FIG. 6-6 Variation of the equation of time in one year. It must also be adjusted for a difference of longitude as shown at the right for Washington, D.C.

is that the real sun is moving eastward faster than the mean sun, as shown by the equation of time. This motion cancels the effect of shortening evenings, and after December 5 the sun sets later than it would otherwise. For the same reason, the sun continues to rise later until about January 5. These dates vary because of a latitude effect, which is greater in Canada and less in Mexico.

TABLE 6-4   The Equation of Time

date	appt. − mean	date	appt. − mean
January 1	− 3^m 21^s	July 1	− 3^m 31^s
February 1	−13 36	August 1	− 6 14
March 1	−12 35	September 1	− 0 11
April 1	− 4 08	October 1	+10 05
May 1	+ 2 53	November 1	+16 19
June 1	+ 2 27	December 1	+11 09

6.7 DETERMINATION OF LONGITUDE

Celestial and terrestrial meridians are concentric circles. Thus the longitude of Washington, for example, can be determined by measuring the angle between either its celestial or terrestrial meridian and the corresponding meridian of Greenwich. Sidereal time is equal to the right ascension of the celestial meridian and can be determined by observing the transit of a star whose right ascension

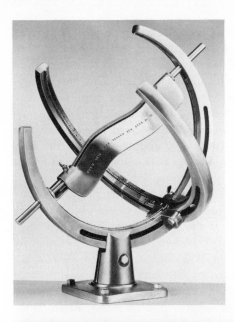

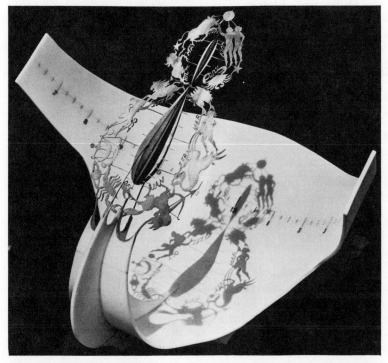

FIG. 6-7 Sundials in which the correction for the equation of time is built into the gnomons. In the dial at the top, the wavy gnomon is rotated until the sunlight passes through the slot and falls on the time scale. The lower dial has an analemma-shaped gnomon surrounded by sculpted signs of the zodiac. It is a calendar as well as a timekeeper. *(Sunquest sundial, designed and constructed by Richard L. Schmoyer; Landisville, Pa.; courtesy Gilroy Roberts; Franklin Center, Pa.)*

is known, as previously described. The difference between the sidereal times at Washington and Greenwich is equal to the difference of their longitudes.

Suppose it was agreed in advance to observe certain stars on the same night as they crossed the meridians of Greenwich and Washington. This would happen when the sidereal clocks at each place read the same—for example, 6^h 41^m for Sirius—but each star would cross the meridian of Greenwich some 5 hours before it reached the meridian of Washington. Suppose a signal is sent to Washington when Sirius is on the meridian at Greenwich, and that 5^h 08^m 16^s later Sirius is on the meridian at Washington. This interval is equal to the longitude of Washington. Any star bright enough to be observed with the instruments at the two observatories can be used to determine the difference of longitude.

The meridian of Washington serves as the standard meridian for the determination of longitudes in the United States. If the time difference between transits of stars is determined for each observatory, the longitude of each is found by adding the time difference to the longitude of Washington. The accuracy of the determination of longitude is therefore exactly the same as the accuracy of determination of time—that is, to about 0.001 second or less.

Longitude is expressed either in hours or degrees and fractions. From Table 6-5, the conversion from time to arc, or the reverse, is possible to whatever accuracy is desired. On maps it is customary to use degrees. In astronomy, longitude is usually expressed in units of time.

TABLE 6-5   Relation between Time and Angle

24 hours = 360° of arc	$1°$ = 4 minutes of time
1 hour = 15°	$1'$ = 4 seconds
1 minute = 15′	$1''$ = 1/15 second
1 second = 15″	

To recapitulate, there are several kinds of day, each dependent on a different timekeeper:

1. *Sidereal day:* the interval between successive transits of the vernal equinox. Length 24^h of sidereal time, 23^h 56^m 4.091^s of mean solar time.
2. *Mean solar day:* the interval between successive transits of the mean sun. Length: 24^h of mean solar time.
3. *Apparent solar day:* the interval between successive transits of the apparent sun. Length: approximately 24^h of mean solar time; variable in length.
4. True period of earth's rotation: 23^h 56^m 4.099^s of mean solar time.[1]

[1] The period of the earth's rotation is with respect to the stars. Sidereal day is rotation of the earth with respect to the vernal equinox. Difference: 0.008^s.

6.8 THE MOTIONS OF THE EARTH

The earth turns about an axis. This motion is called *rotation*. It also moves in an orbit around the sun in accordance with Kepler's laws. The orbit is an ellipse and the sun is at one focus. This motion is called *revolution*. As the earth rotates and revolves, its axis remains nearly parallel to itself. The time of rotation is slightly longer than one sidereal day. The period of revolution, from one position in the orbit back to the same position, is called one *sidereal year*. Its length is 365^d 6^h 9^m 10^s of mean solar time or 365.25636 mean solar days. The sidereal year may also be defined as the length of time between two successive conjunctions of the sun with a star as seen from the earth, or two successive conjunctions of the earth with a star as seen from the sun. (The motions of the stars in one year are negligible.)

The stability of the earth during its revolution is characteristic of rotating bodies, the inertia of their moving parts resisting a change of direction, as well as speed of rotation. However, the orientation of the axis will change if it is subjected to a twisting force, or torque. The toy top is a familiar example. The weight of a top that is leaning away from the vertical exerts a torque tending to topple it. But when it is spinning, the top does not fall; instead, the combined inertial reaction of all the top's whirling parts moves the axis at right angles to the force producing the torque. As a result, the axis moves in a cone around the vertical line passing through the pivot point of the top. This is a conical motion called *precession*. The period of precession depends on the mass and the rate of spin. The top would precess at the same angle indefinitely if friction did not slow it down.

The earth is subjected to a torque, tending to change the direction of its spin axis. This is caused by the gravitational pull of the sun, the moon, and the planets on its equatorial bulge. Figure 6-8 illustrates the origin of the torque resulting from the moon's gravitation; the bulge is greatly exaggerated and shown superposed on the spherical portion of the earth.

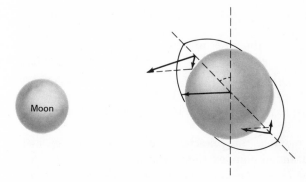

FIG. 6-8 Precessional force of the moon's gravitation on the oblate earth.

The moon tries to pull the earth's equator into the plane of the moon's orbit; the sun tries to pull the equator into the plane of the ecliptic. These two planes are inclined to each other by about 5°. Also the moon's orbit turns around the earth in a period of 18.6 years. The planets are too far away to have any appreciable effect. The result is that the earth's poles turn around the pole of the ecliptic, a line perpendicular to the ecliptic, in a period of 25,900 years. The moon's effect causes an additional periodic irregularity, called *nutation* or nodding of the earth's poles. This appears as a wave of small amplitude on the precessional circle in a period of 18.6 years. The precessional circle is shown on the map in Fig. 3-3, but the nutational effect is too small to show. (See also Fig. 6-9.)

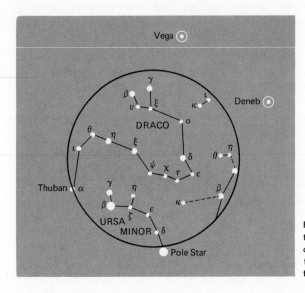

FIG. 6-9 The path of the north celestial pole among the stars due to precession. The pole of the ecliptic is at the center of the circle. The radius of the circle is (about) 23½°.

Since the equatorial plane is tipped about 23½° to the plane of the ecliptic, the celestial pole is also tipped 23½° to the pole of the ecliptic, which is located at right ascension 18 hours and declination +66½°. The precessional circle is 23½° in radius. There is a similar circle of the same size whose pole is at right ascension 6ʰ and declination −66½°.

Precession has been known since the time of Hipparchus. When the Great Pyramid was built in Egypt, the celestial pole was near the star Thuban, α Draconis, a faint star in the Tail of the Dragon. At the time of Hipparchus, Thuban was about 11° from the celestial pole. At the present time, it is about 25° away. It would be better to say that the pole is 25° from Thuban, since it is the pole that moves.

Since the celestial pole is moving, the celestial equator must change also. The vernal equinox and the autumnal equinox slide westward along the ecliptic. This is called the *precession of the equinoxes*. The period is also 25,900 years, in which period of time the equinoxes move completely around the ecliptic at a rate of 50″ per year. The equator changes position with respect to the stars, and the celestial coordinates of all stars change continuously.

Figure 6-10(a) shows the earth in the center of the imaginary celestial sphere. The ecliptic, the celestial equator, and the precessional path of the north celestial pole are shown on the sphere. The vernal and autumnal equinoxes are the two points where the ecliptic and the celestial equator intersect. S_1 and S_2 are two stars located near the equinoxes.

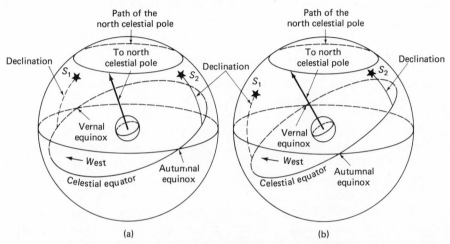

FIG. 6-10   The effect of precession on right ascension and declination near the vernal equinox and the autumnal equinox. The westward movement of the equinoxes and of the north celestial pole is shown in (b).

To illustrate the effect of precession on these coordinates, Fig. 6-10(b) shows the situation after a period of years, when the north celestial pole has moved west because of precession. Stars S_1 and S_2 are still at the same points on the sphere, but because of precession the equinoxes have moved westward along the ecliptic. As a result, the right ascension has increased for each star. Because of the change in the orientation of the equator, the declination of S_1 has increased while that of S_2 has decreased. That is, declinations increase near the vernal equinox and decrease near the autumnal equinox. The coordinates of other stars change also, the amounts and directions depending on the location of the stars.

6.9   THE SIDEREAL AND THE TROPICAL YEAR

The *sidereal year* is the interval between successive passages of the sun through any point on the ecliptic. Its length is 365.25636 mean solar days. The *tropical year* is the length of time between successive passages of the sun through the vernal equinox. But because of precession, the vernal equinox moves westward and the tropical year is shorter than the sidereal year by the time it takes the sun to move along the ecliptic from the vernal equinox to the position of the equinox at the beginning of the following year, about 50″. This takes about 20 minutes of time. Thus the length of the tropical year is only 365.24220 mean solar days. This is the year on which our calendar is based.

To recapitulate: Precession is caused by the rotation of the oblate earth and the pull of the sun, moon, and planets on the equatorial bulge.

The effects of precession are: the changing position of the poles and the resulting westward motion of the equinoxes; the changing right ascensions and declinations of all stars; and the difference in the lengths of the sidereal and tropical years.

6.10 PROOFS OF THE ROTATION OF THE EARTH

We noted in Chapter 4 that the ancient astronomers assumed the earth to be stationary, with seven planets and the stars rotating around it. In the Copernican theory the earth was assumed to be in rotation. Later it was found that under certain conditions bodies in motion with respect to the earth appear to be deflected from the motion predicted by Newton's laws. This deflection is a result of the earth's rotation. It is called the *Coriolis effect*. We will now consider some of the motions where this effect is observed, and which constitute a proof of the earth's rotation.

1. *Deflection of projectiles.* The Coriolis effect is apparent in the trajectories of projectiles that travel long distances in a northerly or southerly direction. The gun and the projectile have eastward velocity because of the rotation of the earth. The projectile retains its eastward velocity as a consequence of its inertia while in flight. If it were fired toward the south from the northern hemisphere, it would move toward a part of the earth moving faster than the eastward component of its velocity. It would therefore appear to deviate toward the right. Similarly, if fired northward it would move toward a more slowly moving part of the earth and would again appear to deviate toward the right. (See Fig. 6-11.) In the southern hemisphere the effect is reversed.

2. *The Foucault pendulum.* A pendulum hung at a terrestrial pole would

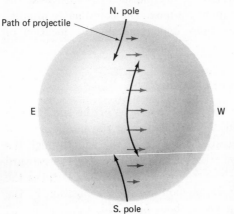

FIG. 6-11   Because of the earth's spherical shape, the speed with which the surface is moving around the axis varies, depending on the latitude. The speed is greatest at the equator and decreases toward the poles, as indicated by the horizontal arrows. This causes an apparent deflection in the trajectory of projectiles fired in a northerly or southerly direction.

continue to swing in the plane in which it was started swinging. The earth would appear to rotate under it. It would seem to an observer that the pendulum was moving, rather than the earth, because he is used to thinking of the earth as a reference. The pendulum would appear to turn toward the right at the north pole and to the left at the south pole.

At the equator there would be no apparent deviation. At other latitudes, the rate would depend on the latitude. At the poles the period of rotation is 23^h 56^m. In the latitude of Minneapolis it is 34 hours and in San Diego, 44.3 hours. The Foucault pendulum was first demonstrated in Paris in 1851 by the French physicist Jean Foucault (1819–1868). It is now demonstrated in many planetariums and museums.

3. *Falling bodies.* A weight dropped from a tower deviates toward the east, since at a higher level it has greater eastward speed than at lower levels and it retains the excess eastward motion during its fall.

4. *The motions of artificial satellites.* Satellites launched from Cape Kennedy make successive passages to the west because of the earth's rotation. These satellites are usually fired with an easterly component to take advantage of the earth's eastward rotation.

6.11 PROOFS OF THE REVOLUTION OF THE EARTH

Tycho Brahe would not accept the Copernican theory of a moving earth. Instead, he proposed the theory that the earth is stationary, that the sun moves around the earth, but that the planets move in orbits around the sun. He reasoned that if the earth were in motion, it should be possible to detect retrograde motions of stars when they are in opposition to the sun. This would be true if the stars were as close as the planet Saturn, the most distant planet known in Tycho's time. The length of Saturn's retrograde loop is about 12°, an angle that could be measured easily. If a star were 12 times farther away than Saturn, its loop would still be 1° long. Tycho could find no evidence of such motion. We know now that the stars are much more distant, the nearest star being some 27,000 times the distance of Saturn.

At the present time it is possible to measure changes of position of the nearest stars with respect to the background stars, but a long-focus telescope is needed. Photographs taken at the proper times are measured under microscopes with special measuring machines, something Tycho did not have. The largest apparent shift of position is only about 1.5″, or only 0.13 mm for a telescope 60 feet long. Angles as small as 0.005″ have been measured.

This apparent motion of the nearest stars is due to the motion of the earth. It can be used to determine the distances of 2000 or more stars. This is one proof that the earth is moving through space and is not stationary. The first *stellar parallax,* as it is called, was measured in 1838.

A second proof of the revolution of the earth was discovered in 1727. It is called the *aberration of starlight.* To illustrate: Suppose it is raining, with the rain coming straight down. An umbrella held upright will keep off the rain. But everybody is familiar with the fact that a moving person must tip his umbrella in the direction in which he is walking or running to keep the rain from his clothes. The greater the speed, the greater must be the angle of tip.

Light travels at a speed of 186,000 miles per second; the earth's speed in its orbit is about 18.5 miles/sec, almost exactly 0.0001 the speed of light. Hence, if an observer wishes to see a star in his telescope, he must tip it a slight amount to compensate for the motion of the earth. The amount of tip is equal to 0.0001 radian or 20.5″, as shown in Fig. 6-12. This is because as light comes down the telescope tube, the lower part of the tube has moved during the time the light travels from the upper to the lower end.

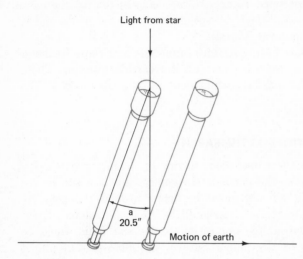

FIG. 6-12   Aberration of starlight. A telescope must be tipped 20.5″ in the direction of the moving earth.

These two methods are entirely different but give the same result. They confirm the assumption that the earth is in motion around the sun.

6.12 THE CALENDAR

Herodotus said that "the Egyptians by their study of astronomy discovered the solar year and were the first to divide it into twelve parts. . . . Their method of calculation is better than the Greeks, for the Greeks to make their seasons work out properly, intercalate[2] a whole month every other year, while the Egyptians make the year consist of 12 months of 30 days each and every year intercalate five additional days, and so complete the regular circle of the seasons."

The Egyptians divided their year into three seasons: Inundation, Winter, and Summer. So it was actually an agricultural year based on the rise of the Nile River and the resulting seasons of growth and nongrowth.

The Hebrew calendar is based on the phases of the moon. Their year has

[2] The word intercalate means to insert (as a day or a month) into the calendar.

12 months of what should be 29½ days each, which come to 354 days in a year. This means that extra months have to be added every two or three years. The months begin when the new moon is first visible in the western sky after sunset. The weeks consist of six days followed by a seventh, the Sabbath. The ancient calendar consisted of four seasons, probably borrowed from the Greeks, since the eastern Mediterranean region knows only two seasons: a hot, dry summer and a cold, wet winter. The lunar calendar is very irregular because of the incommensurability of the phases of the moon with the motions of the sun. Also, the new moon is seen in varying positions with respect to sunset and changes the interval of time at which the moon can be seen after it passes the sun.

When Julius Caesar came to power in Rome, the Roman calendar of 12 months had fallen into a sad state, with spring in December. Caesar with the help of Sosigenes, a Greek in Alexandria, reformed the calendar by making the year 365¼ days in length. That is, he went back to the common year of 365 days, with a leap year every four years, as in the Egyptian calendar. This calendar is called the Julian calendar. It went into effect on January 1, 45 B.C. Table 6-6 gives the origin of the names of the months.

TABLE 6-6   Origin of Names of Months

January:	Janus, the two-faced god
February:	from Latin *Februarius*, feast of purification; month of sacrifices
March:	Mars
April:	(unknown)
May:	Maius Jupiter, the great (god)
June:	a Roman clan
July:	Julius Caesar
August:	Augustus Caesar
September, October, November, December:	7th to 10th months in the old Roman calendar

Since the tropical year (365.2422 days) is a little shorter than the 365¼ days of the Julian calendar, spring came progressively earlier by the calendar. Near the end of the 16th century, spring began on March 11. Pope Gregory XIII in 1582 signed a decree dropping 10 days. The day after October 4, 1582, became October 15. This brought spring back to March 21. The Gregorian calendar was not universally adopted. England adopted it in 1752. Russia adopted a similar calendar after the revolution in 1917.

The Gregorian calendar is based on the tropical year. Each common year has 365 days with a leap year every fourth year, as in the Julian calendar. But leap years are omitted in the century years not divisible by 400. The years 1800, 1900, 2100, and so on are not leap years; but 2000 will be. Thus the length of the average year of the Gregorian calendar is 365.2425 days, as follows:

$$365 + \tfrac{1}{4} - \tfrac{1}{100} + \tfrac{1}{400} = 365 + 0.2500 - 0.0100 + 0.0025 = 365.2425 \text{ days}$$

Therefore the year is too long by 365.2425 − 365.2422 or 0.0003 day, or three days in 10,000 years.

The Russian calendar is the same as the Gregorian calendar until 2800, which will be a leap year in the Gregorian but not in the Russian calendar. The error in the Russian calendar is one day in 44,000 years.

The Gregorian calendar is not entirely satisfactory for several reasons. The months are not of equal length and the year begins on a different day in successive years. Christmas, for example, may come on any day of the week. Easter is a movable feast day and always falls on Sunday. The date of Easter is computed by a complicated formula, approximately as follows:

Easter falls on the first Sunday after the first full moon on or after March 21. The earliest date is therefore March 22 and the latest is April 25. If the first full moon after March 21 falls on Sunday, Easter is the following Sunday.

In the United States, Labor Day is always on Monday and Thanksgiving is always on Thursday. Some other movable holidays have now been fixed on Monday.

The Jewish Passover and other religious festivals are celebrated according to the lunisolar calendar and are therefore on different dates by the Gregorian calendar. For example, both Passover and Yom Kippur come at the time of the full moon.

Astronomers use a perpetual calendar that has many advantages. There are no weeks or weekdays. Each day is numbered and any date is merely the number of the day since January 1, 4713 B.C. This scheme includes all the dates of recorded history. It is therefore possible to calculate the number of days between two events by a simple subtraction. This is very useful in periodic events, such as the study of eclipses and periodically varying stars. This calendar is called the Julian-day calendar. Each day begins at noon Greenwich Mean Time (GMT). Lists of numbers of days are published and are readily accessible to astronomers.

QUESTIONS AND PROBLEMS

Group A

1. Is it possible for an observatory to measure the altitude of Polaris at 6 P.M. and 6 A.M. on the same night and get the same value? Explain.
2. When would a gnomon cast no shadow at noon in latitudes (a) 0°, (b) 23½°N, and (c) 23½°S? *Answer:* (a) March 21 and September 23.
3. If you had a solar clock with a 24-hour dial, starting with 0^h at midnight, during what month would there be a time when the solar clock would agree with the sidereal clock?
4. During what month are the solar days longest? Explain.
5. From Fig. 6-4, find the locality in the United States that has the greatest difference between standard time and local time. Estimate the difference.

6. Sirius is on the meridian of Washington 5^h 08^m 16^s and on the meridian of Anchorage 9^h 59^m 36^s after it crosses the meridian of Greenwich. Calculate the longitude of each. *Answer:* Washington 77° 04'W.

7. Why is the declination of Polaris changing?

8. From Fig. 6-9, estimate when the north celestial. pole will be nearest the bright star Deneb. About how many degrees apart will they be at that time?

9. What would be the disadvantage of having a calendar based on the sidereal year rather than the tropical year?

10. Explain why an object dropped from a high tower on the equator will strike the earth slightly east of the base of the tower.

11. Is the aberration of starlight greater on Mercury or on Mars? Why?

12. Show why the latest possible date of Easter is April 25.

Group B

13. What are the lengths of the longest and shortest shadows cast by a 10-foot pole at local noon (a) at latitude 30°N and (b) at your latitude? *Answer:* (a) 13.5 feet and 1.1 feet.

14. During 1950 an observatory measured the altitude of Polaris on the meridian at 5:30 P.M. and 5:30 A.M. The measured altitudes were 19° 59.1' and 21° 54.9'. Compute the latitude of the observatory and the declination of Polaris. (*Note:* Correct for refraction.)

15. How would time kept by a pendulum clock on Mercury compare with the time it would keep on the earth? Would a wristwatch that keeps time by an oscillating balance wheel driven by a spring be similarly affected? Assume gravity on Mercury to be 0.4 g.

16. If the earth rotated at the same rate but in the opposite direction, (a) how many sidereal days would there be in a sidereal year, (b) what would be the length of a sidereal day in solar time?

17. How do the right ascensions and declinations of stars change, owing to precession, near (a) the solstices (b) the celestial poles?

18. At local apparent noon, what is the (a) Pacific Daylight Time in San Francisco (long. 122½°W) on July 15, (b) Eastern Standard Time in Boston (long. 71°W) on November 20? Give the answer to the nearest minute. *Answer:* (a) 1:15 PDT.

19. Compute the length of a regressional loop of a star 27,000 times the distance of Saturn from the sun.

20. From Table 13-1, (a) what is the orbital velocity of Mars, assuming a circular orbit? (*Hint:* The earth's orbital velocity is 18.5 miles/sec.) (b) What is the aberrational constant for Mars? *Answer:* (a) 15 miles/sec.

21. If calendars were devised for Jupiter, Mars, and Venus, how

many days would there be in a year in each case? See Table 13-1. *Answer:* Venus: about 1.9 days.

22. Assuming the aberrational constant to be 20.5″, compute the distance from the earth to the sun.

7 the nature of light

It is necessary to study the nature of light in order to understand the nature of celestial bodies. Except for the moon, some meteoroids, and some particles of the solar wind, the material of the sun, the stars and other bodies in space cannot be analyzed in the laboratory. But the light they give off can be collected and analyzed. Where it came from and what has happened to it on its long journey through space can be deduced from theory.

We must know the fundamental laws of light in order to understand the principles of optics and the construction of telescopes because it is primarily through telescopes and their auxiliary instruments that we can gain knowledge of the universe.

We must also understand how light behaves under all conditions in space, including the interiors of the sun and the stars. What light is, how it is produced, and how it is related to the atom and the particles of which atoms are composed are fundamental to this understanding.

It is desirable to state the fundamental laws of the behavior of radiation and of the production of the spectrum and how it permits us to determine the nature of radiating bodies. From these laws and the study of light and other radiation we can determine such things as the temperature and density of matter in space, and we can measure the velocities at which celestial bodies move in all parts of the universe. Finally, we can attempt to determine the extent of the universe itself.

7.1 REFLECTION

Light striking a polished surface is reflected; that is, it is turned back by the surface in approximately the direction from which it came. This property is used to form the image of an object at the focus of a reflecting telescope by the use of mirrors.

The *law of reflection* has two parts: (1) The *angle of incidence* (*i*) is equal to the *angle of reflection* (*r*). (2) The incident ray, the reflected ray, and the normal (perpendicular) to the surface at the point of reflection all lie in the same plane. That is, the law states:

$$i = r \qquad \qquad (7\text{-}1)$$

This law is illustrated in Fig. 7-1. *MM'* is a reflecting surface, a plane mirror. *RP* is a ray of light, the incident ray, striking the mirror at some angle *i*, the angle of incidence. *PR'* is the reflected ray. *NP* is a line perpendicular to *MM'* at *P*, the point where the reflection takes place.

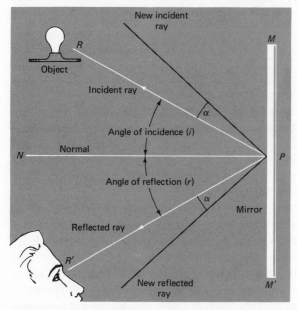

FIG. 7-1   The law of reflection.

The law of reflection was verified experimentally by Alhazen (965–1038), an Arabian scientist, almost 1000 years ago. This law makes it possible to calculate where any ray from an object, such as the electric light in the figure, will travel after it strikes the reflecting surface and therefore where to put the eye of the observer. A luminous object sends out many rays, and more than one ray should be traced. It can be seen that a ray of light emitted at *R* can enter the eye at *R'* without traveling in a straight line between two points. This is the principle of the plane mirror.

It will be noted in the figure that if *i* is increased by the angle *α*, *r* will also be increased by the same angle. Thus

$$i + \alpha = r + \alpha \qquad \qquad (7\text{-}2)$$

Since each ray is increased by the angle *α*, the angle between the incident and reflected rays will be increased by twice the angle (2α). Likewise, if the angle of the mirror is changed instead of the incident ray, the reflected ray will change by twice the angle. This law is important in the construction of the optical instruments of astronomy, such as the reflecting telescope.

The law of reflection is independent of the intensity (brightness) and color of the light as well as the nature and shape of the reflecting surface. The reflected ray is never as intense as the incident ray, since some of the light is lost at the reflecting surface. If the surface is a smooth, curved surface of the proper shape, a beam of parallel light can be brought to a single point, a *focus*. This is a principle that is used in designing a mirror for a reflecting telescope. Each point on a smooth, curved surface can be regarded as a small segment of a plane surface tangent to the curved surface at that point. The law of reflection can be applied separately at each point. If the surface is rough and uneven, the law still holds, but the light is scattered. This scattering is called *diffusion*.

7.2 THE VELOCITY OF LIGHT

There is obviously a difference between the speed of sound and the speed of light. The sound of thunder is heard several seconds after a flash of lightning is seen; the flash of a gun is seen before it is heard. The speed of sound can be timed easily by stationing two men a known distance apart with synchronized clocks that can be used to time the travel of a gunshot report.

Galileo thought the same method could be used for the determination of the velocity of light. He stationed a group of men on a hillside outside the city of Florence in Italy. Another group took a lantern to another hill several miles away. The idea was to uncover one lantern and note the exact time. When the other group saw the light, they immediately uncovered their lantern. The first group then noted the exact time when the returning beam reached them. However, we know now that the time it takes for light to travel a few miles and back is much shorter than the reaction time of the men in uncovering the lanterns. So the method failed.

In 1675 Olaus Roemer (1644–1710), a Danish astronomer, used a more satisfactory method and determined the velocity of light with considerable accuracy. He used observations based on the times of eclipses and occultations of the four large satellites of Jupiter.

The four moons could be seen easily with the telescopes of that day. Roemer compiled a table of their periods of revolution around the planet. He could predict the times each would be eclipsed as it moved into the planet's shadow or occulted when it passed behind the planet's limb. Roemer found that these phenomena occur sooner than expected during part of the year and later than expected during the other part. He correctly inferred that the advance or delay of the occurrence was due to the finite velocity of light. Interestingly enough, Galileo had reached the conclusion that if light is not instantaneous, it must have a high velocity.

Figure 7-2 explains the method. As the earth makes one complete revolution around the sun, Jupiter advances in its orbit by about 30°. The distance between the earth and Jupiter is least when the earth is on the near side of its orbit (at E_1) and increases to a maximum as the earth moves to the far side of its orbit (at E_2). To reach the earth when it is at the maximum distance, the light from the planet and its moons must travel the extra distance across the diameter of the earth's orbit. So if light has a finite velocity, the times of the eclipses should appear later when viewed from the far side than when viewed from the near side. The delay would equal the time required for light to travel across the diameter of the orbit.

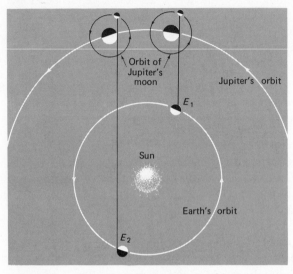

FIG. 7-2   Roemer's method of determining the velocity of light.

Roemer's observations of the times of eclipses and occultations showed that there is a difference of about 16⅔ minutes (1000 seconds) at two dates roughly six months apart. In 1675 the diameter of the earth's orbit was calculated to be 192 million miles (309 million km), which was about 6 million miles too large. Dividing this distance by the number of seconds, the velocity of light came out to be about 192,000 miles (309,000 km) per second. Later results differed slightly. Also it was necessary to allow for the elliptical shape of the earth's orbit and the motion of Jupiter in six months. The modern result, based on the now accepted diameter of 186 million miles is only about 3 percent less than that computed by Roemer, which was amazingly accurate for his time.

Other methods have since been devised to measure the velocity of light with much greater accuracy. The currently accepted value of the velocity of light in a vacuum is 299,792.5 km (about 186,280 miles) per second with an uncertainty of 0.3 km (0.2 miles) per second. The speed is less in material media, such as glass or water. The velocity of light in a vacuum is one of the fundamental quantities in the universe and plays an important role in the theory of relativity. According to this theory, no material object can move faster than the velocity of light.

7.3 WAVE NATURE OF LIGHT

In the early 1800s Thomas Young (1773–1829), an English physicist, and Augustin Fresnel (1788–1827), a French optician and mathematician, performed a series of experiments showing that light exhibits wave characteristics. These experiments dealt with *interference*, a combined effect of light waves from different sources. They found that when a beam of light passes through two pinholes very close together, it will produce a pattern of light and shadow on a screen behind the pinholes.

The behavior of light in this experiment is analogous to that of the water waves shown in Fig. 7-3. The waves, produced by two sticks vibrating in unison on the surface of the water, radiate from each source in ever-widening circular waves. Where the crests of the waves from one source meet those radiated by the other, they combine to produce higher crests. Where the troughs meet they combine to produce deeper troughs. In between these regions of reinforcement are regions where the waves cancel. Here, the crests from one source meet the troughs from the other. These regions of reinforcement and cancellation strike the edge of the water tank in a definite pattern, as can be seen in the photograph.

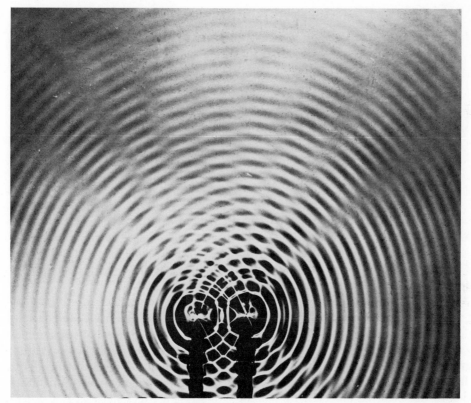

FIG. 7-3   Interference patterns on the surface of water are similar to those produced by the interference of light waves. (D. C. Heath and Company)

This pattern is geometrically similar to the pattern of light and dark areas produced on a screen by light passing through two pinholes. This is strong evidence that light travels in waves similar to water waves. Other experiments have confirmed the wave nature of light.

The lengths of light waves are very small. They can be measured in the physics laboratory by experiments based on the theory of wave motion. The fundamental definitions in wave theory are as follows:

1. *Wavelength* is the distance between the crests or troughs of adjacent waves (see Fig. 7-4).

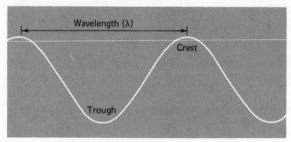

FIG. 7-4 Definition of a wave.

2. *Frequency* is the number of waves passing a given point in a unit of time. For light, the frequency is in cycles per second. The unit of frequency is the hertz (Hz).
3. The *velocity of light* in a vacuum is the velocity with which light travels in kilometers, or miles, per second.
4. The three quantities—velocity, frequency, and wavelength— are related by the following:

Velocity equals frequency times wavelength; that is,

$$V = \nu\lambda \qquad\qquad (7\text{-}3)$$

where V = velocity, ν = frequency, and λ = wavelength.

Another property of light resulting from its wave nature is the manner in which it spreads out when it passes an edge or through a small opening. The result is a very slight bending of the light. This phenomenon is called *diffraction*. The effect is a series of alternate bright and dark fringes, the shape of which depend on the shape of the edge causing the diffraction.

The amount waves are diffracted when passing through an opening is proportional to the wavelength and inversely proportional to the size of the opening. Since light waves are so short, the amount of diffraction is not noticeable unless the opening is very small. Sound waves, for example, are much longer and are greatly diffracted when passing by a corner or through a doorway. If light were diffracted this much, it would be possible to see light around corners. The diffraction of light can be observed by holding two straight edges, such as two pencils, very close together. Between the two will be seen a series of alternating bright and dark bands. They are produced by the interference between the light waves diffracted by each edge.

7.4 THE INTENSITY OF LIGHT

The human eye responds very quickly to light and can detect a very small amount. The photographic plate has the advantage of building up images by long exposures, which the eye cannot do. However, the eye can see faint images the photographic plate would fail to see in the same length of time. But the eye is not a good judge of intensity—the amount of energy per second per unit area—of the light it receives. This ability to differentiate between the brightness of objects is due partly to the eye's sensitivity. The retina is so constructed that its sensitivity increases as the brightness of the light entering the eye decreases.

The eye adjusts to light intensity. When a person leaves a brightly illuminated room and goes out into darkness, the pupil expands until the eye is able to see and distinguish faint objects. For example, only the brightest stars can be seen before the pupil adjusts to the proper size. It may take half an hour for the eye to completely adjust to faint light. In addition to the change of size of the pupil, there are also chemical changes in the retina. Many astronomers claim that the ability to see faint stars can be developed with practice.

Since the determination of the brightness of celestial objects is very important in astronomy, it is necessary to know how light intensity varies with distance. Referring again to the water analogy of the preceding section, it can be seen in Fig. 7-3 that the heights of the wave crests decrease as they radiate from each source in ever-expanding circles. This decrease in the height of each wave occurs because its energy is being spread over increasingly larger areas. Similarly, the energy of a light wave is spread over an increasingly larger sphere as it radiates out from the source. This can be visualized by thinking of a very small light source placed at the center of a sphere with a radius of 1 meter. Each part of the entire surface of the sphere would be illuminated uniformly. If the radius of the sphere were increased to 2 meters, the same amount of light energy would be spread over four (2^2) times as much area, since the area of a sphere is proportional to the square of the radius. This is illustrated in Fig. 7-5.

Each part of a wave travels radially out from the source in a straight line called a *ray*. Four rays are shown in the figure. The area over which the portion of the wave between the rays is spread is shown by squares at distances of 1, 2, and 3 units from the source. It can be seen that, since the rays spread apart both horizontally and vertically, the areas increase as the squares of the distance. This leads to the *law of illumination*, which states: The amount of light reaching an object from a luminous point source decreases as the square of the distance from the light source.

When light from a luminous source reaches a surface, some of it is reflected, some is absorbed by the surface. An airless body, such as the moon, scatters the

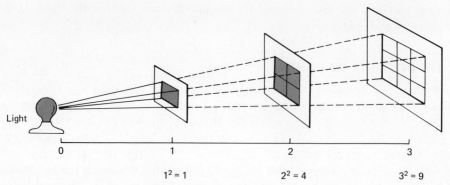

FIG. 7-5 The inverse square law of light. The amount of energy per unit area decreases as the square of the distance from the source.

light. Its *reflecting power* or *albedo* is low because the light is scattered in various directions at random. The amount of reflected light is called the brightness of the object. Its albedo is the ratio of the brightness to the amount of illumination. That is,

$$\text{albedo} = \text{reflecting power} = \frac{\text{reflected light}}{\text{incident light}} = \frac{\text{brightness}}{\text{illumination}}$$

The moon's albedo is only **7** percent. A planet with a dense atmosphere, such as Venus, has a high albedo. Venus' albedo is **77** percent. This can be understood by observing the brilliance of a heavy cloud whose top is illuminated by sunlight. The planets have different albedos depending on the nature of their atmospheres.

7.5 REFRACTION

Light is bent whenever it passes from one transparent medium to another. This phenomenon, called *refraction,* is a result of the difference in the velocity of light in the different materials. It is illustrated in Fig. 7-6, where a ray of light and its associated wave crests are shown passing from air through a piece of glass. Upon entering the glass, the velocity of the waves immediately decreases to two thirds of their velocity in air, and therefore the wavelength decreases. When the rays reenter the air, they immediately regain their former velocity and wavelength. Unless the rays enter the glass perpendicularly, the part of each wave that enters first lags behind the part still in the air, resulting in the bending shown. Likewise, upon leaving, the same part of each wave is the first to reenter the air, gains on the part still in the glass, and thereby bends the ray the other way. It can be seen that the ray is bent toward the perpendicular to the surface when its speed decreases and away from the perpendicular when its speed increases.

The ratio of the velocity of light in a vacuum to its velocity in a transparent medium is called the *index of refraction* of the medium. Table 7-1 gives

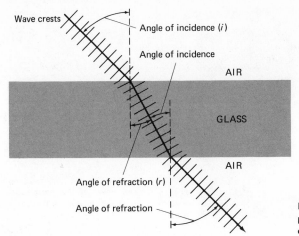

Wave crests

Angle of incidence (*i*)

Angle of incidence

AIR

GLASS

AIR

Angle of refraction (*r*)

Angle of refraction

FIG. 7-6 The refraction of light when passing from one medium to another of different density.

the index of refraction for various transparent substances for ordinary light at room temperatures, except for air, where a temperature of 0°C is used.

Although the index of refraction for air is very nearly equal to that for a vacuum (which is exactly one), there is a significant amount of refraction as light passes from the near vacuum of space into the air around the earth. A star directly overhead is in exactly the same position as it appears to be because its light rays are entering the atmosphere perpendicularly. But a star in any other position looks higher than it really is. The top of the sun appears to be on the horizon at sunrise or sunset, but it is actually below the horizon. This effect makes the days a little longer than they would be if the earth had no atmosphere. At both sunrise and sunset the top of the sun is about 35′ of arc below the horizon when it appears to be touching the horizon. This is greater than the apparent diameter of the sun, which varies between 31′ and 32′.

Refraction at sunrise and sunset increases the length of daylight by about 4 minutes. Refraction must be taken into account by a navigator when the sun or another body is observed for the purpose of determining his position on the earth (see Fig. 7-7).

The atmosphere around the earth is arranged in layers at different heights. It is being stirred constantly by heating and cooling, which produce the winds.

TABLE 7-1 Index of Refraction

Air, dry, 0°C	1.00029
Alcohol, ethyl	1.36
Carbon disulfide	1.63
Carbon tetrachloride	1.46
Diamond	2.42
Glass, crown	1.52
Glass, flint	1.61
Quartz, fused	1.46
Water	1.33

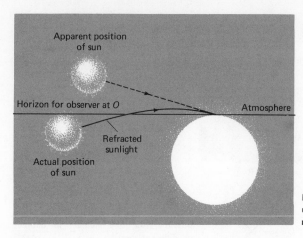

FIG. 7-7   The effect of the earth's atmosphere on the direction of the sun's rays.

The result is variable refraction, making the stars appear to twinkle. Since different colors of light travel through the air with slightly different velocities, the stars seem to change color as they twinkle. This is particularly noticeable with bright stars near the horizon. Variable refraction causes the star images to dance around in the telescope and is very annoying in astronomical observations. Astronomers speak of it as "bad seeing."

7.6 THE ELECTROMAGNETIC SPECTRUM

When white light passes through a glass prism it emerges as a band of color ranging from violet to red, called a *spectrum*. Before Newton's time it was believed that the colors were produced by the interaction of glass and light. Newton found that when the spectrum is passed through a second prism, inverted to the first, the emergent light is white (see Fig. 7-8). This showed that combining the spectral colors produces white light, and Newton correctly concluded that the colors are already present in white light and are separated when passing through the prism.

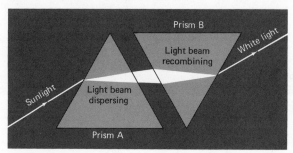

FIG. 7-8   A glass prism will disperse or recombine a beam of light. Two prisms correctly placed in the path of a beam of sunlight will show that white light is a combination of many colored lights.

Later investigation showed that the colors are distinguished by wavelength. Red light has the longest wavelength and violet has the shortest. They are separated upon passing through glass because the index of refraction of glass differs slightly according to wavelength. The indices shown in Table 7-1 are averages. This separation, produced by variation in the index of refraction in different colors, is called *dispersion*. It is troublesome in refracting telescopes, since it interferes with the production of images free from color.

In the middle 1800s James Clerk Maxwell (1831–1879), a Scottish physicist, predicted the existence of electromagnetic waves—combinations of electric and magnetic energy that travel through space at the velocity of light—and concluded that light is one form of such waves. Radio waves were discovered two decades later, verifying Maxwell's prediction. Additional electromagnetic waves have been detected since then. They are similar to light waves in every respect, except that they are of longer or shorter wavelength. The complete electromagnetic spectrum is shown in Fig. 7-9.

Before the discovery of radio waves from cosmic sources in 1931, astronomers were limited to the relatively narrow, optical portion of the spectrum. Practically all the other wavelengths are shielded from the earth by its atmosphere.

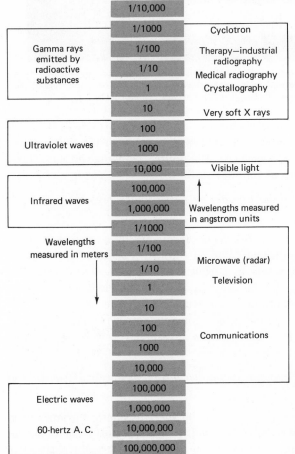

FIG. 7-9 The electromagnetic spectrum. Visible light is only a very small part of the entire spectrum.

Today, cosmic radiation of all parts of the electromagnetic spectrum is being intensively investigated with the aid of space vehicles.

7.7 THE LAWS OF SPECTRUM ANALYSIS

In 1802 William Wollaston (1766–1828), an English chemist, found that when a beam of sunlight was passed through a narrow slit and then through a prism, the solar spectrum was crossed by seven dark lines. He thought that the five strongest lines marked the divisions between the colors in the continuous spectrum. Joseph von Fraunhofer (1787–1826), a German optician, used improved optical equipment and began his study of the solar spectrum in 1814. He counted 600 lines, measured the positions of 324, and assigned letters to the more prominent ones. The dark lines in the solar spectrum are now called Fraunhofer lines (see Fig. 7-10).

FIG. 7-10   The solar spectrum, showing the Fraunhofer lines. The comparison spectrum on each side of the solar spectrum is that of iron. Since it matches many of the lines in the spectrum of the sun, the sun's atmosphere contains atoms of iron. *(Lick Observatory photograph)*

It was some 45 years before the physical significance of these lines was established. In 1859 Gustav Kirchhoff (1824–1887), a German physicist, stated three laws relating the types of spectra to the sources producing them. They may be stated in modern language as follows:

1. *Incandescent (glowing) solids, liquids, or gases under high pressure produce continuous spectra.*

In a *continuous spectrum* the colors blend from one into another without interruption. An electric light bulb when viewed through a spectroscope shows a continuous spectrum. Since all solids give continuous spectra when heated to incandescence, it is not possible in this way to identify the elements that produce them. The reason for a continuous spectrum is that the electrons in the atoms are interfered with by neighboring electrons and are not free to move in their usual orbits, as is possible in gases at low pressure. (See Sections 7.8 and 7.9.)

2. *Incandescent gases at low pressure give discontinuous spectra consisting of bright lines.*

These *bright lines* are images of the slit of the spectroscope in various colors and are very narrow when a narrow slit is used. In gaseous form, each element has its own bright-line spectrum, and each line occupies a definite position in the spectrum. For example, hydrogen shows series of lines from the infrared region into the ultraviolet. Wherever these lines are seen—in the sun, a star, or in space—the element hydrogen can be identified. Salt in the gas flame of a Bunsen burner vaporizes and easily shows the two yellow lines of the element sodium. These lines are so close together that they cannot ordinarily be separated by a small spectroscope.

3. *If the light from an incandescent solid, liquid, or gas under high pressure passes through a cooler gas at low pressure, the spectrum is continuous (from the hot source) and crossed by darker, narrow lines (from the cooler gas). Each dark line occupies the exact place of the bright-line spectrum that would have been produced if the cool gas were heated to incandescence.*

This is called an *absorption spectrum,* because the cool gas absorbs small amounts of energy from the hot source.

Either the bright-line or the absorption spectrum may be used to identify the gas producing it. Since the sun shows an absorption spectrum, it is composed of a hot incandescent gas under high pressure surrounded by an atmosphere composed of a cooler gas under low pressure. Over 60 different elements known on earth have been identified in the sun's atmosphere. This identification would be impossible without the use of the spectroscope and Kirchhoff's laws (see Fig. 7-11).

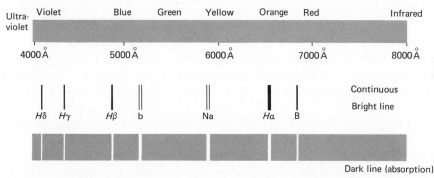

FIG. 7-11   Kirchhoff's three laws of spectrum analysis. Continuous, bright line, and darkline (absorption) spectra are shown. H with Greek letters are lines of Balmer series; Latin letters are Fraunhofer designations.

7.8 ATOMS

Although Kirchhoff's laws established an empirical relationship between spectra and their sources, 50 years elapsed before the physical basis of the second and third laws was found. Maxwell had shown that oscillating electric charges generate electromagnetic waves, and it was known that matter contains electric charges. The continuous spectrum mentioned in Kirchhoff's first law was accounted for by assuming that these charges are oscillating at various frequencies in incandescent solids, liquids, and gases under high pressure. But more had to be learned about the structure of matter before the physical basis of the line spectra was found.

It was once believed that the universe is composed of four fundamental substances: air, water, fire, and earth. These substances were called *elements*. Later it was found, however, that water could be broken down into still more fundamental elements, now called hydrogen and oxygen. The earth was discovered to be composed of many other fundamental substances. The list of elements grew until today we know that there are more than 100.

Materials that are composed of two or more elements are called *compounds;* they can be decomposed into their basic elements by chemical means.

The basic unit of an element is the *atom*. It is defined as the smallest particle of an element that has all the properties of the element. The word atom means indivisible; it was thought that atoms are indestructible until it was discovered in 1902 that atoms of the heavier elements spontaneously change into atoms of other elements. Further experiments led to the development of accelerators and reactors, by means of which it is possible to change all elements.

Atoms are much too small to be seen through a microscope; an average diameter is about 10^{-8} inch. One gram of hydrogen contains 6.0238×10^{23} atoms. The masses of atoms are expressed in relation to each other by means of *atomic weight units*. Hydrogen is the least massive element, with atomic weight equal to 1.008. The atoms of the other elements are heavier than hydrogen, with atomic weights that are approximately an integral (whole) number of units. Carbon is about 12 times heavier than hydrogen, oxygen is about 16, iron is nearly 56 times, and so on. Carbon is taken as the standard for the system of atomic weight units. The atomic weight of an atom comes out closer to a whole number when carbon, instead of hydrogen, is used as the standard. The atomic weight of an atom is, therefore, its mass relative to that of carbon, arbitrarily taken to be exactly 12.

In the early 1900s it was found that most of the mass of an atom is concentrated in a very small volume at the center, called the *nucleus*. More recently, the nucleus has been bombarded with subatomic particles and broken up into still smaller particles. The list of subatomic particles continues to grow. Two of these particles are especially important: a positively charged particle called the *proton* and a neutral particle called the *neutron*. Each has a mass of approximately one atomic mass unit.

Whirling around the nucleus are negatively charged particles called *electrons*. The magnitude of the charge of an electron is equal to that of a proton, but electrons are much lighter, weighing only $\frac{1}{1836}$ as much as a proton. Thus, the mass of the electron is relatively insignificant, and the mass of the atom is approximately equal to the sum of the masses of the protons and neutrons in the nucleus.

Ordinarily, atoms in the free state are neutral, having an equal number of electrons and protons. If an electron is removed or an extra one added, the elec-

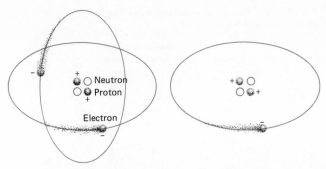

FIG. 7-12 Pictorial models of a neutral and an ionized atom. In the helium atom at the left, the positive charges of the two protons in the nucleus are balanced by the negative charges of the two orbiting electrons. At the right, one of the electrons has been removed, and the resulting helium ion has a net positive charge.

trical balance is upset and the resulting electrically charged atom is called an *ion* (see Fig. 7-12).

Atoms join to form molecules by exchanging or sharing electrons. The chemical identity of an atom is determined by its protons. The simplest atom is hydrogen. It has only one proton and thus only one electron. Helium (see Fig. 7-13) follows with two protons, then lithium with three protons, and so on in succession to the heaviest element. The number of protons in an atom, called its *atomic number,* is unique for each element. The elements are listed in alphabetical order in Table 7-2.

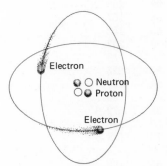

FIG. 7-13 A pictorial model of the helium atom atomic number 2; atomic weight 4. The atom is inert with two electrons in the first shell around the nucleus. Beginning with lithium, additional electrons are located in other shells.

However, the number of neutrons may vary; most elements have subvarieties of atoms with different numbers of neutrons and correspondingly different weights. Since they have the same number of protons for a given atomic number, they are chemically identical although their atomic weights are different. Such atoms are called *isotopes.* Hydrogen, for example, has an isotope of atomic weight 2 (one proton, one neutron) and one of atomic weight 3 (one proton, two neutrons). The former is called deuterium; the latter tritium.

TABLE 7-2   The Elements, Their Symbols, Atomic Numbers, and Approximate Relative Atomic Weights[a]

name of element	symbol	atomic number	atomic weight
Actinium...............	Ac	89	227
Aluminum.............	*Al*	*13*	*27*
Americium.............	Am	95	243
Antimony.............	*Sb*	*51*	*122*
Argon.................	Ar	18	40
Arsenic...............	*As*	*33*	*75*
Astatine...............	At	85	211
Barium...............	*Ba*	*56*	*137*
Berkelium.............	Bk	97	245
Beryllium........	Be	4	9
Bismuth...............	*Bi*	*83*	*209*
Boron.................	B	5	11
Bromine...............	*Br*	*35*	*80*
Cadmium.	Cd	48	112
Calcium...............	*Ca*	*20*	*40*
Californium...........	Cf	98	248
Carbon................	*C*	*6*	*12*
Cerium................	Ce	58	140
Cesium................	Cs	55	133
Chlorine..............	*Cl*	*17*	*35.5*
Chromium.............	*Cr*	*24*	*52*
Cobalt................	*Co*	*27*	*59*
Copper...............	*Cu*	*29*	*63.5*
Curium................	Cm	96	245
Dysprosium...........	Dy	66	162.5
Einsteinium...........	E	99	255
Erbium................	Er	68	167
Europium.............	Eu	63	152
Fermium...............	Fm	100	252
Fluorine...............	*F*	*9*	*19*
Francium..............	Fr	87	223
Gadolinium............	Gd	64	157
Gallium...............	Ga	31	69.7
Germanium............	Ge	32	72.6
Gold..................	*Au*	*79*	*197*
Hafnium..............	Hf	72	178.6
Helium................	He	2	4
Holmium..............	Ho	67	165
Hydrogen.............	*H*	*1*	*1*
Indium................	In	49	115
Iodine................	*I*	*53*	*127*
Iridium................	Ir	77	192
Iron..................	*Fe*	*26*	*56*
Krypton..............	Kr	36	84
Lanthanum............	La	57	139
Lead..................	*Pb*	*82*	*207*
Lithium...............	Li	3	7
Lutetium..............	Lu	71	175
Magnesium...........	*Mg*	*12*	*24*
Manganese...........	*Mn*	*25*	*55*
Mendelevium..........	Mv	101	256

TABLE 7-2 (Continued)

name of element	symbol	atomic number	atomic weight
Mercury...............	*Hg*	*80*	*200.6*
Molybdenum...........	Mo	42	96
Neodymium............	Nd	60	144
Neon..................	Ne	10	20
Neptunium.............	Np	93	237
Nickel................	*Ni*	*28*	*58.7*
Niobium...............	Nb	41	93
Nitrogen..............	*N*	*7*	*14*
Nobelium.............	No	102	254
Osmium...............	Os	76	190
Oxygen...............	*O*	*8*	*16*
Palladium.............	Pd	46	106.7
Phosphorus...........	*P*	*15*	*31*
Platinum..............	*Pt*	*78*	*195*
Plutonium.............	Pu	94	242
Polonium.............	Po	84	210
Potassium............	*K*	*19*	*39*
Praseodymium.........	Pr	59	141
Promethium...........	Pm	61	145
Protactinium..........	Pa	91	231
Radium................	Ra	88	226
Radon................	Rn	86	222
Rhenium..............	Re	75	186
Rhodium..............	Rh	45	103
Rubidium.............	Rb	37	85.5
Ruthenium............	Ru	44	101
Samarium.............	Sm	62	150
Scandium.............	Sc	21	45
Selenium..............	Se	34	79
Silicon...............	*Si*	*14*	*28*
Silver................	*Ag*	*47*	*108*
Sodium...............	*Na*	*11*	*23*
Strontium............	*Sr*	*38*	*87.6*
Sulfur...............	*S*	*16*	*32*
Tantalum.............	Ta	73	181
Technetium...........	Tc	43	99
Tellurium.............	Te	52	127.6
Terbium..............	Tb	65	159
Thallium.............	Tl	81	204
Thorium.............	Th	90	232
Thulium.............	Tm	69	169
Tin..................	*Sn*	*50*	*118.7*
Titanium.............	*Ti*	*22*	*48*
Tungsten.............	*W*	*74*	*184*
Uranium.............	U	92	238
Vanadium.............	V	23	51
Xenon................	Xe	54	131
Ytterbium.............	Yb	70	173
Yttrium..............	Y	39	89
Zinc.................	*Zn*	*30*	*65*
Zirconium............	Zr	40	91

[a] The more important elements are printed in *italic* type.

7.9 ATOMS AND LIGHT

The search for the physical basis of line spectra provided the clues that led to modern atomic theory. Line spectra were puzzling because of the specific pattern of lines for each element. In 1913 Niels Bohr (1885–1962), a Danish physicist, proposed a model for the hydrogen atom that explained the hydrogen spectrum and opened a new era in physics.

He proposed that in the hydrogen atom the electron circles the proton, much as a satellite circles the earth. The force resulting from the electron's circular path is balanced by the electrical attraction between the negative electron and the positive proton, which is analogous to the gravitational attraction between the earth and a satellite.

Bohr made three assumptions to account for the discrete lines in the hydrogen spectrum. One was that light consists of particles, later called *photons*, whose energy is proportional to the frequency of the light. That is, X-ray photons have more energy than radio-wave photons. Photons have all the properties of material particles except that they have no rest mass; they exist only while traveling at the velocity of light. Scientists now accept the dual nature of electromagnetic waves. Under certain conditions they exhibit the properties of particles and under other conditions they exhibit the properties of waves.

The second assumption made by Bohr was that the electron cannot orbit at all distances from the nucleus. Only orbits at certain definite distances, as specified by *quantum numbers*, are stable. This is in sharp contrast with the orbits of earth satellites, which may be at any distance—provided, of course, that they are out of the atmosphere and not significantly affected by the moon. It is as if a satellite could orbit at specific altitudes only—say, 5000, 10,000, and 20,000 miles—and no others. Each orbit is characterized by a specific combination of potential and kinetic energy, analogous to that of an earth satellite as discussed in Chapter 6. In the case of the electron, the energy difference between two orbits is equal to the work required to move the electron from the inner to the outer orbit against the electrical attractive force of the nucleus. Because of the specific energy associated with each orbit, it is customary to refer to them as *energy levels*.

Bohr's third assumption was that the electron can change from one energy level to another in a quantum jump by absorbing or emitting an amount of energy exactly equal to the difference between the two levels. Energy is emitted in the form of a photon. Since its frequency is directly proportional to its energy, the photon's frequency, and thus its wavelength, is precisely determined by the difference between the two energy levels. Energy can be absorbed in several ways: (1) from the impact of another electron, as when a spark passes through hydrogen gas; (2) from the collision of hydrogen atoms, as when the gas is at high temperature; and (3) from a photon of exactly the right energy to boost it to one of the higher levels, the reverse of photon emission. When an electron is elevated to a higher level, it almost immediately returns to the lowest level, emitting a photon.

Not only did the Bohr model account for the two spectral series of hydrogen that had been observed; it also predicted a new series in the infrared and one in the ultraviolet. These were soon found. The various series occur because the electron does not always return from a higher level to the lowest in one jump. It may make the transition in two or more jumps, stopping at intermediate levels on the way. In this case, a photon with the corresponding energy is emitted at each jump. The origin of the first four spectral series of hydrogen is illustrated by the energy-level diagram of Fig. 7-14. A fifth series, called the Pfund series, has also been found.

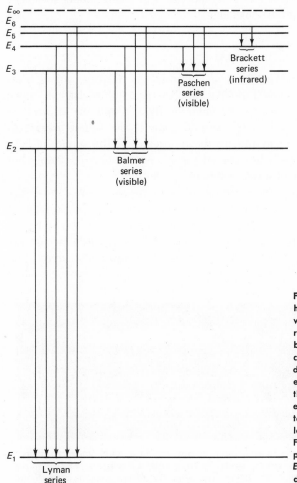

FIG. 7-14 Energy-level diagram for hydrogen showing the origin of the various spectral series. Horizontal lines represent the energies of the Bohr orbits. As the energies of the levels increase, the differences between them decrease. E_∞ represents the energy of escape from the nucleus, which ionizes the atom. Vertical arrows represent electron jumps. Each series corresponds to jumps terminating on a particular level, and is named after its discoverer. For example, the Lyman series comprises those jumps terminating on level E_1. Since these have the greatest energy differences, this series has the shortest wavelengths.

The first series to be discovered, the Balmer series, is in the visible and ultraviolet part of the spectrum. A second series, the Lyman series, is in the ultraviolet; a third series, in the infrared, is the Paschen series. The fourth series, the

Brackett series, is produced by transitions from the outer energy levels to the fourth level, and has lines in the infrared. The figure shows the transition levels of all these four series. All lines in the hydrogen spectrum occur in both emission and absorption.

The Bohr model was extended to include all elements. Since each element has a different nuclear charge, the differences between energy levels vary from element to element, and each has a unique set of spectral lines.

When atoms are packed closely together, as they are in solids, liquids, or highly compressed gases, the energy levels are greatly modified, resulting in a continuous ranges of transitions. Hence these substances are characterized by continuous spectra rather than discrete line spectra. The Bohr model has been extensively modified in light of later knowledge.

7.10 THE DOPPLER EFFECT

Another very useful law of the behavior of light is the *Doppler effect*, first stated by Christian Doppler (1803–1853), an Austrian physicist, in 1842. Doppler showed that when the source of sound or light is moving toward an observer, more waves will reach him per second than if the source were stationary. This will have the effect of raising the frequency of the sound or light waves. Likewise, the frequency will be lowered for a receding source. It does not matter whether the source or the observer is moving, or both.

In the spectrograph, a light source moving toward the slit causes the entire spectrum to shift slightly toward the shorter, blue side. The advantage of having a comparison spectrum alongside that of the spectrum of the source is that the slight shift of the spectrum can be measured easily and accurately compared with the position of lines in the spectrum of a stationary source (see Fig. 7-15).

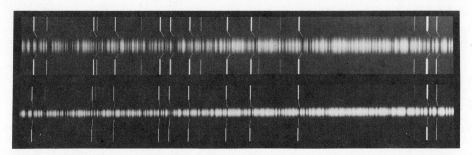

FIG. 7-15   Spectra of two stars, showing Doppler shift toward the red in each. Upper: Aldebaran (Alpha Tauri) is receding at 45 miles/second. Lower: Arcturus (Alpha Bootis) is receding at 12 miles/second. The slanting lines connect the stellar lines with the comparison lines, which are placed on each side of the stellar spectra. *(Photographs from the Hale Observatories)*

If the source is moving away from the slit, the lines are shifted toward the longer, red end of the spectrum. The amount of the shift in each case is proportional to the velocity of approach or recession, and the effect is the same for both bright-line and absorption spectra.

The Doppler effect may be stated as follows:

$$\frac{\text{amount of shift}}{\text{wavelength of spectral line}} = \frac{\text{relative velocity of source and observer}}{\text{velocity of light}}$$

or, expressed as a mathematical equation,

$$\frac{\Delta\lambda}{\lambda} = \frac{V}{c} \tag{7-4}$$

This is an adequate approximation for moderate velocities. For velocities approaching that of light, it becomes

$$\sqrt{\frac{1 + (V/c)}{1 - (V/c)}} = \frac{\lambda + \Delta\lambda}{\lambda} = 1 + z \tag{7-5}$$

where, in both equations, V = relative velocity, c = velocity of light, λ = wavelength of line, $\Delta\lambda$ = amount of shift; and $z = \Delta\lambda/\lambda$.

The Doppler effect is used for the study of many kinds of motion in the universe. The approach, recession, and rotation of all visible objects in space lend themselves to measurement by this important proportion, as will be shown in later chapters.

For example, the Doppler effect is a third method used to demonstrate the revolution of the earth around the sun. The velocity of the earth can be calculated by using the constant of aberration (Section 6-11) and is found to be 18.5 miles/sec. Suppose a spectrum is taken of a star on the ecliptic 90° east of the sun. The Doppler-effect velocity should equal that of the star, which may be unknown, plus the velocity of the earth. Six months later the earth would be moving away from the star and the relative velocity would be the velocity of the star minus the velocity of the earth. Half the difference is equal to the velocity of the earth in its orbit. By making many measures of many stars, we arrive at a fairly accurate velocity of the earth, 18.5 miles/sec. We may use it to determine the distance to the sun, as can be done by using the aberration constant. Again, this method is not sufficiently accurate, but it serves as a demonstration of the earth's motion around the sun.

QUESTIONS AND PROBLEMS

Group A

1. (a) Draw a diagram showing that when the rays of light from a light bulb are reflected by a flat mirror, the reflected rays diverge. (b) Draw a mirror that would cause the light rays to converge upon reflection.
2. (a) Assuming that the hills used by Galileo and his assistants in their attempt to measure the velocity of light were 10 miles apart, how long did it take the light to travel this distance? (b) How could they have determined that their reaction times were an unsurmountable source of error?
3. Different colors of light travel at different speeds in the same medium. How could Roemer's method for measuring the

velocity of light be used to determine whether or not light of different colors also travel at different speeds in space?

4. What properties of the eye could cause an error in comparing the relative brightness of stars seen at different times?

5. How much are the centers of the sun and moon displaced by refraction (a) when rising and (b) when 5° above the horizon?

6. Does refraction cause a star to rise faster or slower than it would if the earth had no atmosphere?

7. Show from the data in Table 6-2 that the sun and moon should appear flattened when rising.

8. Calculate the velocity of light in flint glass from its index of refraction.

9. For a given type of glass, the index of refraction is slightly greater for red light than for violet light. Draw a diagram showing white light entering a prism and the violet and red rays leaving.

10. Why does removing an electron from an atom make it positively charged?

11. Ordinary photographic film is least sensitive to red light, hence the use of red safe lights in darkrooms. Explain, using the photon theory of light.

12. The spectral lines of hydrogen emitted by a recently discovered quasi-stellar source are observed to be 16 percent longer than the same lines emitted by hydrogen in a laboratory. What is the relative velocity of the quasi-stellar source with respect to the earth? *Answer:* Away, about 30,000 miles/sec.

Group B

13. What length mirror does a 6-foot person need to see his full length? Make a diagram showing the person standing in front of the mirror and draw the rays emanating from his feet and the top of his head that are reflected by the mirror to his eye.

14. Compute the distance from the earth to the sun, using Roemer's light time of 1000 seconds.

15. What is the frequency of light of wavelength (a) 6×10^{-5} cm, and (b) 4×10^{-5} cm. *Answer:* (a) 5×10^{14} waves/sec.

16. Jupiter is approximately 5 times and Saturn 10 times farther from the sun than the earth is. How much more solar energy does the earth receive per square foot of surface area than (a) Jupiter and (b) Saturn? *Answer:* (a) 25 times.

17. Compare the brightness of the sun as seen from (a) Venus, (b) Mars, (c) Pluto, and (d) the nearest star, in terms of its brightness as seen from the earth. *Answer:* (c) approximately 1/1600.

18. Compute the angle of refraction of a beam of light entering

water from air, if the angle of incidence is (a) 30°, (b) 45°.
(c) Compute the angles for a diamond.

19. Natural chlorine gas is a mixture of two isotopes, one of atomic mass 35, the other of atomic mass 37. What must be the ratio of the amounts present to give the natural mixture an average atomic mass of 35.5?

20. The diameter of an atomic nucleus is about 10^{-13} inch. (a) What fraction is this of the atom's diameter? (b) What fraction of an atom's volume is empty space (neglecting the space occupied by the electrons)?

8 optics and the telescope

8.1 INVENTION OF THE TELESCOPE

It is not known who discovered the magnifying power of a lens. This property must have been known before the time of the Arabian scientists, because we know that Alhazen used small, glass hemispheres as early as about A.D. 1000. Shortly after that time, lenses were used to correct defective vision.

The manufacture of spectacles became an important industry, especially in Holland, where it appears that a Dutch spectacle maker, Hans Lippershey (1560–1619), discovered the principle of the telescope in the early part of the 17th century. Lippershey was famous for his telescopes in 1608, when the discovery came to the attention of Galileo in Italy. The government of Holland awarded Lippershey 900 florins (probably about \$360) for his invention, which had been used for military purposes. Galileo immediately constructed a telescope that magnified only three diameters, but he soon built an improved model with a magnification of 33 diameters. Turning the telescope toward the sky for the first time, Galileo in 1609 and 1610 made the astronomical discoveries for which he is famous. Two of his telescopes are exhibited in the museum of Florence, Italy.

8.2 THE REFRACTING TELESCOPE

There are four major types of telescope: the refractor, the reflector, the Schmidt camera, and the radio telescope. They all operate on the principle of refraction or reflection, and each is designed for a particular use.

The refracting telescope uses lenses that bring the light from a distant object into a focus by refraction. The principal lens, the *objective*, collects the parallel rays of light from an object and brings them to a *focus*, a point where they may be examined by the eye of the observer with a second lens or combination of lenses, the *eyepiece*. Or they may be recorded on a light-sensitive film placed in the focal plane of the objective. In the latter case, no eyepiece is used.

The distance from the center of the objective to the focus is the *focal length, F*. The diameter of the objective, *A*, is the aperture. In the visual refractor, the usual focal length selected is about 15 times the aperture. The *focal ratio, F/A*, is therefore about 15. The figure varies for different telescopes and is sometimes as great as 20.

The function of the eyepiece is to magnify the image formed by the objective. There are two principal types of eyepiece (see Fig. 8-1). The positive eyepiece consists of two plano-convex lenses (that is, flat on one side and curved outward on the other) of the same focal length, placed with the curved sides facing each other. The first lens is called the field lens; the other is the eye lens. The image is formed at the focus of the telescope objective just in front of the field lens. This eyepiece can be used as a hand magnifier.

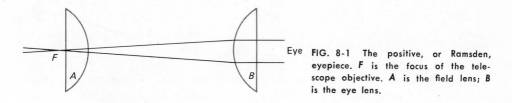

FIG. 8-1 The positive, or Ramsden, eyepiece. **F** is the focus of the telescope objective. **A** is the field lens; **B** is the eye lens.

In a negative eyepiece the field lens is larger and has a focal length about 3 times that of the eye lens. They are placed as shown in Fig. 8-2. The image falls between them. The metal tube holding the two components may be provided with a cross of fine lines on which the image falls. Since the purpose of this cross is to locate the center of the field of view, it is placed on the axis of the telescope and is illuminated with a faint light whose intensity can be changed according to the brightness of the image. The focal length of the eyepiece, *f*, is the distance from the image to the center of the field lens. A negative eyepiece cannot be used as a hand magnifier.

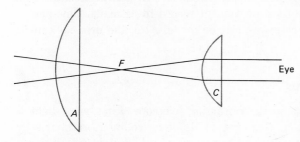

FIG. 8-2 The negative, or Huygenian, eyepiece. **F** is the focus of the telescope objective. **A** is the field lens; **C** is the eye lens.

The *magnifying power* of a telescope is the number of times the telescope enlarges the object compared to its size as seen by the eye alone and depends partly on the focal length of the objective and partly on the focal length of the eyepiece used. It can be calculated as the ratio of the two focal lengths; thus

$$M.P. = \frac{F}{f} \tag{8-1}$$

This can be shown as follows (see Fig. 8-3). Tracing the light rays from the two ends, MM', of an object in the sky, the angle subtended by the object MM' is equal to that subtended by the image, mm', as seen from the center of the objective. Call this angle α. The angle β of the image as seen from the eyepiece is larger than α because the eyepiece is closer to the image. That is, the focal length of the objective is always longer than the focal length of the eyepiece. The ratio of the two angles is the magnifying power of the telescope. Notice also that the image is inverted by the objective.

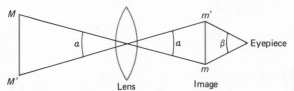

FIG. 8-3   Magnifying power of a telescope.

Objectives are expensive because of their large size. Eyepieces are small and relatively inexpensive. It is therefore the custom of observatories to keep on hand a number of eyepieces of different focal lengths to be used with a given objective, depending on the magnification desired, the object to be observed, and the steadiness of the air. The function of a telescope is to collect the light, bring it to a focus, and enlarge the image.

Images of the planets and other objects are large enough in angular diameter to be enlarged by the telescope. However, stars are so far away that, even in the absence of distortion, their images appear to be points in even the largest telescopes. Because of scattering of light by the air and diffraction in the telescope, images of bright stars may appear to be of appreciable size, especially on nights when the air is unsteady. The telescope under the best conditions will not magnify a stellar image, but it will increase the apparent distance between stars. That is, the telescope magnifies the apparent size of a field of stars. The use of a photographic plate enables a permanent record to be made. This photograph can be enlarged or contact prints can be made. Also, the original can be stored for future study.

8.3 THE TELESCOPE OBJECTIVE

There are three principal defects in the refracting telescope. First, since light is brought to a focus by refraction, the light of different colors is refracted by different amounts. Therefore the component colors of white light are brought to different focal points. This effect, called *chromatic aberration,* can be very annoying. However, it can be corrected by the use of an achromatic objective. The reflecting telescope, in which light is reflected only, is not troubled by chromatic aberration.

The *achromatic objective* is composed of two lenses. It would be better to have more than two, but this adds to the cost and, because of absorption, more pieces of glass reduce the available light. So usually only two lenses are used. In expensive cameras three or four units are used in the lens (see Fig. 8-4).

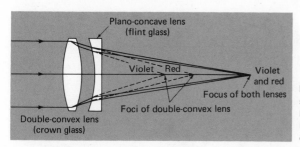

FIG. 8-4 An achromatic objective, showing the paths of violet and red light through a single double-convex lens and after correction by a plano-concave lens.

The first lens of an achromatic objective is a double-convex lens and the second lens is a plano-concave lens. The first lens converges the light toward a focus. In doing so it bends the violet light more than the red and intermediate colors, thus dispersing the beam into component colors. The second lens bends the light in the opposite direction, tending to diverge the beam. Since this second lens also bends the violet light more than the other colors, the effect is to combine them, thus removing at least some of the chromatic aberration. The amount of convergence or divergence caused by each lens depends on its shape and on its index of refraction, a property of the type of glass. By the use of two lenses of different types of glass, it is possible to design the objective so that the convergence exceeds the divergence, bringing the light to a focus, while the dispersion into colors by the converging lens is canceled by the diverging lens.

The converging lens is usually made of crown glass, a type commonly used for windowpanes. The diverging lens is usually made of flint glass. Because the degree of dispersion varies according to color, it is not possible to match the lenses for light of all colors. The refractors of the 19th century and early 20th century were so designed that the yellow and green light were brought to the same focus. These telescopes were designed to be used visually. The red and blue light, to which the human eye is not quite so sensitive, do not come to the same focus as the yellow and green. This produces a yellow-green image, surrounded by a purple fringe that is not troublesome for any objects except the brightest ones. The images of the sun and moon have very noticeable fringes.

In modern telescopes the two components of the objective are separated, but in small lenses, such as those used in cameras, they are cemented together. In the time of Hevelius (1611–1657) attempts were made to correct for chromatic aberration by making very long telescopes. The long-focus objectives produced less bending of the light and thus less dispersion. These telescopes must have been very difficult to handle (Fig. 8-5).

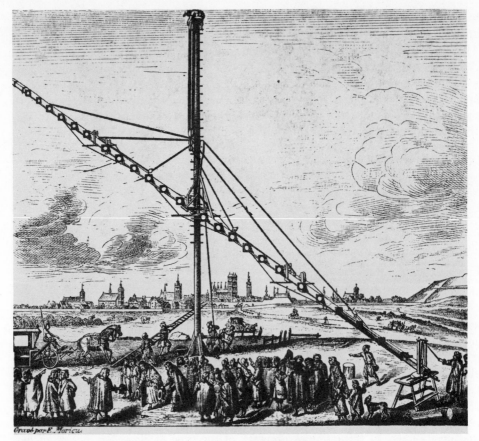

FIG. 8-5 Contemporary sketch of the long-focus telescope used by the astronomer Hevelius in the 17th century. *(The Bettman Archive, Inc.)*

For photography with a visual telescope, it is necessary to use a filter that transmits only the rays that are in focus and cuts out all that are not in focus.

Photographic telescopes are built today with smaller focal ratios, thereby permitting shorter exposure times for a given aperture. If the focal ratio is less than eight ($f/8$), the instrument is called an *astrograph*.

A second defect of a telescope is *spherical aberration*. As the name suggests, this defect is most troublesome if the surfaces of the objective are sections of spheres. Aberration means a wandering away, or deviation from a standard. In a lens with spherical aberration, the light rays through the edge of the objective do not come to the same focus as those that pass through the center. This aberration can be corrected by making the surfaces not quite spherical. We shall see later how spherical aberration is corrected in the Schmidt camera. Since an achromatic lens has four surfaces of curvature, two for each component, aberration is difficult to correct. The surfaces must be computed for each size of objective separately. Such corrections are time-consuming and of course add to the cost.

A third defect is *coma*. This is an aberration produced when the rays strike the lens from an angle distorting the images. Technically, this means that objects at an angle with the telescope axis are not in focus, but have images that look like the short tails of comets, hence the name. This aberration limits the field of sharp images and is impossible to correct in large telescopes. Coma may be observed in some of the photographs in later chapters in this book.

8.4 LIGHT-GATHERING POWER AND RESOLVING POWER

The *light-gathering power* of a telescope is a measure of the amount of light the telescope collects. This measure must be compared to some standard, but is more easily used as a comparison between telescopes of different size. Light-gathering power depends on the area of the objective that collects the light. The area of a circle is proportional to the square of the diameter of the circle. Hence, the light-gathering power of a telescope is proportional to the square of the aperture of the objective. Expressed as a formula,

$$\text{relative light-gathering power} = \frac{\text{area of lens 1}}{\text{area of lens 2}} = \frac{a_1^2}{a_2^2} \qquad (8\text{-}2)$$

As an example, the light-gathering power of a telescope with objective 10 inches in diameter is 10^2, or 100 times that of a 1-inch telescope ($1^2 = 1$).

This equation also permits the calculation of the brightness of a star as seen with the unaided eye compared with its apparent brightness through a telescope. In this computation the size of the pupil of the eye can only be estimated, since it varies with the amount of light entering. A good approximation in the dark is about $\frac{1}{4}$ inch. For example, the ratio of brightness of a star seen through a telescope of aperture 10 inches compared with its brightness seen by a $\frac{1}{4}$-inch pupil is 100 divided by $\frac{1}{16}$, or 1600.

Resolving power is the ability of a telescope to form distinguishable images of objects that are separated by very small angles. To the unaided eye, most stars look like single objects. Through a telescope, however, many of these apparently single stars are seen to be two or more stars very close together. The ability of a telescope to "split" such multiple stars is called the resolving power of the telescope. Resolving power is expressed as the angle between two stars just barely separated in the telescope under the most favorable atmospheric conditions. The following formula is used:

$$\text{resolving power} = \frac{4.56''}{a} \qquad (8\text{-}3)$$

where a is the diameter of the objective in inches and $4.56''$ (seconds of arc) is the angle between two barely separable images seen in a 1-inch telescope. The number can be computed and involves the wavelength of the light received. We may consider this formula to have been obtained by observation alone, and not by theory. Such a formula is said to be empirical.

Although the eye is an optical instrument, its resolving power cannot be determined by the formula, because the retina of the eye is made up of rods and cones, which are the light-sensitive elements. Because of the distances between the elements, the eye is limited in its ability to resolve objects close together. The resolving power of the average human eye is between 40″ and 75″.

As an illustration of the use of this formula, the largest refracting telescope, the 40-inch telescope at the Yerkes Observatory, can resolve two stars only 0.11″ apart. In other words two mountains on the moon 0.11″ apart can just barely be seen separately. Two features on the moon less than 240 yards (220 meters) apart cannot be seen separately with the 40-inch telescope (Fig. 8-6). With the 200-inch telescope, the minimum separation is about half the length of a football field, 50 yards (46 meters).

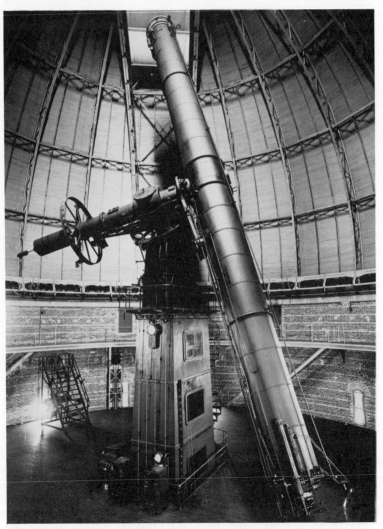

FIG. 8-6   The 40-inch telescope of the Yerkes Observatory. *(Yerkes Observatory photograph)*

Assuming, as before, that the diameter of the pupil of the eye is ¼ inch, the eye alone should be able to see distinctly two stars 18″ apart. This is the theoretical resolving power of the eye as calculated from formula 8-3. However, the double-double star, ε Lyrae, mentioned in Section 3.4 is composed of two pairs of stars 207″ apart. They can be separated with difficulty, if at all, by the unaided eye.

8.5   THE REFLECTING TELESCOPE

We have seen that a parallel beam of light may be brought to a focus by refraction through a double-convex lens. A parallel beam may also be brought to a focus by reflection, a principle used in reflecting telescopes. Instead of passing light through a refracting medium, the reflecting telescope brings the beam to a focus by the use of a concave mirror. In this case, the telescope is built so the light passes down the telescope tube, strikes a mirror, which changes its direction by reflection, and forms an image at the focus located in the upper end of the tube (Fig. 8-7).

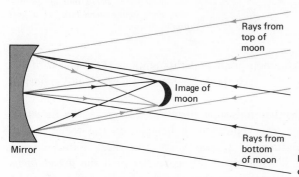

FIG. 8-7   Formation of image by a concave mirror.

If the surface of the mirror is a section of a concave sphere, not all the rays will come to the same focus. This is spherical aberration, as in the spherical lens. If the surface is a section of a paraboloid, all the rays in a parallel beam will meet in a common focus. Thus spherical aberration is eliminated. The mathematical principle is as follows:

If a right circular cone is cut by a plane parallel to one side of the cone, the intersection is a parabola. Another definition of the parabola is that it is a plane figure so drawn that any point on the parabola is equally distant from a point inside, the focus, and a straight line, the directrix. In Fig. 8-8, CD is the directrix and O is the focus. For any point R on the parabola, $OR = RS$. If a parabola is rotated about its axis, AOB, a three-dimensional figure is generated. This figure is called a paraboloid of revolution.

From the mathematical theory of a parabola, any line parallel to the axis of the parabola (or paraboloid) will be reflected at equal angles to the perpendicular at the point of reflection. All rays of a parallel beam of light coming from the

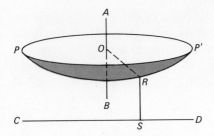

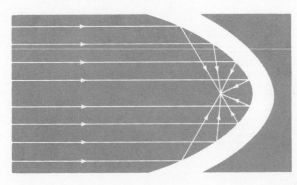

FIG. 8-8   Definition of a parabola or paraboloid (upper figure). Every point. R, on the parabola is equidistant from O, the focus, and the directrix, CD. The lower figure shows that all parallel rays come to the same focus of a parabolic mirror.

direction of the axis meet in a common point and this point is the focus O of the parabola (paraboloid). Hence in a parabolic reflector there is no spherical aberration.

Since the mirror of the reflecting telescope forms an image in the direction from which the light is coming, an eyepiece may be used in the focus. Alternatively, a photographic plate may be placed there (see Fig. 8-10a) and a photograph of the object made. This form of telescope is called the *direct-focus reflecting telescope*. It should be pointed out that the eyepiece or the plate must be held in position at the upper end of the tube, with the result that the beam of light is obstructed and some light is lost before the beam reaches the mirror. This is not a serious difficulty, however, since only a small percentage of the light is lost. A "hole" is not produced in the center of the image, as might be expected. Only the largest telescopes are used in the direct focus.

The 200-inch telescope has a cage at the upper end of the tube where the observer sits to make his observations. In spite of the large size of the cage, the 200-inch mirror is so big that only about 15 percent of the light is lost by the obstruction (Fig. 8-9).

In the reflecting telescope, since the light of all colors is reflected equally, there is no chromatic aberration. However, reflectors are troubled by coma. The 200-inch telescope (which has a mirror 200 inches in diameter) has a usable field of only 10 minutes of arc, although some correction can be made by means of a lens of small diameter placed in the beam of light in front of the focus.

In smaller reflecting telescopes it is obviously impossible to use an eyepiece

FIG. 8-9   The 200-inch Hale telescope on Mount Palomar. *(Photograph from the Hale Observatories)*

in the focus, since all the light would be obstructed by the head of the observer. It is necessary, therefore, to deflect the light before it reaches the focus. This is done by placing a plane (flat) mirror in the emerging beam at a distance from the focus such that this secondary mirror is just filled by the light (see Fig. 8-10b). If this mirror is set at an angle of 45° to the axis of the telescope with the reflecting surface facing the objective mirror, the beam will be deviated 90°, since the angle of reflection is equal to the angle of incidence. The light is then brought, through a hole in the tube to a focus outside the telescope, where it can be viewed with an eyepiece. The observer thus looks at right angles to the direction of the object, as shown in the figure. This form of reflecting telescope was first used by Newton and is called a *Newtonian reflector,* or Newtonian telescope.

In large telescopes, such as the 100-inch telescope, there is an observing platform that can be raised or lowered for the convenience of the astronomer in reaching the focus. This platform runs on a track attached to the observatory dome. The observer may be as high as 50 feet above the floor. A photographic plate is usually placed in the focal plane instead of an eyepiece.

Sir William Herschel (1738–1822) and others used the direct focus of their telescopes by tipping the mirror slightly so that the image was formed at one side of the open end of the telescope tube, where an eyepiece was used. Since

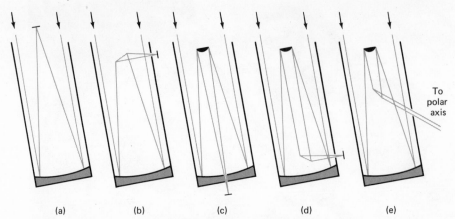

To
polar
axis

FIG. 8-10   Five forms of reflecting telescopes: (a) direct focus; (b) Newtonian, with flat secondary mirror; (c) and (d) Cassegrain and modified Cassegrain, with convex secondary; (e) Coude
with beam directed down polar axis. Telescopes (d) and (e) each have a third flat mirror.

this technique introduced a slight amount of coma, because the observations were
made off the axis of the telescope, the Newtonian form is to be preferred. Many
amateur astronomers own and use the Newtonian type of reflector. The small
secondary, diagonal mirror usually blocks out about 10 percent of the incoming
light. This loss can be compensated for by increasing the exposure time of a
photograph by 10 percent, but no compensation is possible for visual observations
—except to get a larger telescope.

Another type of reflecting telescope is the *Cassegrain reflector*. In this form
(see Fig. 8-10c) the secondary mirror is not flat, but is convex toward the
primary mirror (the objective). It is not placed at an angle of 45° as in the
Newtonian form, but is at right angles to the beam of light. The reflecting
surface faces the primary mirror and reflects the light back again toward the
objective below. If used alone, the convex mirror would diverge the light. But,
as in the case of the second component of the achromatic lens, the combination
of a converging and a diverging beam produces an image at any selected distance
below the secondary mirror. Usually the beam is allowed to pass through a hole
in the primary mirror and the image is formed at a convenient location at the
lower end of the telescope tube. Here it can be viewed from the floor. Also, the
handling of auxiliary apparatus is much easier than from the upper end of the
telescope.

The 100-inch and 60-inch telescopes at the Mount Wilson Observatory do
not have holes in their primary mirrors. But they are used in a modified Cassegrain form by placing a third mirror just above the primary mirror in each
telescope (Fig. 8-10d). This mirror is set at 45°. It deflects the light at right
angles, where it can be examined at the side of the lower end of the telescope.
In modern forms of the reflecting telescope, the primary mirror may be spherical
and the secondary some other form, such as an elliptical mirror.

The 100-inch telescope is also equipped with mirrors that deflect the beam of
light down the polar axis—the axis about which the telescope is turned automati-

cally to compensate for the rotation of the earth (Fig. 8-10e). This form is called the *Coude focus*. It has the advantage of permitting the placing of heavy auxiliary equipment in a stationary place, such as a separate room under the observing floor, where temperature control is possible.

8.6 TELESCOPE MIRRORS

At the time of Herschel, telescope mirrors were made of speculum metal, a combination of copper and tin. Herschel cast his own mirrors. The difficulty with speculum metal was that the proportions of copper and tin (about one third tin, two thirds copper, and a little arsenic, antimony, or zinc added for whiteness) had to be almost perfectly correct or the metal was too hard or too soft. Also, if the whiteness deteriorated with time, the only way to improve the mirror was to start all over with a new casting. The term speculum is still used by some amateurs to mean the mirrors of their telescopes, even if the mirrors are made of glass and coated with metal.

Later, plate glass was used because it could be ground and polished to the desired shape, usually a paraboloid, and then coated with silver, which became the reflecting surface. Notice that the silver is not deposited on the back of the glass as in the usual mirrors used in the home, but is on the front surface. If the silver tarnishes, as it does in the presence of sulfur, it can be removed easily and a new surface deposited by a chemical process. This troublesome and relatively expensive procedure is necessary. Still later, aluminum was used in place of silver, since it lasts longer and has even greater reflectivity in the violet region of the spectrum. However, it must be deposited on the glass by exploding[1] aluminum foil in a vacuum. This requires special equipment, including a vacuum tank and vacuum pumps. The aluminizing of the 200-inch telescope mirror is an involved process, but it is necessary only about every five years.

The 200-inch telescope mirrors are made of Pyrex glass, a modification of the glass used in baking dishes in the kitchen, which has a low coefficient of heat expansion. Still more recently, heat-resistant glass has been substituted. Experiments with fused quartz, which had failed before the casting of the disk for the 200-inch telescope was attempted, have now proved successful. The cost of this material is almost prohibitive. There were also plans for casting mirrors of aluminum instead of using aluminum for coating of glass mirrors. Now that techniques for making quartz mirrors have been perfected, these plans have been superseded by the use of quartz, which has a still lower coefficient of expansion.

8.7 THE SCHMIDT CAMERA

In 1930 another type of reflecting telescope was invented by Bernhard Schmidt (1879–1935), a German optician.

The Schmidt telescope uses a spherical mirror, which is relatively easy to

---

[1] Aluminum foil is placed at precalculated positions inside the chamber and is exploded by a sudden charge of electricity. The aluminum is deposited over the mirror and also over the inside of the chamber.

make but produces spherical aberration. To correct this, Schmidt used a special thin plate of glass, called a correcting lens, placed in the upper end of the telescope tube. This lens is so shaped that it corrects for spherical aberration by bending the entering rays by the proper amount before they strike the spherical mirror. It does not have to be as large as the mirror, but is quite difficult to design correctly. It is so thin that no chromatic aberration is introduced. However, the field of focus of the telescope is not flat, but curved. Therefore it is necessary to use a curved film, which is placed between the mirror and the correcting plate. This type of telescope, which perhaps should be called a camera, since it cannot be used with an eyepiece, can be built with very short focal length (see Fig. 8-11).

The large Schmidt camera on Mount Palomar has a spherical mirror of 72-inch diameter and correcting plate of 48-inch diameter. Its focal ratio is 2.5, which makes it very fast, and it is therefore an excellent instrument for photography. The Palomar Sky Atlas was made with this camera. The photographic plates are 14 inches square and the star images are sharp out to the edges. Times of exposure run from 10 minutes to one hour, depending on the speed of the emulsions used on the plates and the color filter through which the photographs are taken. The field of the telescope is 6.6°, so the entire sky visible from Mount Palomar can be covered in less than 100 hours. It is used in conjunction with the 200-inch telescope. It photographs the general area of the sky in which an interesting object is located. The larger telescope is used for fine details.

A new type of reflecting telescope is now being built, called the Schmidt-Cassegrain telescope. It uses a Cassegrain mirror and a Schmidt correcting lens, but has a secondary reflection from a small mirror attached to the back of the Schmidt lens. The light is then reflected back through a hole in the center of the Cassegrain mirror and comes to a focus below the mirror, where it can be photographed or observed with an eyepiece. These telescopes are available with apertures from 5 inches to 14 inches.

8.8 IMAGE FORMATION

Because of the optical properties of lenses and mirrors, there is a minimum magnification that should be used with a telescope. This minimum is four times the diameter of the objective expressed in inches. For example, if a telescope is 15 inches in diameter, the smallest magnification that should be used is 60. The reason is that with any smaller magnification some of the light from the eyepiece will not enter the eye of the observer, but will be lost outside the pupil.

The maximum magnification should not be greater than 50 times the diameter of the objective. This is because any greater magnification will produce blurred images, rather than points. On nights when the seeing is poor—that is, when the atmosphere is turbulent and the images unsteady—a smaller magnification than normal should be used. Most observatories have a supply of eyepieces to fit various atmospheric conditions, and they can be changed according to the judgment of the observer.

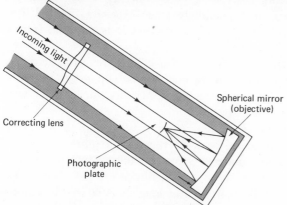

FIG. 8-11 The 48-inch Schmidt camera at the Mount Palomar Observatory (top), and a diagram showing the optical principles of a Schmidt camera (bottom). (*Photograph from the Hale Observatories*)

Telescopes collect light and form images that can be viewed with eyepieces or photographed and otherwise examined with special equipment. A telescope permits objects to be seen that are too faint to be seen with the unaided eye. It separates objects too close together to be seen otherwise. The reason for building larger and larger telescopes is to take advantage of these properties of light-gathering power and resolving power. A telescope on the moon or on a space platform would not have an atmosphere to contend with. The image would always be steady and the highest powers could be used at all times. Also, telescopes located above the earth's atmosphere can be used to photograph and study the ultraviolet part of the spectrum totally absorbed by the air.

Radio telescopes gather and study radiation from space that is invisible to the eye and to the optical telescope.

8.9 REFRACTOR OR REFLECTOR?

What are the advantages and disadvantages of refractors and reflectors? The refractor's lens is always ready for use after cleaning with alcohol and distilled water. The long focal length makes it useful where a steady image under high magnification is desirable. However, the refractor is troubled by chromatic aberration; the reflector is not. Both types do have spherical aberration unless carefully corrected by proper shaping of the surfaces of the objective. The achromatic objective requires the figuring (design) of four and sometimes more surfaces. The reflector has only one surface on each of its mirrors.

Light passes through a lens, but is reflected from the coated surface of a mirror. Thus the lens must be made of high-quality glass without bubbles large enough to affect the optical quality. The long-focus refractor is not always suitable for photography. Because of its large focal ratio, the image formed on the photographic plate is comparatively dim, and thus long exposure times are required. The short-focus refractors now being built concentrate the light into small, bright images that require correspondingly short exposure times. Ultraviolet light will not pass through the glass of a lens, but it is reflected from a metal surface.

The reflector's surface must be covered with a highly reflective coating. Aluminum is most widely used today. The mirror can be supported from the back; the lens must be held by supports at the edges. Mirrors can be made with short focal lengths and are fast for photography.

For the same diameter, a reflector has a shorter tube than a refractor, especially when the Cassegrain form is used. It therefore requires a smaller dome than a refractor of comparable diameter. Cost of construction per unit of diameter is therefore almost entirely in favor of reflecting telescopes. The largest refractor is 40 inches in diameter and about 60 feet long. It is housed in a dome 90 feet in diameter. The largest reflector presently in operation is the Hale telescope on Mount Palomar. Its diameter is 200 inches; it is 55 feet long; its dome is 137 feet in diameter. (A 236-inch reflector is nearing completion in the Soviet Union.)

8.10 RADIO TELESCOPES

The radio telescope is a relatively new form of telescope. It is used to detect and measure radiation from the stars and from matter between the stars that was not previously detectable. Such telescopes have now added a great deal of information about the makeup of stars and of the location and composition of interstellar matter. This branch of astronomy has become very important, and many large receivers of radio waves from space have been built.

Radio waves from space were discovered accidentally by Karl Jansky (1905–1950) in 1931. He built a receiver that could be rotated on a circular track in order to investigate the causes of static radio interference. To his surprise, he found that certain noises were coming from a particular point in the sky and that the celestial sources from which they came moved across the sky at the rate of the diurnal motion of the stars.

Radio waves are part of the electromagnetic spectrum but have longer wavelengths than the visual waves studied by optical telescopes. Because they are not affected by atmospheric conditions, radio telescopes can be operated in

FIG. 8-12   Grote Reber and the first steerable radio telescope. Reber built the 31-foot instrument in Wheaton, Ill., in 1937. (National Radio Astronomy Observatory)

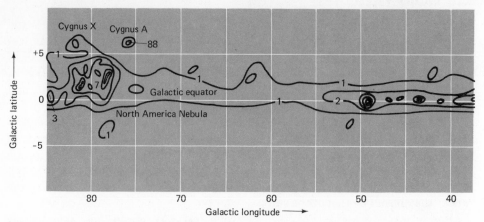

FIG. 8-13   A radio map of the Milky Way shows intense radiations along the galactic equator. Some of the sources are known to lie outside the Milky Way system. The area indicated as Cygnus A is an example of an extragalactic source. Only a small part of the sky is shown here.

the daytime as well as at night; also clouds and haze do not interfere significantly with the reception of radio waves.

Radio telescopes are built on a large scale. Grote Reber (1911–   ) in 1937 set up a receiver in Wheaton, Illinois, a suburb of Chicago (Fig. 8-12). He was the first astronomer to follow up Jansky's experiments. He had trouble avoiding electric waves generated by ignitions of passing cars, but he did succeed in mapping the directions of radio-wave sources from space. The heating of his equipment by the sun on clear days was another source of trouble. In 1943 Reber discovered radio waves from the sun by tuning his receiver to a wavelength of 1.87 meters. These waves come from the solar corona.

A map of radio frequencies from the Milky Way is shown in Fig. 8-13. The numbers on the lines indicate the strengths of the signals. The lowest strength is marked 1, increasing to 2 and 3. Note the strong signals from the galactic center, the area from which Jansky first discovered radio signals from space. Notice also the high intensity of the signals coming from Cygnus A, one of the first radio sources discovered.

It should be emphasized that radio waves are not sound waves; they are electromagnetic waves. They vary in length from about 8 mm to about 17 meters. These are the wavelengths of the spectrum that can penetrate the earth's atmosphere, and must be detected by receivers tuned to those wavelengths. They have very low intensities and may be amplified and recorded on a chart (see Fig. 8-14).

Radio waves can be reflected and refracted like light waves. The radio telescope, therefore, consists primarily of a large parabolic dish that collects the waves and brings them to a focus. This dish may be a solid surface or a mesh of wires stretched in the shape of a paraboloid of revolution. At the focus is an antenna in which the radio signal produces a current that is amplified in

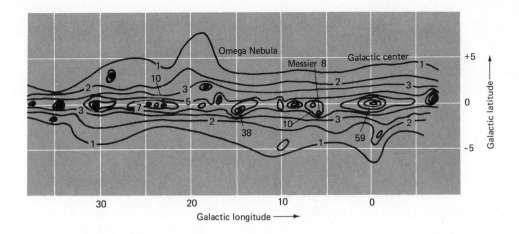

a receiver. The amplifier is similar to one in a home radio set, except that the signal is usually not converted to sound. If a signal from space is received, it appears as a peak above the line formed by background noise, as in Fig. 8-14.

When radar signals are sent to the moon and reflected back to the earth, they return in about 2.5 seconds and are recorded as "pips." The time between the sending of a wave and its reception on the earth can be used to determine the distance to the moon, since the velocity of radio and radar waves is the same as that of light waves.

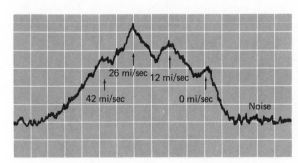

FIG. 8-14 Drawing of a radio signal from space, adapted from an actual tracing from a source on the galactic equator. The velocities have been determined from measures of the Doppler shifts of the peaks above the background noise.

A similar method using a laser has also been developed. By the use of a laser it is possible to send a very narrow beam of light to great distances. In 1969 American astronauts Armstrong and Aldrin placed a screen on the landing site on the moon to receive and reflect back to earth laser beams from American observatories. The purpose of this experiment was to determine very accurately the exact distance to the moon, and the exact location and possible motion of the continents on the earth. These measures all depend on an accurate determination of the location of the landing site on the moon. This laser reflector and others placed in different locations permit a determination of the distance with an accuracy of 1 foot.

Since the radio signals from space are so weak, radio telescopes must be very large and the receivers must be very sensitive. The resolving power of radio telescopes depends on wavelength, as with optical telescopes. The formula is

$$\alpha = 2.52'' \times 10^5 \frac{\lambda}{a} \tag{8-4}$$

where α is the resolving power in seconds of arc, λ is the wavelength to which the receiver is tuned, and a is the aperture of the dish, both in the same units of length. Since the wavelengths are so long, the resolving power is poor; that is, the smallest detectable angle between radio sources is large. See also equation (8-3). A 50-foot dish tuned to 21 cm has a resolving power of about 1.0°. Since the resolving power is inversely proportional to the aperture, as can be seen from equation (8-4), it is necessary to build telescopes with large diameters to improve the resolution.

The largest steerable radio telescope is 250 feet in diameter. It is located at Jodrell Bank, near Manchester, England (see Fig. 1-7). An earlier instrument at the same observatory was composed of a mesh of wires strung near the ground in the shape of a paraboloid and with its antenna at the top of a pole placed vertically above the dish. The instrument could be pointed only to a region of the sky near the zenith. By tilting the pole slightly, it could be pointed to any area near the zenith.

A more recent, large radio telescope is in Puerto Rico and is operated by Cornell University. Its surface is spherical rather than parabolic because the dish is stationary, and the beam can be swung over a wider range of angles. It is used principally for planetary studies.

Recently, the resolving power of radio telescopes has been greatly increased by utilizing telescopes several thousand miles apart. The widely separated telescopes receive signals from the same celestial source simultaneously. The signals are first recorded on magnetic tape at each site with a very accurate frequency standard so that they may be accurately correlated. Comparing the signals provides a resolution much greater than that of either telescope when used alone.

A radar telescope is a special kind of radio telescope that sends out a powerful pulsed beam directed at celestial objects and receives the echo pulses when they are reflected back to earth. It was with such an instrument that the United States Army Signal Corps in 1946 bounced a signal off the moon. Radar is also used to detect and measure the velocities of meteors, since the signals are reflected from the meteor trains. A radar telescope was also used to determine the distance to Venus and progress is being made toward identifying some of the larger surface features. Radar signals have even been reflected from the sun (see Fig. 8-15).

FIG. 8-15   A radar telescope combination of transmitter and receiver. This 160-foot instrument at the Stanford Research Institute is used to explore the sun, moon, planets, and interplanetary gases. The receiving equipment is mounted at the peak of the tripod, one leg of which is partially hidden in the photograph. *(SRI Stanford University)*

8.11 TELESCOPE MOUNTINGS

There are two principal methods of mounting telescopes. They are the altazimuth and the equatorial mountings. In order to reach all parts of the sky, the telescope must be able to point in any direction. This is accomplished by the use of two axes that are rigidly connected and about which the telescope can be moved in two independent directions. The *altazimuth* mounting has a vertical axis about which the telescope can turn in a horizontal plane. A second axis, which is horizontal, permits the telescope to turn in a vertical direction. That is, the two motions are at right angles to each other. This is the principle of the surveyor's transit. To keep a star in its field, the telescope must keep moving slowly in the proper direction, using its two axes. (For example, see Fig. 8-16.) This cannot be done by a simple mechanical device. However, modern telescopes, especially those in space, are pointed toward a bright star after a search with a photoelectric cell, which then holds the telescope pointing toward that or another star that

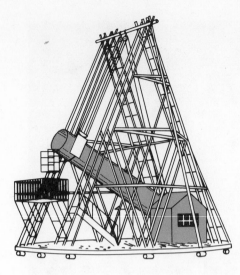

FIG. 8-16 Sir William Herschel's largest telescope. Forty-eight inches in diameter and 40 feet
long, it was mounted as an altazimuth telescope
and operated by man power.

is being observed, by keeping the light of the star centered in the field of view
of the photocell.

The usual procedure in astronomical observatories is to point one axis, the
polar axis, toward the celestial poles. If the telescope is rotated about this axis
by a clock mechanism at the same rate at which the earth rotates, but in the
opposite direction, a star will always remain in the center of the field of view.

To find a star, it is desirable to have a second axis, perpendicular to the
polar axis, about which the telescope can be moved north or south. This second
axis is the *declination axis*, which usually carries a circle divided into degrees
and fractions of degrees reading zero when the telescope is pointed towards any
point on the celestial equator, and 90° at each pole. The *declination circle* reads
the declination of the star to which the telescope points. Thus, when the mounting
is turned about the polar axis, the telescope follows any star as it moves along
its diurnal circle. This form of mounting is called an *equatorial mounting*.

There are several ways of mounting an equatorial telescope. The older forms
used a central column placed on a structure of stone or cement that extended
into the ground. This was for stability and to prevent vibrations inside the
observatory from being transferred to the telescope. The central column was not
in contact with the building at any point.

On top of the column was the mounting head, which contained the polar
axis. This short axis was rotated by a clock mechanism or by an electric motor.
The declination axis was attached at right angles to the polar axis. The telescope
was suspended from one end of this axis; since its position brought the telescope
off the center of gravity, it had to be counterbalanced by a heavy weight at the
opposite end of the axis. The clock was mounted either inside or outside the
central column. The 40-inch telescope at the Yerkes Observatory is of this type
(see Fig. 8-6).

A better, more modern way of mounting a large telescope is used for the
100-inch telescope. It has two vertical piers placed some distance apart north

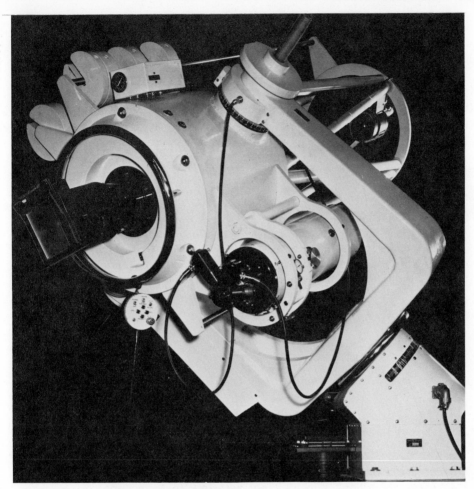

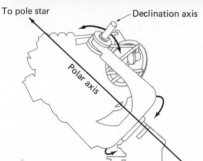

To pole star

Declination axis

Polar axis

FIG. 8-17 The equatorial fork-type telescope mounting, showing the polar and declination axes. The declination axis is graduated in degrees north and south, the polar axis in hours east and west. (*The Ealing Corporation photograph*)

and south. Each pier supports part of the weight of the moving telescope and its two axes. The polar axis rests on the two piers. The telescope is rotated about this axis by a motor beneath the floor. The 200-inch telescope is similarly mounted. But there is one notable improvement over the 100-inch installation. In the latter case, the telescope cannot be pointed directly at the pole, since the axis is in the way. In the 200-inch mounting the upper end of the polar axis is cut in the form of a horseshoe. The telescope can be depressed inside this U-shaped axis and the stars near the pole can be reached for observation.

Another type of modern mounting is the fork type, shown in Fig. 8-17. Here the two axes and the telescope tube are mounted on a single pier. The 236-inch Russian reflector will be the first large optical telescope to have an altazimuth mounting.

QUESTIONS AND PROBLEMS

Group A

1. Figure 8-18 shows three parallel light rays and their associated waves entering a convex lens. (a) Copy the diagram and complete it, showing how the rays pass through and converge to a focus. (b) Draw a similar diagram for a concave lens, showing how it causes the rays to diverge.

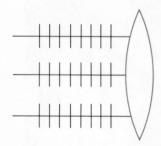

FIG. 8-18   Convex lens and parallel rays.

2. What are two ways of correcting a reflecting telescope for spherical aberration? Why is it not necessary to correct for chromatic aberration?
3. The diameter of the moon is 2160 miles, corresponding to an angular diameter of approximately 30'. What is the size of the smallest feature that can be resolved by the average human eye?
4. Discuss the ways in which the diameter and focal length of a telescope objective affect its ability to detect faint objects.
5. The Hale 200-inch telescope has a focal ratio 3.3. (a) What is its focal length? (b) Compute its magnifying power, if an eyepiece of ½-inch focal length is used. Compute its resolving power. (c) Compare the light-gathering power with that of the 100-inch telescope. *Answer:* (a) 55 feet. (b) 1320; 0.0228" (seconds of arc).
6. In some photographs, symmetrical "spikes" radiate from the stellar images. (See, for example, Neptune in Fig. 13-9.) Is this because stars are shaped that way, or is it due to some characteristic of some telescopes? Explain.

7. How does the Schmidt camera differ in construction from an ordinary reflecting telescope? What is the major advantage of a Schmidt over other optical telescopes?

8. What are the maximum and minimum magnifications that should be used in (a) a 6-inch telescope and (b) a 15-inch telescope?

9. What are the advantages of placing a telescope on a space platform?

10. How would the image be affected if the lower half of the lens in Fig. 8-7 were covered? (*Hint:* What would be the effect of removing the light rays that pass through the lower half of the lens?)

11. What attributes of a telescope determine its (a) light-gathering power, (b) resolving power, and (c) magnifying power?

12. Compare optical reflectors with refractors as to (a) aberrations, (b) loss of incident light, (c) maximum feasible aperture, (d) use in photography, and (e) maintenance.

13. Radio astronomy is well developed in England and Holland. What probably contributed to the early interest in radio astronomy in these countries?

14. How large an earth-based optical telescope would be needed to see craters 3 feet in diameter on the moon?

Group B

15. The 36-inch Lick refractor has a focal length of 60 feet. What is its (a) focal ratio, (b) magnifying power with a ½-inch eyepiece, and (c) resolving power? (d) Compare its light-gathering power with that of the 200-inch telescope. *Answer:* (d) 0.032.

16. (a) How many times greater would the diameter of a telescope objective need to be to increase its light-gathering power by a factor of 25? (b) How would this affect the resolving power and magnification? *Answer:* (a) 5.

17. If the secondary mirror in the Cassegrain form of a 100-inch telescope has a circular shape, what percentage of the incoming light does it obscure if it has a diameter of (a) 25 inches and (b) 20 inches? *Answer:* (a) 6.25 percent.

18. If the mirrors of the 100-inch telescope each reflect 90 percent of the incident light, what percent of the original light is reflected by the last mirror in the modified Cassegrain form?

19. Compute the effective focal length of the 200-inch telescope used as a Cassegrain. Assume that the second mirror is 20 inches in diameter and 35 feet from the eyepiece and that the light exactly fills it.

20. How large would the dish of a radio telescope need to be to have the same resolving power as the 200-inch telescope? Assume the radio telescope is tuned to 21-cm radiation and the optical telescope to 4000 angstroms wavelength. *Answer:* 1675 miles.

9 auxiliary instruments of astronomy

At first, astronomical observations were made without the use of instruments. We have mentioned the count of the number of days in the year by observing sunrises and sunsets. It is agreed that the beginning of the Egyptian year was based on the times of rise of the Nile River. The rise of the Nile also coincided, during early Egyptian history, with the first visibility of the bright star Sirius in the dawn. This is known as the heliacal rising of Sirius, a star that was personified as the goddess Sothis.

The Jewish year began when the new moon was first seen in the twilight.

The gnomon was also used to determine the length of the year and to determine the north and south motions of the sun in the sky. The changes in the position of its shadow were used to determine time.

9.1 ANCIENT INSTRUMENTS

Later the Egyptians also used their pyramids as observing stations for the determination of time. In the Great Pyramid there was a Grand Gallery, which was oriented north and south. Times of passage of stars across this gallery, which was open to the sky before the pyramid was finished, served as a clock at night. During daylight, the Egyptians used a vertical shaft called an obelisk. It cast a shadow that served as a sundial to mark the hours. These were among the first instruments of astronomy.

An Egyptian obelisk, now in New York City, and the massive structure at Stonehenge, England, are shown in Fig. 9-1. Very early in the history of England, two circles of large stones (dated between 2000 and 1500? B.C.) were used at Stonehenge for religious and, almost certainly, for astronomical purposes.

FIG. 9-1 Ancient timekeepers included the Egyptian obelisk and the prehistoric arrangements of stones, called Stonehenge, in England. (Photograph of obelisk taken in 1918.) *(Courtesy of the Metropolitan Museum of Art; C. M. Huffer)*

Summer began when the sun could be seen to rise approximately over the Heel Stone as seen from the Altar Stone (Fig. 9-2). The Heel Stone was a large, vertical stone placed several hundred feet northeast of the Altar Stone, which was in the center of the two rings of stones. This marked the date of the summer solstice. Other alignments of stones marked the various positions of sunrise, sunset, moonrise, and moonset. Thus the alignment of stones served as an astronomical observatory.

FIG. 9-2   The summer solstice at Stonehenge. Framed in an archway for an observer standing at the center of the structure, the Heel Stone marks the point on the horizon where the sun is about to rise. (S. G. Perrin)

By the end of the 16th century Tycho Brahe was observing the sky from an observatory on the island of Hven off the coast of Denmark. Among his instruments were the sextant and wall quadrant, which were described in Chapter 1. The sextant was movable in two directions, permitting the measurement of the altitude and azimuth of a celestial object. The wall quadrant was fixed in position and permitted the determination of altitude only. The arc of the circle of this quadrant was graduated accurately in degrees, and Tycho invented a way of measuring fractions of degrees by a device that was the forerunner of the modern vernier. Tycho also had a clock (shown in Fig. 1-2) and could determine the altitude at a given time.

With these instruments Tycho measured the positions of the planets among the stars. His observations were the most accurate that had been made up to that time. They served Johannes Kepler as a basis for his three laws of motion of bodies in the solar system. Tycho also constructed a celestial sphere on which he plotted the positions of about 1000 stars. His observations marked the beginning of modern astronomy of position. From them Kepler was able to compute the distances of the known planets, particularly of Mars, in terms of the distance from the sun to the earth, which was not known at that time.

All instruments before the time of Galileo were, of course, without optical assistance. When Galileo adapted the telescope to the study of astronomy, it became possible to use it also with the auxiliary instruments then in use, such as the sextant and the quadrant.

9.2 TRANSITS AND SEXTANTS

In the surveyor's transit, a telescope is mounted on two axes with graduated circles reading to less than 1′ of arc. There is a fundamental plate that can be rotated about a vertical axis for measuring azimuth. The telescope turns in altitude about a horizontal axis. This instrument is similar to Tycho's sextant equipped with a telescope (Fig. 9-3).

The mariner's sextant has a small telescope mounted on a frame that can be held in the hand. It is used to measure the angle between an astronomical body and the sea horizon. Methods have been developed by which a skillful navigator can determine his position from a moving ship with an accuracy of about 2 miles. The sextant has been modified for use in the air. This modification makes use of a bubble in a fluid that marks the horizontal. The position of an aircraft can be determined to within about 10 miles with a bubble octant.

The astronomical transit is a telescope mounted on one axis placed in a horizontal east-west position. The telescope is thus so directed that it can be used only for stars on the meridian. It is therefore similar to Tycho's wall quadrant, except that it is equipped with a telescope and very accurately graduated circles with verniers[1] (Fig. 9-4). With this instrument the latitude of an observatory can be determined to about 0.1″ of arc, or about 10 feet.

An even more accurate instrument is the meridian circle, a special form of transit usually equipped with a larger telescope and still more refinements for reading angles. It is used in a north-south position (on the meridian) and the positions of stars can be determined with the greatest precision—about 0.01″ of arc. With the astronomical transit and the meridian circle the sun, moon, and planets are kept under observation and their orbits determined from these observations with great accuracy. Work of this nature is one of the principal tasks of the United States Naval Observatory.

9.3 ASTRONOMICAL PHOTOGRAPHY

Photography was developed during the 19th century. The first process—called the daguerreotype process after its inventor, L. J. M. Daguerre (1789–1851)—used sensitized metal plates. The first photograph of the moon was made in

[1] In the vernier, a second scale below the main scale may have 9 or 11 divisions in the same interval occupied by 10 divisions on the main scale. If, as in this drawing, the vernier scale is marked in units of 1′, the coincidence of two marks can be read in units of the vernier scale. Note that the zero (0′) on the lower scale is between 0° 10′ and 0° 20′, and the coincidence is at 5′. Hence the reading here is 0° 15′. (See Fig. 9-4b.)

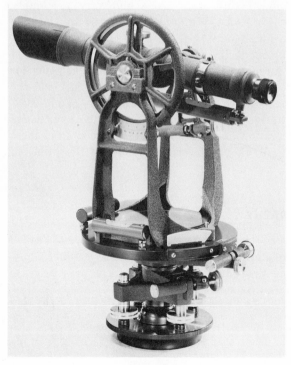

FIG. 9-3 The surveyor's transit (top); and the mariner's sextant (bottom). (Courtesy of Keuffel & Esser Co; Department of Defense)

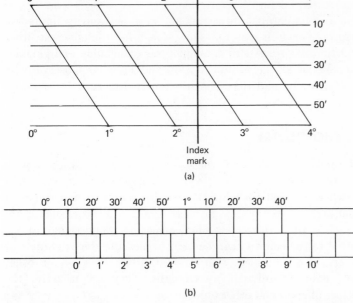

(a)

(b)

FIG. 9-4 Tycho Brahe's device for measuring angles. The index mark indicates the direction of the object being observed. In this drawing the angle
reads 2°/24'.

1840 by this technique. A later development was a light-sensitive emulsion, used
at first on glass plates and at present also on flexible film. The first astronomical
photograph of a star was made at the Harvard Observatory by this improved
process about 1850.

The use of photography makes it possible to record an entire section of the
sky at one time instead of observing each star separately, as with the transit
and meridian circle. A photograph shows the relative positions and brightness
of the stars, including those below the limit of visual observation. The photographic emulsion on glass plates as used at the telescope is placed at the focus.
The exposure times can be varied according to the judgment of the astronomer,
depending on the purpose for which the photograph is being made.

For example, in the Palomar survey of the sky made with the 48-inch
Schmidt camera, exposures of 10 minutes were made with a yellow filter in
front of the plate and then one of about 50 minutes through a red filter. Figure
3-9 reproduces a pair of such photographs (in black and white). They are posi-
tives made from the original negatives.

Photographs such as these give the positions of stars with respect to each
other or to some central star whose exact position can be determined with the
meridian circle. Also the brightness of stars can be measured by the size of the
images from either the original negative or the positive copy. The brighter
stars have the brighter and larger images. The colors of the stars can be found
by comparing the brightness as measured on the yellow and red photographs.

The original plates have been copied and printed by ordinary photographic processes and have been stored for future reference.

Before 1959 almost all photographs in astronomy were necessarily limited to black and white because color film was too slow except for the sun and other bright objects. In 1959 the first successful color pictures of nebulae and galaxies were made by William Miller at the Mount Wilson and Palomar Observatories with the 200-inch Hale telescope and the 48-inch Schmidt camera.

9.4 ASTRONOMICAL PHOTOMETRY

The instrument used to measure the brightness of an object is called a *photometer*. Modern photometers use photoelectric cells. The cell generates a small current of electricity when exposed to light. It must be amplified and can then be measured by a sensitive meter. Such an instrument, a *photoelectric photometer,* is easily attached to a telescope. The current is proportional to the amount of light admitted to the cell and can be used to measure the brightness of a star. Photometers are also equipped with colored filters of glass or other material that transmit light of different wavelengths. They are used for the measurement of the colors of stars and other objects.

Formerly the eye was used as a light meter. This *visual photometry* has been largely superseded by photographic photometry, which is widely used for faint sources. Photographs can be made of many stars at the same time, whereas the photoelectric photometer is limited to one object at a time. Although most photocells are sensitive only in the blue region of the spectrum, red-sensitive cells have been developed. Also, cells with lead sulfide as the sensitive surface permit measurement of radiation in the infrared.

In the early 1960s a device called the *image-converter* tube was developed and successfully put into operation. This tube uses a photoelectric surface at the focus of the telescope. Light striking the surface releases electrons that are focused on a sensitive photographic plate. These electrons act on the emulsion, which is then developed into a photograph. Ordinarily it takes about 50 times more light intensity to produce a photograph directly than it does with the image tube. Thus the telescope is being made more efficient than formerly by an increase in the effectiveness of its equipment rather than in the size of the telescope.

There are several ways of adapting photography to the problems of photometry. If a photograph of a portion of the sky is made with the images of stars in focus on the plate, the blackness of the image on the negative is proportional to the brightness of the star. The silver grains are affected at greater distances from the centers of the images of bright stars than of fainter ones. Hence the stars show as round, black spots and the sizes are proportional to the brightness of the stars. Figure 9-5 shows this effect. However, in this photograph the stars are white because the reproduction is a positive made from the negative. The magnitude of a star can thus be determined by comparing the size of its image with that of a star of known magnitude.

FIG. 9-5   A photograph of a star cluster showing the effect of a star's magnitude on the size of its image. *(Yerkes Observatory photograph)*

Visual photometry for many years used Polaris as the standard of magnitude. It was selected because it could be seen throughout the night. It always remains in a nearly fixed position and is always in reach of observers in the northern hemisphere. All other stars could be compared with Polaris by a special double telescope at the Harvard Observatory. Unfortunately, it was shown by accurate photoelectric observations that the standard is not constant in brightness, but varies by about 4 percent. Thereafter, the average magnitude of Polaris was used as the standard.

For photographic photometry a series of stars, the Polar Sequence near the north celestial pole, is now used. This sequence includes Polaris. The stars in other parts of the sky are compared with those in the Polar Sequence, taken with the same equipment and with the same exposure times. This photometry is accurate to about 1 percent. Secondary standards have also been set up in selected areas of the sky. These stars have been carefully compared with the Polar Sequence. A similar series has also been set up in the southern hemisphere.

9.5 THE SPECTROGRAPH

A spectroscope is an instrument for producing a spectrum and permitting it to be examined visually. If such an instrument is equipped with a plateholder on which the spectrum is photographed instead of viewed through an eyepiece, it is called a *spectrograph*. This instrument is provided with a narrow adjustable slit through which the light of a luminous object is allowed to pass. The slit is placed in the focus of the telescope objective. The light diverges after passing through the slit and then passes through a lens that makes the beam parallel. This lens is referred to as the collimating lens, or collimator.

The light next passes though a prism made of glass or some other refracting material, where it is dispersed into its component colors. Finally it passes through a small telescope, composed of an objective lens and an eyepiece. The eyepiece may be replaced by a plateholder and the spectrum permanently recorded by photography. Figure 9-6 shows the optical principles of the prism spectrograph.

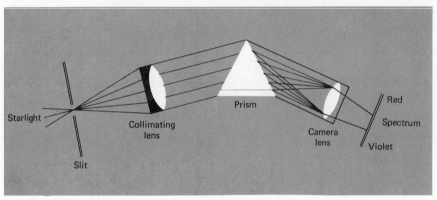

FIG. 9-6 Optical principles of a single-prism spectrograph.

The prism may be replaced by a grating, which also disperses light into a spectrum. A grating consists of fine lines very close together (about 0.0001 inch apart) on glass or metal. When the light passes through the small openings between the lines of a glass grating, it is diffracted as described in Section 7.3. The grating separates the light into its component colors. The grating has two advantages over the prism. First, the dispersion is greater; that is, the spectrum lines are farther apart and can be measured with greater accuracy. Second, in the grating spectrum the dispersion is the same for all colors, whereas in the prism spectrum the dispersion changes continually from red to violet.

Several hours of exposure time are sometimes required to obtain a good photograph of the spectrum of a star. Since the temperature drops in the observatory during the course of an exposure, it is necessary to keep the spectrograph at a constant temperature to prevent a change in the refraction caused by the expansion and contraction of the optical parts. For that reason the entire spectrograph is housed in an insulated case equipped with an electric heater controlled by a thermostat.

It is also desirable to compare the spectrum of a source under observation with the spectra of various elements, each of which has its own set of spectral lines. This comparison with a known source makes it possible to determine whether or not a particular element is present in the atmosphere of the star, to measure the Doppler effect (which determines the amount of radial motion of a star toward or away from the earth), and to discover many other effects of interest to the astrophysicist. The light from the comparison source is allowed to pass through the two ends of the slit to form two spectra, one on each side of the spectrum under investigation. Having the spectrum of star and source side by side provides a very accurate method of measurement (see Fig. 9-7).

FIG. 9-7 Spectrum of the star WW Aurigae. The comparison lines above and below the spectrum of the star make it possible to identify specific elements in the star's atmosphere. *(Courtesy Arthur Young, California State University, San Diego)*

Astronomical spectrographs are built with one, two, or more prisms or with gratings with as many as 30,000 rulings per inch. Individual lines that cannot be separated by a series of prisms can often be separated with a grating spectrograph. With modern gratings it is possible to observe fainter stars than could be observed before.

For many years, research was done to develop methods by which the spectra of stars could be obtained with spectrographs without slits. Some degree of success was achieved, but this technique did not give accurate velocities.

By placing a thin prism in front of the telescope objective, it is possible to photograph a field of many stars at the same time. Instead of a round image for each star as in direct photography, the starlight is stretched out into a low-dispersion spectrum that can be studied under a microscope. The prisms are made with sides inclined to each other at an angle of about 5°. This instrument is called an objective prism spectrograph. Figure 9-8 shows the spectra of a field of stars taken with a spectrograph of this type.

Since in a well-designed telescope the image of a star is a point, its spectrum would be too narrow to permit the lines to be seen and measured accurately. In the spectrograph it is customary to widen the spectrum by allowing the star to move back and forth on the slit until the proper exposure of the widened spectrum is made. The width of the spectrum can be controlled by adjusting the length of the slit. In the objective prism spectrograph, the spectrum is widened by moving the plate in the direction of the lines. The amount of widening is more easily controlled by moving the plate itself than by moving the telescope. The widening of spectra requires increasing the exposure times, the length of exposure depending on the brightness of the star and the amount of widening desired.

An even more recent instrument, the *photoelectric scanner,* combines a grating spectrograph and a photoelectric photometer. Here the spectrum is passed

FIG. 9-8   A photograph made by the use of a transmission grating. By placing a diffraction grating in front of the telescope objective, the spectra of a number of stars can be photographed at the same time. (*Yerkes Observatory photograph*)

through a slit to eliminate all except a short region of the spectrum. As the grating is rotated by a small motor, each part of the spectrum in turn falls on a photomultiplier and is then recorded as a line on a meter. The lines in the spectrum show as dips in the tracing. The width and depth of each line can then be measured directly without the use of photography. This method, though somewhat slow, is becoming important in astrophysical studies of the brighter stars. It is obvious that only one star can be observed at a time. Figure 9-9 shows an artist's drawing of a spectrum scan. The scanner is being used in the orbiting astronomical observatory, OAO-2.

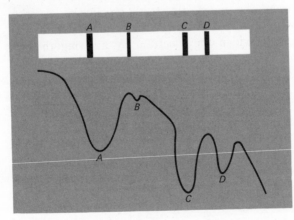

FIG. 9-9   Artists drawing of a photo-electric scanner tracing of a spectrogram. A is a strong, but broad and diffuse, absorption line; B a weak, sharp line; C a stronger, somewhat diffuse line; and D a still weaker, slightly broadened line.

9.6 PHOTOELECTRIC PHOTOMETRY

Photoelectric photometry has become one of the most important branches of astronomy. The photocell is the basic instrument for the measurement of stellar magnitudes and colors, although a considerable amount of photometry is also done by photography. The magnitudes and colors of the standard stars mentioned previously have been accurately measured and used as comparisons for the photometry of fainter stars and other types of celestial objects.

In 1873 it was found that the element selenium changes its electrical resistance when radiated with light. This effect was first used to measure the intensity of starlight in England in 1891 and later in the United States. The selenium cell was used by Joel Stebbins (1878–1966) at the University of Illinois in 1907 when he measured the changing light of the moon at various phases. The selenium cell was also adapted to the measurement of sunlight and starlight at the Illinois observatory. It was used for measuring the light of the solar corona at the times of eclipses and for the measurement of variations of starlight during eclipses of stars, the first having been the star Algol in the constellation Perseus (see Map 1). This was the first eclipsing star to be discovered and the first star for which a complete light curve was observed electrically (see Chapter 18).

The *photoelectric effect* was discovered in the latter part of the 19th century. When light is allowed to fall on the alkali metals, such as rubidium, sodium, and potassium, a very small current of electricity is produced. This current is proportional to the amount of incident light.

The photoelectric is much more sensitive than the selenium cell. Photoelectric cells made at the University of Illinois were used in a photometer with an electrometer to measure the current—that is, the rate at which electrons were released from the potassium when exposed to light. The first photoelectric photometer at the University of Wisconsin used an electrometer that hung vertically, as shown in Fig. 9-10. The difference of brightness of two stars could be computed by comparing the photoelectric current produced by each star. The photocell is capable of detecting small changes of less than 1 percent in the light of a variable star.

The photoelectric current is extremely small, and for many years efforts to amplify it were unsuccessful. Amplification has now been accomplished in several ways. The first success achieved at the University of Wisconsin in 1932 was by the use of a vacuum tube, the GE FP-54, which could be operated by a low-voltage storage battery. By placing the vacuum tube and the photocell in a partial vacuum, Albert Whitford (1905–   ) amplified the photocurrent about 20 times, thus permitting the measurement of stars as faint as tenth magnitude with a 15-inch telescope. Seventh magnitude had been possible before amplification.

The invention of the 1P21 photomultiplier tube during World War II further increased the usefulness of the photoelectric effect. This tube (shown in Fig. 9-11) contains a sensitive element that produces the photocurrent. A succession

FIG. 9-10   The first photoelectric photometer of the Washburn Observatory, University of Wisconsin, 1923, showing oblong light-tight box with Kunz photocell and hanging string electrometer. *(University of Wisconsin)*

of nine plates at high electric potential amplifies the current about one million times. The tube works best when refrigerated with dry ice, and photometers are built that allow this to be done easily. The signals from the 1P21 are further amplified by a complex circuit and finally are recorded on a moving drum,

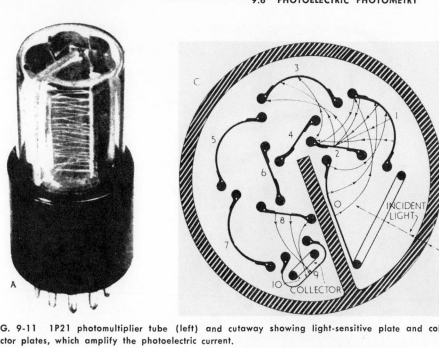

FIG. 9-11 1P21 photomultiplier tube (left) and cutaway showing light-sensitive plate and collector plates, which amplify the photoelectric current.

where the deflections can be measured and converted to differences of magnitude at the convenience of the astronomer. This system is in use in many observatories (Fig. 9-12).

The method of recording and computing the magnitude of the amplified photocurrent has now been further improved. By the use of a tape recording device and a very accurate clock, reading to hundredths of a second, it is possible to record all the essential steps in the observation. For example, the tape punch reads the output of a voltmeter, duplicating the deflections on the scale of the recording meter. It records the exact time and the photometer settings that are used in making the observations. It lists the number of the color filter used and the symbols for the variable star and the comparison stars. The astronomer can also reject the last observation if in his opinion something had spoiled it, such as a cloud passing over the star.

These data are fed into a computer along with the deflections on the recording meter, which still must be read by the astronomer or his assistant. The computer then calculates the effect of changing altitude of the stars on the amount of light absorbed by the earth's atmosphere. It does this by solving a set of equations obtained from a series of observations of special stars scattered over selected areas of the sky. If the star is an eclipsing binary, the time of observation and the phase computed from the time of the last eclipse are included in the computation. The observations are reduced to the magnitude scale, referred if desired to the brightness scale of the standard stars.

This equipment of the observatory at the California State University at San

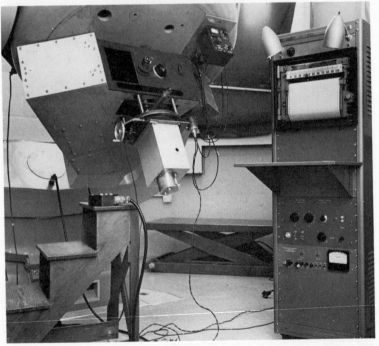

FIG. 9-12 The photoelectric photometer attached to the 36-inch Cassegrain telescope of the University of Wisconsin. (Houck, from University of Wisconsin)

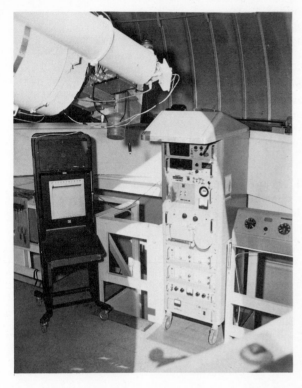

FIG. 9-13 Photoelectric equipment of the Mt. Laguna Station of the California State University at San Diego. (Courtesy Burt Nelson and R. Roome, California State University, San Diego)

Diego is shown in Fig. 9-13. The next step will be remote control, so the astronomer need not be present while his observations are made. Early attempts at remote control at the Kitt Peak National Observatory were not satisfactory and have been temporarily discontinued. However, a similar technique is being used for observations from the orbiting observatory, OAO-2 (Fig. 22-7).

QUESTIONS AND PROBLEMS

Group A

1. What modern instruments are similar to Tycho's (a) sextant and (b) wall quadrant?
2. What is the derivation of the terms "sextant" and "bubble octant"?
3. Why might the altitude of a star measured with a mariner's sextant be slightly larger than when measured with a bubble octant?
4. Why is it not possible to use a bubble octant or a mariner's sextant in a spacecraft?
5. List the advantages and disadvantages of photography over visual observations in astronomy.

6. What two ways of measuring the brightness of stars are more accurate than visual observation?
7. In photographic studies of star brightness, the Polar Sequence is used as a reference. Why is it desirable to photograph it each time a comparison is made, rather than use the same photograph over and over?
8. Gratings with low dispersion are used to obtain spectra of very faint stars. Why cannot gratings with greater dispersion, which are more accurate, be used?
9. Trace briefly the development of photometry.

Group B

10. Verify the statement that 0.1″ of arc along the earth's surface corresponds to a distance of about 10 feet. (Assume the earth to be a sphere 25,000 miles in circumference.)

10 the moon

10.1 A GIANT LEAP FOR MANKIND

The moon was probably the most discussed object in the universe during the middle 1960s. Would we be able to land on it? Who would be first? What would it be like when we got there? Is the surface hard or soft? Can it support the weight of a spaceship? Is the surface covered with dust? If so, how thick is the dust layer?

These perplexing questions were answered with the landing of the American astronauts, Armstrong and Aldrin, from Apollo 11 on July 20, 1969. They were the first humans to walk on the moon's surface. They brought back to earth specimens of rocks and surface material to be analyzed in our laboratories. The landing of the module proved that the surface, at least in the Sea of Tranquillity, is hard and can support the weight of a spaceship. The dust is not as deep as some astronomers had predicted.

10.2 THE MOON'S DISTANCE

In order to plan for a landing on the moon, the question of its distance from the earth had to be answered accurately. This is a fundamental necessity, since all questions of the size of the moon and its surface features depend on a knowledge of its distance. The approximate distance had been known for many years, even centuries. The orbital flight and the location of the landing site had to be predetermined at the space laboratories on the earth. The orbit was computed with an accuracy of a mile or two, and several space missions had been sent to photograph the moon's surface from close up before the actual mission of

Apollo 11 could be attempted. The first flights to and around the moon could not be calculated with the accuracy needed because of gravitational irregularities. These led to the discovery of mass condensations, called *mascons*, below the lunar surface, which because of their additional gravitational attraction had led to a false computation of the orbit.

We now propose to discuss both the older and the newer methods of determining the lunar distance, its size, its mass, and thus its density.

The older methods depended on the determination of the moon's parallax. The parallax method is worth discussing, both from a historical viewpoint and because the parallax method is still basic for determining the distances of stars.

The *parallax* of the moon is defined as the angle formed (subtended) at the center of the moon by one half the diameter of the earth, as shown in Fig. 10-1.

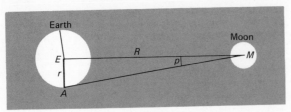

FIG. 10-1 The parallax of the moon.

The problem of finding the moon's distance is similar to the determination of the distance across a body of water where the distance cannot be measured directly. In that problem a base line is set up, its length measured accurately, and the angles to a point whose distance is to be found are measured from both ends of the base line. In Fig. 10-2 the base line is *AB* and the angles *BAC* and *ABC* are measured from the ends of the line *AB*. The sides of the triangle, *AC* and *BC*, are then computed.

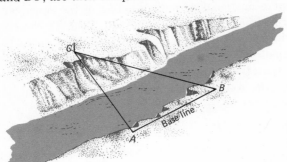

FIG. 10-2 Determination of the distance to an inaccessible point by triangulation.

If the length of the base line is comparable to that of the sides, we must use trigonometry to obtain an accurate solution. But if the base line is very short compared with the sides, the lengths of the sides can be found by using only radians. The earth's radius is very small compared with the distance from the earth to the moon, so radians can be used to solve the long slender triangle in Fig. 10-2 and will be used in this discussion.

If the parallax p is small, less than about 4°, the arc of the circle is so nearly equal to the chord that the small difference in length may be neglected.

Hence the radius of the earth may be used as an approximation of the length of the arc.

In the appendix, the following formula is developed:

$$\text{angle (in radians)} = \frac{\text{length of base line}}{\text{length of radius}} \qquad (10\text{-}1)$$

Adapting equation (10-1) to the problem of the moon's distance, $p =$ the parallax, EM or $R =$ the distance to the moon, and $AE = r$, the radius of the earth.

Rewriting equation (10-1), we have

$$R = \frac{r}{p} \qquad (10\text{-}2)$$

where p is in radians, or

$$R = 57.3° \times \frac{r}{p°}$$

$$= 3437.8' \times \frac{r}{p'}$$

$$= 206,265'' \times \frac{r}{p''} \qquad (10\text{-}3)$$

where the first form is to be used if p is in degrees, the second if p is in minutes, and the third if p is in seconds. These equations are not restricted to the parallax of the moon; they may be used for any distances and parallaxes in the solar system.

If A and B are two points on the earth, assumed for simplicity to be at opposite ends of a diameter, the distance AB is known. If the angle p can be measured, the distance to the moon can be calculated. In practice, it is not possible to observe the moon simultaneously from two ends of a diameter. But if A and B are as far apart as convenient, the distance between them can be calculated from their latitudes and longitudes.

As can be seen from Fig. 10-3, the angle between two points, S_1 and S_2, on the celestial sphere is also equal to twice the parallax. Because the sphere is infinitely large, one half the angle between S_1 and S_2 as seen from the earth and from the moon is essentially the same. It averages 57.04′, but varies constantly because of the changing distance of the moon.

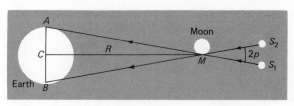

FIG. 10-3   The parallax method of determining the moon's distance.

Assuming the average parallax and the equatorial radius of the earth, 3963 miles, the distance from the center of the earth to the center of the moon may be calculated from equation (10-3), as follows:

$$R = \frac{3438' \times 3963 \text{ miles}}{57.04'} = 239{,}000 \text{ miles}$$

Using the most accurate data available, we obtain the average distance to the moon as 238,857 miles with an uncertainty of 1.3 miles (384,403 km ± 2.1 km). This distance is called the *mean distance*. It was checked by a radar echo from the moon in 1946 and by light from a laser beam in 1969. Since the distance by radar and by laser beam are from surface to surface, the distance from center to center must be calculated by knowing the radii of the moon and the earth. The accuracy by laser is about 1 foot.

EXAMPLE: On January 23, 1964, the horizontal parallax of the moon was given in the *American Ephemeris and Nautical Almanac* as 59' 15.083". Compute the distance between the centers of the earth and moon at that time.
SOLUTION: Substituting in equation (10-3),

$$R = 3438' \times \frac{3963 \text{ miles}}{59.25'} = 229{,}500 \text{ miles}$$

10.3 THE MOON'S DIAMETER

Knowing the moon's distance, and by measuring the moon's apparent angular diameter, we may adapt equations (10-3) to the determination of the linear diameter (Fig. 10-4), as follows:

$$AB = D = \frac{R \times d'}{3438'} = \frac{R \times d''}{206{,}265''} \tag{10-4}$$

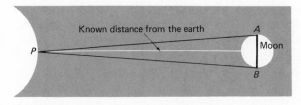

FIG. 10-4   Determination of diameter of the moon.

The moon's diameter subtends an angle of about one-half degree on the sky as seen from the surface of the earth. Its mean angular diameter is 31.09'. Hence the diameter

$$D = \frac{238{,}857 \text{ miles} \times 31.09'}{3438'} = 2160 \text{ miles}$$

Since the parallax and distance of the moon vary, its angular diameter is also variable. This can be shown by measuring the apparent diameter with the surveyor's transit or with the marine sextant.

The moon is a relatively large satellite; no other planet has a satellite that is as large in comparison with the parent body. The two bodies are often called the *earth-moon system*. The two bodies move together through space, revolving around a common center of mass, called the *barycenter*. The barycenter moves in an elliptical orbit around the sun, and the earth and moon revolve around it. The earth departs from the orbit of the barycenter by nearly 3000 miles. The determination of this distance makes it possible to calculate the mass of the moon.

10.4 THE MOON'S MASS

The product of the earth's mass and its distance from the barycenter is equal to the product of the moon's mass and its distance from the barycenter. This relationship defines the barycenter and may be expressed by the following equation:

$$M_E d_E = M_M d_M \tag{10-5}$$

where M_E = mass of the earth, M_M = mass of the moon, d_E = distance from center of the earth to the barycenter, and d_M = distance from center of the moon to the barycenter.

The distance between the center of the earth and the barycenter has been carefully determined at the United States Naval Observatory. The average distance, d_E, is 2903 miles. Hence the barycenter lies inside the earth.

The average distance of the center of mass of the moon from the barycenter is 238,857 miles − 2903 miles = 235,954 miles. Substituting in equation (10-5):

$$M_E \times 2903 \text{ miles} = M_M \times 235,954 \text{ miles}$$

From this equation,

$$M_E = \frac{235,954 \text{ miles}}{2903 \text{ miles}} \times M_M = 81.28 M_M$$

A recalculation in 1966 gave this ratio as 81.30; thus the mass of the moon is 1/81.30 or 1.23 percent of the mass of the earth. Using this ratio and the known mass of the earth, we find the mass of the moon to be 8.10×10^{19} tons. From this and the volume calculated from the moon's diameter, its density is 3.343 g/cm³, which is comparable to the density of rocks in the earth's crust.

In addition to density, the surface gravity and velocity of escape can also be computed from its mass and diameter. The surface gravity of the moon is 0.1653 or about one sixth that of the earth; and the velocity of escape is 0.212 or about one fifth that from the surface of the earth; that is, 0.1653×9.80 mi/sec² = 1.62 m/sec², and 0.212×6.95 mi/sec = 1.47 mi/sec (or 2.36 km/sec), respectively.

10.5 THE MOON'S APPARENT PATH

The apparent path of the moon among the stars as seen from the earth is a great circle roughly following the ecliptic. The moon's motion is direct (from west to east) from night to night, and of course appears to rise and set as other bodies do because of the rotation of the earth. The period of time between successive conjunctions with a star, the *sidereal month,* is about 27⅓ days. It is somewhat variable because of the attractions by the sun, the bulge at the earth's equator, and the planets. The average sidereal month is 27.32166 mean solar days and the amount of variation is as much as **7** hours.

As shown in Fig. 10-5, the path of the moon crosses the ecliptic in two points, called *nodes.* At the ascending node, the moon moves from south to north of the ecliptic and at the descending node from north to south. Because of attractions by other bodies, the nodes are not stationary, but move completely around the ecliptic from east to west in about 18.6 years. This motion is called the *regression of the nodes.*

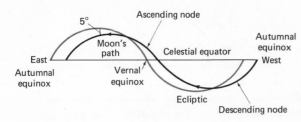

FIG. 10-5 The path of the moon and the ecliptic, showing the ascending and descending nodes and inclination of 5°.

Since the sun is also moving along the ecliptic in an easterly direction, the intervals between new moons, when the sun and moon are in conjunction, are longer than the sidereal month. This month of the phases, called the *synodic month,* averages about 29½ days in length. A more accurate value is $29^d 12^h 44^m 2.8^s$, or 29.53059 days.

The inclination of the moon's path to the ecliptic is about 5°, but it is also variable from 4° 59′ to 5° 18′ because of the external attractions. Its average is 5° 08′.

10.6 THE MOON'S ORBIT

The changes in parallax and apparent diameter of the moon are due to the variation in the distance. Both parallax and apparent diameter are inversely proportional to the distance. Hence, relative distances can be plotted on a convenient scale, inversely proportional to either parallax or apparent diameter. From the changes in distance and direction, it can be deduced that the moon's orbit is an ellipse with the earth at one focus. Also, since gravitation holds the two bodies together, it can be deduced mathematically that Kepler's laws, restated for satellites, hold for the moon and all artificial satellites in orbit around the earth.

The period of the moon in its orbit around the earth is one sidereal month. The eccentricity of the orbit is about 0.055, but is variable because of external attractions, called *perturbations.* Perturbations are disturbances of the normal elliptical motion. The principal perturbations of the orbits of the moon and the earth satellites are due to the sun and the bulge at the earth's equator. Others are numerous and difficult to compute.

Perturbations affect the eccentricities of the orbits causing them to vary by small amounts, but the averages remain about the same. For the moon, the average eccentricity e is 0.0549. Perturbations also cause the regression of the nodes and the changes in the inclination of the orbit to the plane of the ecliptic. The average inclination i is $5.145° = 5° \, 8.7'$.

The point of closest approach of the moon to the earth is called *perigee;* that of greatest recession is *apogee.* The line joining these two points is the major axis of the ellipse, or the *line of apsides.* The center of the major axis is also the center of the ellipse (Fig. 10-6).

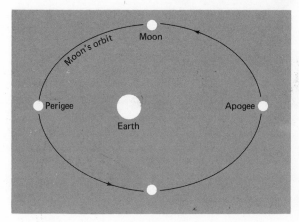

FIG. 10-6 The elliptical orbit of the moon around the earth.

Because of perturbations, the moon's line of apsides rotates from west to east—that is, in the opposite direction to the regression of the nodes. This is called the *advance of the perigee.* The perigee point makes a complete circle in 8.85 years. Because of this advance of over 3° per month, the moon reaches perigee at intervals of 27.55455 days. This is known as the *anomalistic month.*

The distance of the earth from the center of the ellipse is found by multiplying the mean distance by the eccentricity. It is 13,100 miles. Therefore the average perigee distance is 225,757 miles and the average apogee distance is 251,957 miles. But because of perturbations, the smallest actual perigee distance is 221,463 miles and the largest apogee distance is 252,710 miles.

As Fig. 10-7 shows, the moon's orbit is everywhere concave toward the sun. This can only be seen by drawing the path of the earth-moon system to the same scale as the distance to the sun.

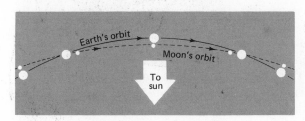

FIG. 10-7 Since the sun's gravitational pull is greater than that of the earth or the moon, both bodies follow orbits that are continually concave toward the sun.

10.7 PHASES AND ELONGATIONS

The *phases* of the moon are produced by the changing positions of the three bodies, as shown in Fig. 10-8. At new moon, the elongation of the moon can vary between 0° and 5° because of the inclination of the orbit.

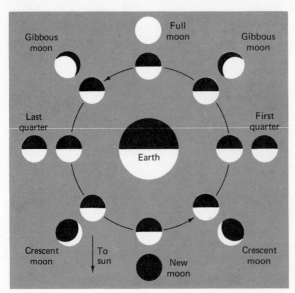

FIG. 10-8 Production of phases of the moon.

The age of the new moon (*conjunction*) is zero and increases through the 29½ days of the synodic period. After new moon the elongation increases, and the moon shows a crescent phase until it is 90° from the sun, at *quadrature*. It is then at first quarter and the moon's age is about seven days, or one fourth of its synodic period. After quadrature the moon is *gibbous,* and we can see more than one half of its illuminated surface. At full moon (*opposition*), age 15 days, the elongation is between 175° and 180°. Elongation is always measured as the shortest angular distance between two bodies. So the full moon can never have an elongation greater than 180°. After full moon the phases occur in reverse order, until at age 29½ days the moon is new again.

The edge of the moon is called the *limb.* There are both bright and dark limbs, except when the moon is new or full. The line between the bright and dark areas, the sunrise or sunset line, is the *terminator.* It is roughly the arc of a circle, but is very irregular because of the rough surface features. Just after new moon, when the crescent is very narrow, the dark surface can be seen faintly illuminated. This is sometimes called "the old moon in the new moon's arms." The reason is that from the moon the earth would appear fully illuminated— a full earth. Light reflected from the earth's atmosphere returns toward the moon. Since the reflecting power of the moon, its *albedo,* is only 7 percent and the earth's albedo is about 40 percent, the earth at full phase would appear 77 times brighter from the moon than the full moon seen from the earth. Thus the light reflected from the earth is bright enough to faintly illuminate the dark part of the moon.

The average interval between moonrises or moonsets is about 24^h 50^m. However, because the declination of the moon ranges between $-28\frac{1}{2}°$ and $+28\frac{1}{2}°$ (see Fig. 10-9), the times of moonrise are strongly affected. They also depend on the latitude of the observer. At the north pole when the moon is above the celestial equator, as it is half the time, it is always above the horizon. If below the equator, the moon never rises. As one moves from the pole, there comes a time when the moon just barely rises or sets. From the equator, it is above the horizon for exactly 12 hours at a time, except for its daily eastward motion.

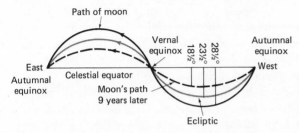

FIG. 10-9 The changing position of the moon's path with the position of the nodes. Note particularly the change of declination.

At intermediate latitudes the most noticeable change in the intervals between successive moonrises is at full moon in autumn. Since the moon is at opposition, it is near the vernal equinox at the full phase in autumn. As it rises at sunset, its northward motion carries it above the equator and it rises earlier. The eastward delay and the northward speed-up in time of rising nearly cancel each other. There are several nights when the moon rises only a few minutes later than on the previous night. The full moon nearest the autumnal equinox is the *harvest moon*. One month later, it is the *hunter's moon*. They are much more noticeable at high latitudes than nearer the equator. (See Fig. 10-10 for an explanation of the harvest and hunter's moon.)

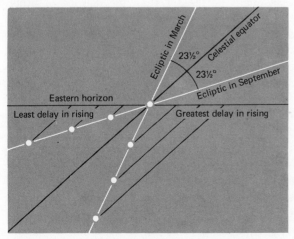

FIG. 10-10 Explanation of the harvest and hunter's moon. The moon's path is near the ecliptic, which is nearly parallel to the horizon in September and again in October.

The ascending node and the vernal equinox coincided in 1969 and will again coincide in 1988 when the moon's declination will range from $-28.5°$ to $+28.5°$. The harvest and hunter's moons are most conspicuous during those years. In 1978 the ascending node will coincide with the autumnal equinox and the moon's declination will range from only $-18.5°$ to $+18.5°$. The harvest and hunter's moons will be less noticeable then.

10.8 THE MOON'S ROTATION

It is quite obvious that the moon does not rotate rapidly. The same surface features are always visible except for a few near the limb, which when viewed through a telescope can be seen to disappear and reappear during a month's time. This is because the moon always presents approximately the same face to the earth. However, if the moon could be viewed from a distant planet, all parts could be seen each month. So the moon does rotate and the period of rotation, one sidereal month, is equal to the period of revolution.

This situation has been produced by tidal action by the earth during the billions of years the two bodies have been in existence. It is believed that when the moon was less solid than it is at present, the earth raised tides in the mass of the moon similar to the tides in the earth's oceans. When the moon's material was plastic, there was a great deal of friction in the moving masses inside the moon. This friction slowed the moon's original rotation until the period of rotation equaled the period of revolution around the earth. The internal lunar tides remained fixed when the material solidified. There remained a bulge in the figure of the moon directed toward and away from the earth. The maximum and minimum diameters differ by less than 2 miles.

Accurate tracking of the Orbiter satellites revealed that the center of mass of the moon, which governed the satellite orbits, is 1.6 miles closer to the center of the earth than is the center of the moon's shape. It was then predicted that the near side of the moon might consist of material heavier than average and that the density of the lunar seas might be greater than the density of other parts of the moon.

This prediction was proved to be correct by the discovery of *mascons* (mass concentrations) below the lunar surface. Attention was called to them, as mentioned in Section 10.2, by irregular gravity measures detected by the lunar orbiters. Ten mascons had been discovered and verified by the middle of 1969 (see Fig. 10-11). As the figure shows, the mascons are mostly under the lunar seas. They appear to be of the order of 30 to 120 miles (50 to 200 km) in extent and some 30 miles below the surface. One mascon, possibly double, has been found near the center of the moon to the left of Sinus Medii in the drawing of Fig. 10-14. It is possible that a body the size of a small planet forms the major portion of a mascon and may have formed the sea under which it lies.

The far side of the moon has been called the dark side. This is not true. When the moon is new, the dark side faces the earth and the far side is completely illuminated. The duration of sunlight on any part of the moon is half a synodic month, or about 15 days. The length of time a star is above the moon's horizon is half a sidereal month, or about 14 days.

FIG. 10-11 A lunar gravity map. Analysis of lunar satellite orbits has revealed the presence of mass concentrations (mascons) under the circular maria, such as Mare Imbrium, Mare Serenitatis, and Mare Crisium (top left to right). The irregular maria do not show mascons. (Courtesy Jet Propulsion Laboratory and Physics Today)

10.9 LUNAR LIBRATIONS

The disappearance and reappearance of features near the moon's limbs can be explained by four effects, called *librations*. They are as follows:

1. *Diurnal librations.* When the moon rises, an observer on the earth can see over the top limb. As the earth rotates, he is

carried to a position where he can see beyond the other limb, which he sees as the upper limb of the setting moon. The total effect is equal to the moon's parallax, or about 1° at both moonrise and moonset (Fig. 10-12).

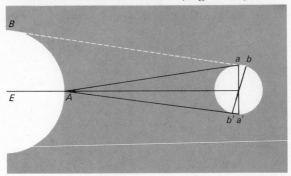

FIG. 10-12   Diurnal libration of the moon; *aa'* is the apparent diameter of the moon for an observer at *A*; *bb'* is the lunar diameter seen from *B*.

2. *Latitudinal librations.* The moon's equator is tipped about 6½° to the plane of the orbit. This permits alternately 6½° of the northern and southern limbs to be visible from some part of the earth each month.

3. *Longitudinal librations.* The moon rotates uniformly on its axis, because there is no known force to cause the rotation to slow down or speed up. The moon's revolution is not uniform, as stated by Kepler's second law. Hence the rotation and revolution get out of step. At perigee revolution is ahead, and a few degrees of the following limb become visible. At apogee the preceding limb becomes visible by a similar amount, about 8° at each limb.

4. *Physical librations.* There is a slight shift of the moon, produced by its bulge in the earth's direction. These liberations move the features of the moon by about 1 mile as seen from the earth.

The expressions "see over the top limb" or "see beyond the other limb" mean that an observer may see a different limb from that seen by an observer in line between the centers of the earth and the moon. Thus an observer at *B* watching the rising or setting moon sees beyond the limb, *aa'*, that would be visible to an observer at *A*, where the moon is on the meridian. Also, an observer at a north or south latitude would see a different limb from one seen by an observer on the equator.

The result is that 41 percent of the moon's surface can never be seen from the earth. Forty-one percent is always visible, and the remaining 18 percent is visible at times from some place on the earth because of librations. This area is along the limb and is very irregular in shape.

10.10 THE MOON'S ATMOSPHERE

The moon has little or no atmosphere. The velocity of escape from the lunar surface is so low that, if the moon ever had an atmosphere, it has now all escaped except for a very thin trace from the solar wind—atoms held temporarily captive by the moon's gravitation. The velocity of escape is only 1.475 miles/sec, as we have shown. This is very near the velocity of molecules in an atmosphere of oxygen and nitrogen at moderate temperature, such as we have on the earth. In this case, at some time each atom or molecule of the original atmosphere must have reached the escape velocity and left the moon's gravitational field. It has been calculated that if the average molecular velocity were as high as 0.5 mile/sec the moon's atmosphere would have escaped in a few weeks.

In 1958 a Russian astronomer, N. A. Kozyrev, reported that he had seen carbon gas escaping from the region of the central peak in the crater Alphonsus. This would indicate that some molecules are coming from the cracks in the floor of the moon. Kozyrev failed to mention that black areas are visible near the crater walls and that they have depressions in them. Ranger 9, which landed in Alphonsus, relayed back to the earth photographs of the crater including the dark areas, but failed to detect any trace of carbon gas. Figure 10-13 shows the crater, the central peak, and the black areas. (See also Fig. 1-9, a Ranger photograph of Alphonsus.)

The temperature of the lunar surface can be estimated from the earth by the use of a thermocouple. If two strips of different metals are welded together at both ends and if one joint is heated while the other is kept cold, a difference of electric potential (voltage) is produced. When one joint of a thermocouple is placed at the focus of a telescope and moonlight is allowed to fall on it, the amount of radiation can be measured, since it is proportional to the difference in potential, and the temperature of the moon's surface can be calculated. A thermocouple placed at the focus of the 100-inch telescope was also used to measure the very small amount of heat from the stars.

The results for a region of the moon directly below the sun show a high temperature of 110°C (230°F), slightly above the boiling point of water. The dark side measures −173°C (−280°F). Away from the subsolar point a beam of sunlight covers a greater area and the temperature is not quite so high. The temperature at the lunar poles is uncomfortably, but not unbearably, high. The vehicles that landed carried instruments for measuring the temperature and for sending the resulting data back to the earth. The high surface temperature results from the fact that the sun is above the lunar horizon for two weeks at a time.

The lack of a lunar atmosphere is particularly striking during a total eclipse of the moon. At those times, sunlight is intercepted by the earth; the lunar temperature drops very rapidly, then rises immediately when the moon moves out of the earth's shadow. The rise in temperature may be as much as 200°C (360°F) in 1 hour. Under the extreme changes, rocks would be shattered by alternate expansion and contraction unless protected by a layer of dust. Before Surveyor I

FIG. 10-13   The surface of the moon around the crater Alphonsus (white arrow). The central peak and black areas near the walls are easily visible. (*Lick Observatory photograph*)

landed in 1966 the thickness of the dust layer was entirely unknown. Former estimates varied from an inch or two to 1 mile.

Because of the absence of atmosphere, the moon has no water and no clouds. The extremes of temperature result from rapid heat radiation after sunset. The moon has no twilight and no auroras. The sky is black, and one should be able to see the stars even in daytime by shielding the eyes from direct sunlight. Blue sky on the earth is caused by scattering of blue light mostly by air molecules and dust in the air.

On the moon there is no sound, since sound requires a medium for transmission. Meteors are not visible, although it is certain that both large and small meteoroids hit the moon all the time. Ranger 7 photographs showed the effects of meteoric bombardment on the small surface craters (Fig. 1-8). Their sides appear to be eroded, and, there being no water or wind to erode them, they must have been struck by small meteoroids or by ultraviolet radiation, which would not get through an atmosphere.

10.11   THE MOON'S SURFACE

There are several types of features on the lunar surface, all of which, except the smallest, can be observed from the earth. The following is a list of features: (See Figs. 10-14 and 10-15).

1. *Seas.* The relatively smooth, dark, and large areas representing seas are conspicuous even in small instruments.
2. *Craters.* Hundreds of thousands of depressions in the lunar surface are called craters. They vary from 146 miles to 3 feet in diameter.
3. *Mountains and mountain ranges.* Through the telescope the moon's mountains look much like the mountains on earth.
4. *Rays.* Rays are bright streaks that radiate from some craters. They are as long as 2000 miles.
5. *Rills (or rilles).* Rills are cracks in the surface, probably produced by faulting.
6. *Domes and calderas.* Swelling of the surface, followed by subsidence, results in domes and calderas. Occasionally a crater is seen on the top of a mountain peak.

Galileo was the first to observe the dark areas, which he called seas, or *maria* in Latin. There are about 20 maria, which cover nearly half the surface facing the earth. The largest is the Mare Imbrium, the Sea of Showers, some 700 miles across. Others have similar fanciful names, such as Tranquillity, Serenity, Fertility, Crises, Clouds, Nectar, and the Bay of Rainbows. Most of the seas were named by Hevelius, a Polish astronomer, who published a book on selenography. (Selene was the goddess of the moon.)

The seas are not as smooth as they look at first glance, but are pitted with craters and surrounded by mountain ranges. Their dark appearance is due to

FIG. 10-14 Photograph of the western hemisphere of the moon and drawing (on opposite page) showing principal features. *(Lick Observatory photograph)*

their low albedo, a reflecting power of only 2 percent compared to 7 percent for the average and 40 percent for Aristarchus, the brightest spot on the moon. The curvature of the moon is so great that the walls around the seas are not visible from the centers. For example, the highest peak in the lunar Apennines is Mt. Huygens, which is 3.4 miles high. This peak cannot be seen from a distance greater than 86 miles, which is only about one eighth the distance across the Mare Imbrium.

The rays are nearly straight lines of higher albedo extending outward from certain craters. The most conspicuous ray system radiates from the crater Tycho, near the lunar south pole. During full moon these rays are easily seen, and it is possible to trace one across the Sea of Serenity almost to the northern limb. Allowing for the moon's curvature, the length of this ray is more than 2000

SOUTH

Moretus
Watt
Cuvier
Steinheil
Faraday
Metius
Janssen
MARE AUSTRALE
Fabricius
Stöfler
Rheita
Valley
Maurolycus
Rheita
Aliacensis
Piccolomini
Apianus
Werner
Playfair
Fracastorius

MARE
NECTARIS
Catharina
Abulfeda
Mädler
Cyrillus
Albategnius
Langrenus
MARE
Torricelli
Theophilus
Pickering
Hipparchus
MARE SMYTHII
Delambre
Horrocks
Messier
Ritter
Godin
Triesnecker
WEST
Sabine
Maskelyne
Agrippa
MARE
Hyginus Cleft
Boscovitch
TRANQUILLITATIS
Julius Caeser
Ariadaeus
Cleft
Proclus
Plinius
Manilius
MARE MARGINUS
MARE
Macrobius
CRISUM
MARE
Bessel
Le Monnier
Linne
Autolycus
Posidonius
Aristillus
SERENETATIS

Caucasus Mts.
LACUS SOMNIORUM
Cassini
Eudoxus
Piton
Atlas
Mortis
Aristoteles
Endymion
Hercules
Alpine Valley

MARE FRIGORIS
MARE
HUMBOLDTIANUM
Gartner

NORTH

SEAS
Craters and walled plains
Mountains and other formations

miles. It can be traced on Fig. 10-14, the western hemisphere of the moon.

Another easily seen system of rays extends radially from the crater Copernicus (see Fig. 10-15). Others radiate from Kepler, Herodotus, Aristarchus, and others. Ranger 7 landed in Mare Nubium (Sea of Clouds) in a relatively smooth area on a ray from Tycho and possibly one from Copernicus. This region was selected as a test site for manned spacecraft landings because of its apparently smooth surface. Ranger 8 landed in the Sea of Tranquillity and found conditions there very similar to those in the Sea of Clouds. The name of the latter sea was changed after the Ranger 7 landing to Mare Cognitum (the Known Sea).

Surrounding many of the seas are ranges named for mountain ranges on the earth, such as the Apennines, Alps, and Caucasus. Other mountain ranges are

FIG. 10-15   Photograph of the eastern hemisphere of the moon and drawing (on opposite page) showing principal features. (*Lick Observatory photograph*)

located in the southern hemisphere. The highest are the Doerfel Mountains and the Leibnitz range, which reach heights of some 30,000 feet. Other mountains are single, such as Pico and Piton, rising out of the Mare Imbrium. The height of Piton can be calculated from measurements on a photograph taken near the quarter-moon, since its shadow is sharp and easily measured (see Fig. 10-15).

The lunar features on the far side of the moon have been studied on photographs made by both Russian and American lunar probes. (See photographs and maps in this chapter.) The first map of the far side was published by the Russians, who named some of the seas and craters. The American spaceships have photographed the far side, as well as the near side, in more detail. Many of the lunar features have been named by the Americans. To become official

SOUTH

Clavius
Maginus
Longomontanus
Tycho
Schickard
Walter
Aliacensis
Pitatus
Werner
Purbach
Thebit
Straight Wall
MARE HUMORUM
Alphonsus
Arzachel
Gassendi
Albategnius
Bullialdus
MARE
Ptolemaeus
Hipparchus
Herschel
Latronne
Mösting
Flammarion
NUBIUM
OCEANUS
Grimaldi
EAST
SINUS
Copernicus
MEDII
Kepler
Eratosthenes
Aponaine Mts.
MARE
PROCELLARUM
Pytheas
Timocharis
Lambert
Herodotus
Autolycus
Aristillus
Archimedes
Cassini
Piton
IMBRIUM
Pico
Jura Mts.
Plato
MARE FRIGORIS
Fontinella

NORTH

SEAS
Craters and walled plains
Mountains and other formations

these names must be approved by the International Astronomical Union. (For a complete list of names on the far side, see *Sky and Telescope* for November 1970.)

10.12 THE NATURE OF THE SURFACE

The first soft landing on the moon was made by the Soviet Luna 9 on February 3, 1966, which transmitted pictures for three days. It was followed by the American Surveyor I on June 2, 1966 (Fig. 10-16). Its pictures showed much better resolution and gave a better understanding of the lunar surface. One of the most significant pictures was a narrow-angle photograph showing one of the vehicle's footpads implanted in the lunar soil (Fig. 10-17). The pad sank only about 2 inches into the surface, indicating that the lunar soil is underlaid by a

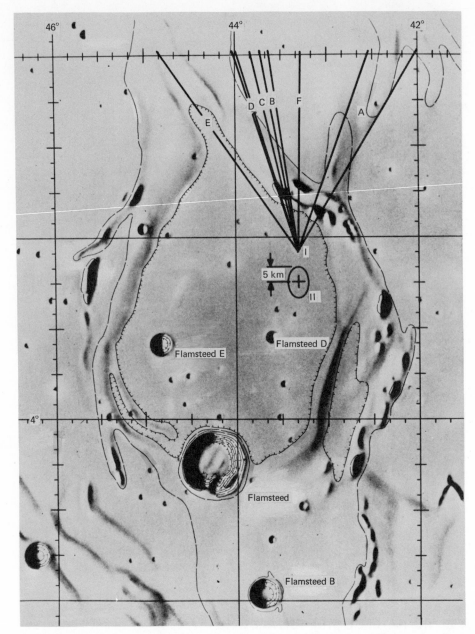

FIG. 10-16 Lunar chart showing the landing site of Surveyor 1. Area 1 designates the site as deduced from horizon features in sections A through F, while area 11 was deduced from trajectory tracking data. (NASA)

hard material that can withstand a pressure of at least 10 lb/in². This is covered by a layer of loose granular material similar to terrestrial soil.

In the next year and a half, four more surveyors were successful in landing,

FIG. 10-17 Surveyor 1 photo of the moon with a portion of the spacecraft landing gear at the upper right. The construction of the foot pads made it possible to obtain data about the land-bearing capacity of the lunar surface. (NASA)

analyzing the surface material, and transmitting this data, plus pictures, to the earth. During this time, five lunar orbiters photographed over 99 percent of the lunar surface and televised the pictures to the earth. The general character of the surface, as shown by these pictures and those of the Apollo missions, is that of a rolling terrain of plowed-up soil, covered with craters ranging from inches to many miles in diameter.

The Apollo 11 mission of July 20, 1969, carried out several scientific experiments mentioned in Chapter 1. The command module, piloted by Michael Collins, continued to orbit the moon while the lunar module carried Neil Armstrong and Edwin Aldrin to a safe landing on the surface and afterward to a successful re-docking with the command module. Armstrong and Aldrin wore special suits that contained oxygen, water, cooling and heating equipment, and other materials essential for survival on the moon for about 2½ hours. Armstrong was on the surface for just over 2 hours and Aldrin about 15 minutes less.

Walking on the surface was not as difficult as had been predicted. The layer of dust was only a few inches thick at most. The original landing site was found to be filled with boulders and was passed over by skillful navigation of the landing module. The actual landing site was covered with scattered rocks, some 50 pounds of which were collected into sealed bags and brought back to the earth.

Several experiments were set up. One was a panel to catch any material from the solar wind. This panel was removed at the end of the lunar walk and brought back to earth for a determination of the composition of the solar wind. A lunar seismometer was set up and left in place to record and send back to earth any lunar tremors that might have been produced by lunar quakes, landslides, or the possible fall of meteoroids onto the surface. It was more sensitive than the usual seismometers used by geologists on the earth.

All lunar materials were carefully placed in sealed, airtight containers to avoid possible contamination. The astronauts and their lunar material were placed in quarantine in a special building for debriefing, physical examinations, and complete isolation for three weeks to be sure no contaminant from the moon was carried back to earth. At the end of three weeks all were released and no contaminant was found. Lunar dust was injected into mice and studied for reaction to any lunar virus. No reaction was found.

10.13 LUNAR CRATERS

The type of lunar feature that occurs in greatest numbers is the crater. More than 30,000 have been counted on photographs taken from terrestrial observatories. The typical crater is a pit, frequently with a peak in the center. The largest crater is Clavius, 146 miles in diameter. There are millions visible in close-up photographs, the smallest about 3 feet across and 18 inches deep.

Craters are named for philosophers and astronomers of the past, a custom started by Riccioli, an Italian astronomer, in 1651. A glance at the charts facing the two photographs of the quarter moon in this chapter will give the student a good idea of those represented. More detailed lunar maps have been published, including one by NASA of the far side. The new craters are named for astronomers of the 20th century who are no longer living.

The floor of the typical crater is below the average level of the surface, sometimes several thousand feet lower. The walls are piled up above the average level. It has been estimated that the volume of material in the walls is about the same as the volume of the crater, indicating that the material was somehow dug out of the crater and deposited around the rim. Running outward from some of the prominent craters are systems of rays, as mentioned previously. Where did this material come from and why is it brighter than that of the crater floor? The answer seems to be related to the problem of crater formation.

Rills are cracks in the lunar floor. The usual rill is about half a mile wide, of unknown depth, and many miles in length. They are sometimes straight (for example, the Great Wall, about 60 miles long), but many are curved and run across other types of features. Ranger 9 photographed beautiful rills on the floor of the crater Alphonsus (Fig. 1-9) with small craters strung along the rims.

Rills are apparently faults in the lunar surface. It is possible that when the moon cooled it contracted and the rills were formed by slippage during the process.

Surveyor I landed on the Oceanus Procellarum (Ocean of Tempests) near the east limb and inside a large partially buried crater, which has the smaller crater Flamsteed on its southern rim. Photographs of the horizon show low mountains on the crater rim about 7.5 mi distant. They project up to 300 feet above the ordinary, near horizon. The ancient crater has been almost buried beneath the covering material of the ocean. This crater is about 60 miles in diameter. About 20 small craters, ranging from about 10 feet to 300 feet in diameter, are in sight of the cameras on board the spacecraft. One crater, about 35 feet away, is 10 feet wide and 2 feet deep.

The close-up photograph (Fig. 10-18) shows that the surface is covered with granular material down to at least $\frac{1}{50}$ inch in diameter, with large boulders 20 inches long. These boulders are pitted and cracked. The result of a brief study indicates that the surface consists of a hard material, with a weaker material on top having a depth of about 1 inch. The coarse blocks are scattered at random. There is no appreciable layer of dust in this region. Most of the smaller visible craters are shallow and have low, rounded rims or seem to be rimless.

FIG. 10-18 One of 4000 photos taken from Surveyor 1 during its first five days on the moon. The distance along the horizon (from upper left to lower right) is about 50 yards. (NASA)

10.14 FORMATION OF LUNAR FEATURES

The question of the formation of craters and other lunar features has never been answered satisfactorily. One theory is that the moon was a molten mass that solidified, producing the craters by volcanic action. However, since the moon's density is about the same as that of rock on the earth and since it is believed that there is no dense core, as there is inside the earth, the central temperature is probably not over 2000°C, and the interior is solid rock. If the moon was once a hot, molten mass, it is difficult to see how it could have cooled to its present temperature.

There is no overall lunar magnetic field like the earth's, which is thought to result from electric currents in a molten core. The lack of such a pattern of magnetism indicates that the moon does not now have a molten core. However, the Apollo astronauts have measured weak magnetism—less than 1 percent of the earth's. The fact that it varies from place to place indicates that it is caused by local sources. These sources must have acquired their magnetism in the past. Possibly the moon at that time had a molten core and a strong magnetic field.

Seismometers set up by the Apollo astronauts have provided additional clues as to the nature of the lunar interior. The seismic events that have been recorded were small. Some, which occurred within three days of perigee, were apparently shallow moonquakes. Others were apparently caused by the impact of meteoroids. The low level of seismic activity, relative to the earth, suggests that the outer shell of the moon is more stable than the earth's.

Another, and probably more widely accepted, theory is that the craters were formed by bombardment of meteoroids coming from orbits around the sun, which would strike the moon with velocities as high as 45 miles/sec or even higher. The kinetic energy of motion would be transformed mainly into heat, raising the temperature sufficiently to melt the surface material. The sudden rise in temperature, possibly to as much as 1,000,000°K, would vaporize at least a portion of the lunar rock and result in an explosion that would be most efficient in a direction perpendicular to the surface. Since the surface gravity is only one sixth of the earth's surface gravity, the explosion would be 6 times as effective as on the earth, and, because of the lack of atmosphere, the material would be thrown to great distances. Experiments with bullets fired at a lead plate give results similar to the depths and shapes of the lunar craters, including central peaks.

Some of the kinetic energy would be dissipated into shock waves, which may have caused great destruction of already existing craters and mountains on the opposite side of the moon. Such effects have not yet been fully investigated. The seismographs left by the Apollo astronauts are helping to solve this problem.

The fact that craters are seen on the walls of other craters strengthens this theory. But the objection has been raised that no new craters are now being formed. However, if the moon is 5 billion years old and if there are 1 million craters visible from the earth, the rate of production is only one every 5000 years, and the moon has been under observation with telescopes for less than one tenth of that time.

The rays consist of material that may have been thrown out of craters when they were formed. This would explain why the rays run in nearly straight lines, which they would not do if they were formed by material coming from cracks in the lunar surface. But why only a few craters have ray systems is not known.

The meteoric theory is also strengthened by the belief that there were probably more meteoroids in space several billion years ago than there are now.

Alphonsus might have been formed by a dome pushed up by gases from inside the moon, a theory strengthened by Kozyrev's observations of carbon fumes. As the dome settled, the floor shrank and cracked, then was hit later by meteoroids.

One theory of the formation of the seas suggests that grazing collisions with low-velocity particles produced lava by melting. Another scientist has proposed the theory that the seas were formed by impact with comets, which released chemical energy in addition to kinetic energy. Still another thinks they resulted from internal basalt formed when the fluid interior released gaseous elements. The craters could have formed later. A close examination of the Mare Imbrium shows original craters almost completely filled with the heated material at the time the sea was formed.

Probably a better theory is that the moon was originally covered with craters such as those in its southern hemisphere and that a very large fall of meteoroids or large comets melted the surface, causing the material to flow into the existing craters and even into some of the neighboring seas. The smooth floor of the Mare Imbrium could have been produced as a result of the solidification of this melted lava and was later hit by other meteoroids to form the craters of smaller size visible today. The maria were not all formed at the same time. Rocks brought back from the Ocean of Storms are several hundred million years younger than those from the Sea of Tranquillity.

At present it is safest to conclude that the lunar features were not formed by a single kind of catastrophe, but that both impacts from outside and forces inside the moon have been at work for billions of years. Almost certainly, craters like Tycho and Copernicus and their ray systems were formed by impact. They also appear to be younger than others outside the system of seas. Alphonsus and possibly even Clavius may have been formed by the collapse of domes.

A map and photographs of the far side of the moon are shown in Fig. 10-19.

10.15 MOON ROCK STUDIES

More than 500 scientists from nine countries studied intensively some 50 pounds of lunar soil and rocks picked up on the surface by the astronauts from the Apollo 11 mission. The studies included analyses of the chemical composition, general and special mineralogy studies, tests for the determination of the age of the Sea of Tranquillity, investigations of magnetic and electrical properties, and search for organic material.

The samples from Tranquillity Base were the first extraterrestrial objects,

FIG. 10-19   The moon's far side. The three charts (see opposite page) are based on American Lunar Orbiter photos, one of which is shown above. The photo was taken from an altitude of about 900 miles. The prominent crater is about 150 miles in diameter. (NASA)

other than meteoroids, that scientists had been able to study in terrestrial laboratories. There were some surprising results. It was recognized early, before the intensive studies were started, that the lunar rocks were pitted and contained glassy beads. It was assumed that these beads were formed by the cooling process after the material had been heated to a high temperature in a very dry atmosphere at very low pressure. Sufficiently high temperatures would occur directly below the impact of a meteoroid and the glowing drops would be scattered in all directions mixing with the soil.

In general the rocks were basaltic, many consisting of small angular fragments cemented together, apparently by the heat and pressure generated by meteoroid impacts. The soil samples were a mixture of fragments of rocks, iron meteorites, and glass. Most of the rock fragments apparently are small pieces of larger rocks that were probably once part of the bedrock underneath. It is thought that these fragments came originally from the lunar highlands, some 25 miles south of Tranquillity Base.

The chemical contents confirm the findings from the Surveyor studies, except for the minor variations to be expected in view of the different locations of the samples. Oxygen is the most abundant element, being in compound with iron and silicon and other elements. There are traces of many other elements, including the rare earths and uranium and its derivatives.

The textures of the igneous rocks—those produced by the action of heat and usually implying fusion—are similar to those on the earth. Chemically, however, they differ significantly from terrestrial rocks and also from meteoroids. Although oxygen was the most abundant element, as it is on the earth, its con-

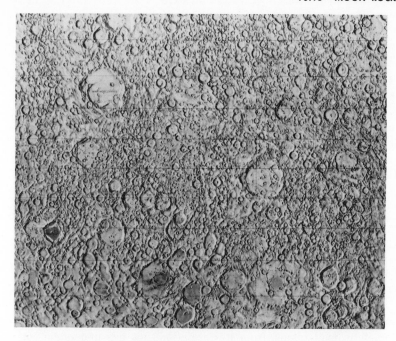

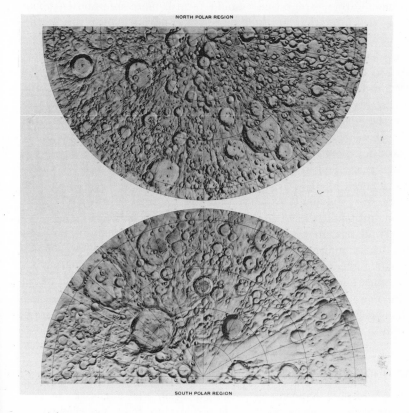

NORTH POLAR REGION

SOUTH POLAR REGION

centration was considerably less than in common terrestrial igneous rocks. Free metallic iron, which is extremly rare in rocks on the earth, was common in the lunar rocks. The silicate materials were unusually transparent and clear, indicating crystallization in a very dry system.

Tests for carbon in lunar material showed carbon in compound with oxygen, mostly carbon monoxide, including some gas bubbles in the glass beads. The abundance of carbon was about 200 parts per million. The result indicates that most of the fine particles had been heated and that the carbon had combined with oxygen from mineral oxides at that time.

The only positive results of tests for hydrocarbons, compounds of carbon and hydrogen such as those found in oil and gas on the earth, were attributed to pollution by gases from the landing retro-rockets of the landing craft.

Radioactive dating, based on the relative concentration of radioactive elements and the products into which they decay, gives an age for the lunar soil of 4.6 billion years. This is the same age obtained from radioactive dating of meteorites and is assumed to be the age of the solar system. The ages of crystalline rocks indicate that they were formed several million years later.

Apparently the soil samples provide an effective average age of the lunar crust. The relative youth of the rocks and maria indicates that the moon has not been a completely dead planet from its formation, but has undergone considerable evolution. Future explorations should furnish more information about the era prior to the formation of the lunar rocks, during which the earth's record has been obliterated.

Additional evidence of crustal changes is provided by gases that were emitted by the hot surface of the sun (the solar wind) and trapped in the rocks, and by the microscopic tracks produced by the high-energy impact of nuclei from the galaxy (cosmic rays). The concentrations of these gases and nuclei in samples obtained below the surface indicate that the material must at one time have been exposed to the surface. Although there is no atmosphere on the moon, the lunar rocks showed evidence of erosion at about 10^{-8} to 10^{-7} inch, about one atomic layer, per year. This resulted from the high velocity impacts of micrometeoroids as evidenced by the microscopic pits on the surface of rocks and soil fragments. There was abundant evidence that the material in the crust must have been mixed by bombardment of very small and very large particles.

The lower abundance of certain elements compared with that of the same elements on the earth indicates that the lunar material separated from a primeval nebula at a temperature of 1000°C or higher. Among these elements are potassium and other alkali metals. Titanium and others, including some of the rare earths, are more abundant than on the earth. The soil is enriched with nickel and other precious metals. This is consistent with observed abundances of meteoric material in the lunar soil.

10.16 LUNAR AND SOLAR TIDES

In the description of the shape of the moon, mention was made of the tides raised by the earth on the moon and the fact that they have solidified into a permanent bulge in the direction of the earth. The moon also raises tides in the earth, which, though very small, have been measured. But the most obvious tides are those raised in the sea by a combined gravitational pull of the moon and the sun.

Tides are caused by differential forces. That is, because of a difference in distance, the force exerted by the moon and sun on the oceans is different from that on the earth itself. The gravitational pull is inversely proportional to the square of the distance, according to Newton's law. But the differential tidal force varies inversely as the cube of the distance.

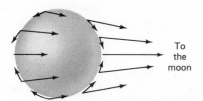
To
the
moon

Moon FIG. 10-20   Relative magnitude of the moon's gravitational force on different parts of the earth.

Figure 10-20 shows the relative magnitudes of the moon's gravitational force on different parts of the earth. On the near side (toward the moon), the gravitational force per unit mass of the ocean is greater than the inertial reaction on the solid earth, and the ocean bulges toward the moon. On the far side, the inertial reaction on the earth is less than the force on the water, and the ocean bulges away from the earth. That is, the earth is pulled away from the water.

The earth rotates under the two tides, so they appear to move westward at a rate of about 1000 mph at the equator, assuming that the moon is directly above the equator and that the earth is completely covered with water. Because of the land masses, however, the actual computation of the tides is very complicated and can be made only by high-speed computers for each port individually. On the average the high tides, followed by low tides, return at intervals of $12^h 25^m$. This is half of the average interval between transits of the moon across any meridian. This coincidence was responsible for the discovery that the tides are produced by the moon.

The sun also produces tides on the earth. Although the tidal force of the sun's attraction on the earth is much greater than that of the moon, it has less effect on the tides. Because of its greater distance from the earth, the sun's force per unit mass does not vary as much from one side of the earth to the other as does the moon's force. As a result, the sun's tide-raising force is not quite half that of the moon's.

When the moon is new or full (that is, at syzygy), the two tides reinforce each other. This is called a *spring tide*. At the times of spring tides, the high tides are unusually high and are followed by unusually low tides. At quarter moon (quadrature), the solar and lunar tides tend to cancel each other. The moon's tide being slightly greater, there is a shallow high tide followed by a moderate low tide. This is called a *neap tide*. Since the distances to the moon and sun are involved in the tide-raising forces, it is easy to see that the highest spring tides occur when the earth is at perihelion in January and when the moon is at perigee.

As the tides ebb and flow, there is friction between the water and the ocean floor. The narrow Bay of Fundy in eastern Canada has tides that are sometimes as high as 50 feet. The friction is probably greatest in the Bering Straits. The effect of these forces is to slow the rotation of the earth. The amount can be calculated from the times of eclipses centuries ago. It has been deduced that the length of the day has increased by about 0.0016 second per century, equivalent to a rate of decrease of energy of some 2 billion horsepower.

To allow for the increased length of the day, all clocks keeping solar time were set back one second at the end of June 30, 1972. This day became the longest day in history, having 86,401 seconds instead of the usual 86,400 seconds!

The final result of the exchange of energy between the earth and the moon will be that the earth's rotation will eventually be stopped entirely and the earth and the moon will keep the same hemispheres directed toward each other. The moon will slow down in its revolution around the earth until the month and the day are equal to an estimated 47 of our present days. The moon will meanwhile recede from the earth.

After the lunar tides stop, the solar tides will still be acting. They will bring the moon back toward the earth and its distance will keep on decreasing until finally the moon will come so near the earth that the tidal forces will shatter the moon and the earth will acquire a ring system like that of Saturn. These changes will require an immense amount of time.

QUESTIONS AND PROBLEMS

Group A

1. Trace the apparent path of the moon by noting its position in the constellations each night. Mark the position on the star maps in this book. Extend your observations through one cycle, if possible, and summarize your findings about positions and phases.

2. If an astronaut is on the moon when it is at an average distance from the earth, how much time is required for him to receive a radio signal from Houston?

3. If the resolving power of your eye is 1′, how far away is a car when you are just able to resolve the headlights, if they are 5 feet apart? *Answer:* about 3.25 miles.

4. The earth has about 13.4 times as much surface area as the moon. Using their albedos, show that the earth at full phase reflects 77 times more light than the full moon. (*Hint:* The total amount of light reflected is proportional to the area and the albedo.)

5. (a) If you observed the moon to be 30° above the western horizon at sunset, where would it be the following night at sunset? Would it be brighter or dimmer? (b) Repeat for 30°

above the eastern horizon at sunrise. (c) Repeat for sunset, 30° above the eastern horizon.

6. Why is the albedo of the earth so much greater than that of the moon?

7. Show why the dark side of the moon receives more light from the earth at crescent phase than at gibbous phase by making a diagram of the positions of the sun, moon, and earth at those phases.

8. What is the phase of the moon when on the meridian at (a) 6 P.M.; (b) midnight; and (c) 3 A.M.? *Answer:* (a) first quarter.

9. Show from the synodic period that the moon rises approximately 50 minutes later each night.

10. What phases would the earth exhibit to an observer on the moon? Would the earth rise and set?

11. Draw a diagram showing that the moon must make one rotation with respect to the stars during each revolution in order to keep the same face toward the earth.

12. Give reasons for the extremes of temperature on the moon.

13. What could cause changes on the surface of the moon?

14. At what phases of the moon are the craters Archimedes, Kepler, and Alphonsus observable?

15. How do the surface features of the far side of the moon differ from those on the near side?

16. Measure the lengths of the Alpine Valley and the diameter of the crater Archimedes on the moon photographs and compute their dimensions in miles. *Answer:* Alpine Valley, about 80 miles.

17. Judging from their appearance in the photographs, would you estimate the crater Tycho or Hipparchus to be the older? Why?

18. Summarize the arguments for and against the two main theories on the origin of the lunar craters.

19. A spring tide occurs at a given port at 1 P.M. on August 1. (a) When will the next low tide occur? (b) On what day will the next neap tide occur? (c) On what day will the next spring tide occur?

Group B

20. (a) What is the apparent angular diameter of the earth when observed from the moon? (b) If Venus were observed from the moon and showed an angular diameter of 1' of arc, how far away would it be?

21. Assuming the distance from the moon to the earth to be the same as now, compute the mass of the moon if the barycenter were located 21,700 miles from the center of the earth.

22. Compute the surface gravity of the earth if it were (a)

compressed to half its present diameter, and (b) expanded to three times its present diameter.

*23. Calculate Saturn's (a) velocity of escape, (b) surface gravity, and (c) average density. Assume the mass to be 100 times and the diameter 9 times those of the earth. Check the answers with Table 13-1. (They will be slightly different because of the approximations.)

24. Compare the kinetic energy of a meteoroid with a velocity of 45 miles/sec with that of a rifle bullet with muzzle velocity of 0.5 miles/sec, if their masses are the same. (See Chapter 4 for the formula for kinetic energy.)

25. The ratio of the solar tide to the lunar tide is given in this chapter. Show that the moon is responsible for 69 percent of the total tide-raising force.

26. Restate Kepler's three laws of planetary motion for satellites around the earth or other planets.

27. Using the sidereal period of the moon and its mean distance, calculate by Kepler's law restated in Question 26 the distance from the earth's surface to a satellite whose period is one day.

28. From the data of Question 1, determine the inclination of the moon's orbit to the ecliptic and the longitude of the node.

11 the sun

The sun is a star; but since the earth is at just the right distance for it to support the kind of life that exists on the earth, it is the most important star to us. It is also the only star near enough to be examined in detail with our telescopes. In fact it is so close that the earth is in the solar atmosphere. Since the sun is a star, a good start can be made in stellar astronomy by a careful study of all the solar details.

11.1 THE SUN'S DISTANCE AND SIZE

The mean distance to the sun is a fundamental unit that must be known before we can determine the dimensions of solar features, and, going still further out into space, investigate the physical nature of stars and the parts of the universe to which they belong.

The relative distances of the planets from the sun can be found with the aid of Kepler's third law. The sidereal periods can be calculated from the synodic periods (see Chapter 4), which are readily determined from observations extending over several years. The times of oppositions of Mars have been recorded since the time of Tycho Brahe, from which its synodic period is 780 days. The sidereal period, 687 days, is derived from the relation between the sidereal periods of Mars and the earth. From Kepler's third law, it follows that Mars is 1.524 times farther from the sun than the earth is.

This distance is expressed as a ratio rather than a distance in miles or kilometers. All such planetary distances can be computed from their sidereal periods in terms of the earth's distance. They can all be converted to miles after one of the distances inside the solar system has been measured. The distance from the earth to the sun has been adopted as the standard of distance inside the solar system. All distances in space depend ultimately on this standard.

The problem of determining the distance from the earth to the sun has been a very difficult one for centuries. But we now have a good value, accurate probably to within a few hundred miles. The reason for the difficulty is that the parallax method formerly used for the distance to the moon cannot be used because no bright stars can be seen on the sky background behind the sun. Also the angle is quite small, less than 9″ of arc.

However, other methods are available now that we have radar and space vehicles. In 1900 a parallax of 8.8″ was adopted by international agreement as the best value at that time. Since 1900 the sun's distance has been expressed in terms of the distance deduced from that parallax. The distance of 149,500,000 km, or about 92,890,000 miles, was called the *astronomical unit*. The 1968 *American Ephemeris and Nautical Almanac* gives the value of the astronomical unit as 149,600,000 km, or about 93 million miles.

After the distance to the sun has been determined, its diameter can be found from its apparent mean angular diameter of 32′ of arc. The linear diameter is 864,000 miles (1,390,000 km) with a probable error of about 100 miles.

11.2 THE SUN'S MASS

The sun's mass can be determined from gravitational formulas and is found to be 1.99×10^{30} kg, or 332,930 times the mass of the earth.

Dividing the mass in grams by the volume in cubic centimeters, the average density of the sun is 1.41 g/cm³, which is smaller than the earth's average of 5.51 g/cm³ and the moon's of 3.34 g/cm³.

11.3 THE SUN'S RADIANT ENERGY

The earth obtains all its heat and light energy from the sun, except for a few sources that cannot be traced directly to it. There are, for example, energy sources from nuclear reactions, volcanoes and hot springs, and lunar tides. Other sources, coal and oil deposits, can be traced to plant and animal life that grew in sunlight. Water power is possible because solar energy lifts the water, which then falls from higher to lower levels. Even the wind, which drives sailboats and turns windmills, comes from a circulation of air due to solar heating. And solar tides furnish part of the tidal power, which we have not yet learned to harness satisfactorily.

The determination of the sun's temperature depends on a measure of the amount of energy received at the earth's mean distance. The space between the earth and the sun is so empty that it can be assumed that very little solar energy has been absorbed on the way. Thus it is possible to calculate the amount of energy a square centimeter of the solar surface radiates into space.

One difficulty in the measurement of solar energy is that it is absorbed by the earth's atmosphere. Until recently, it was necessary to set up radiation-measuring instruments at various altitudes and calculate what it would be outside the atmosphere. Now it is possible to send instruments into space, make measurements, and send back the results by radio.

The instrument used is the pyrheliometer, which measures the rate of increase in temperature of a solid or liquid exposed to sunlight. It is of course necessary to be sure that only the radiation from the sun strikes the pyrheliometer and that none is lost by reradiation during the experiment.

The amount of solar radiation received outside the earth's atmosphere per second per square centimeter at the mean distance is called the *solar constant*. It is not quite constant, but varies slightly, usually not over 0.4 percent. Its value obtained by a long series of measurements is 1.96 small calories per minute per square centimeter of surface at right angles to the direction of the sun. Expressed in joules, the solar constant is 1.37×10^{-1} j/sec/cm^2.

It is safe to assume that the sun radiates uniformly in all directions; otherwise there would be cooler and hotter regions in space. There is no evidence that this is the case. The total amount of energy radiated by the sun can be calculated by thinking of the sun as surrounded by a sphere with radius equal to the mean distance of the earth from the sun. Each part of the sphere would be at the same distance from the sun and would receive the same amount of radiation. All the energy radiated by the sun would strike this sphere. The total energy per second would be equal to the intensity per square centimeter, multiplied by the number of square centimeters in the surface of the sphere.

Using this method, it can be calculated that the solar radiation is equivalent to 63,200 kilowatts of continual radiation per square meter of the sun's surface.

11.4 RADIATION LAWS AND THE SUN'S TEMPERATURE

Knowing the total radiation per square centimeter of the sun's surface, we can calculate its temperature from radiation laws formulated in the 19th century. It was found that all heated solids and liquids radiate energy in a similar manner. The amount of radiation emitted at a given temperature depends on physical characteristics, such as gloss and color, rather than chemical composition. Kirchhoff, who formulated the laws of spectrum analysis, discovered that at any given temperature the radiating power of a surface is directly proportional to its absorbing power. Black, the best absorber, is also the best emitter. Although there is no surface that absorbs all incident radiation, it is convenient to base the radiation laws on a hypothetically perfect absorber, called a *blackbody*. The results for real surfaces and with specific physical characteristics can then be corrected to blackbody radiation.

Blackbodies emit electromagnetic radiation even when cool, but the waves are in the infrared portion of the spectrum and are too long to be seen by the eye. When a body is heated, the first radiation apparent to the eye is a dull, red glow. As the temperature rises, the color changes to white and then to blue. The observed color at any temperature is a mixture of many wavelengths. The distribution over the spectrum for several temperatures is shown in Fig. 11-1.

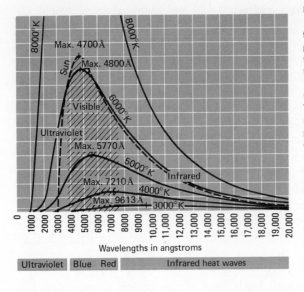

FIG. 11-1 Distribution of energy (heat) along the spectrum. The distances from the base line to various points in the temperature curves show the amounts of heat radiated at various temperatures. These quantities of heat (and the areas) are proportional to the fourth power of the temperatures. This is Stefan's law. Note that the maximum points in the curves move progressively toward the short-wave end as the temperature rises. Wien's law gives the relation between temperature and the wavelength of maximum energy. The dotted line is the heat curve of the sun's spectrum as observed from the surface of the earth. It most nearly matches the 6000°K temperature curve. Note that much of the sun's ultraviolet radiation is missing. The air absorbs everything beyond wavelength 2950 Å.

These curves are based on the theoretical blackbody. Their general shapes were first obtained in the 19th century from measurements of radiating surfaces that closely approximated blackbody radiation. It will be noticed that the sun's radiation curve is close to that of a blackbody at 6000°K.

Initially, attempts to derive a law from existing theory that completely describe the shapes of these curves failed, but two laws that describe certain characteristics were derived. Stefan's law states:

The total energy radiated by a blackbody is proportional to the fourth power of the temperature.

The formula is:

$$E = aT^4 \tag{11-1}$$

where T is the temperature in degrees Kelvin (absolute temperature) and a is a constant that equates the temperature to the total energy, expressed in joules per second: $a = 5.72 \times 10^{-12}$.

The total energy radiated at all wavelengths at a given temperature is proportional to the area under the energy curve for that temperature. It is evident from Fig. 11-1 that this area increases rapidly as the temperature increases. The total energy at one temperature can be compared to that at another temperature by taking the fourth power of the ratio of the two temperatures. For example,

$$\frac{\text{total energy at } 6000°\text{K}}{\text{total energy at } 3000°\text{K}} = \left(\frac{6000°\text{K}}{3000°\text{K}}\right)^4 = \left(\frac{2}{1}\right)^4 = 16$$

The temperature of the visible layer of the sun can be found from Stefan's law by substituting the total energy radiated per square centimeter and solving for T. The total radiation, 6.32×10^3 j/sec/cm^2 or 6.32×10^7 watts/m^2 has been computed from the data and by the method described above. We find

$$T^4 = \frac{6.32 \times 10^3}{5.72 \times 10^{-12}} = 1.11 \times 10^{15}$$

from which $T = 5770°$K.

It is also apparent from the radiation curves that as the temperature increases, a greater portion of the energy is radiated at shorter wavelengths. This is described by a law derived by Wilhelm Wien (1864–1928), a German physicist, as:

The wavelength for which the radiation is most intense, λ_{max}, *is inversely proportional to the absolute temperature.*

The higher the temperature, the more energy is concentrated in the shorter, bluer wavelengths. Expressed as a formula:

$$\lambda_{max} = \frac{0.289}{T} \qquad (11\text{-}2)$$

where λ_{max} is in centimeters and T in degrees Kelvin.

This law affords another means of calculating the sun's temperature. The value of λ_{max} from measures of the solar energy is in the blue region at 4700 Å. Substituting in equation (11-2): Since 4700 Å $= 4.7 \times 10^{-5}$ cm,

$$T = \frac{0.289}{\lambda_{max}} = \frac{0.289}{4.7 \times 10^{-5}} = 6150°\text{K}$$

Since Stefan's and Wien's laws are based on different data, it is not surprising that the two values of T differ. It is sufficiently accurate to accept 6000°K as the solar temperature. This is based on radiated energy and is an average, effective observed temperature.

11.5 OBSERVING THE SUN

The sun was once considered to be a perfect heavenly body without blemish. But at times, when sunlight was sufficiently dimmed by mist, fog, or dust so one could look directly at it without damage to the eyes, dark areas were seen on its surface. Galileo saw them with his telescope but did not recognize them for what they really are. Herschel thought they were holes in the bright atmosphere and that he could see the darker surface underneath. Some astronomers drew them as depressions in the solar surface. It is now possible to observe sunspots with small telescopes and watch them develop and move across the visible disk of the sun. They are also being observed intensively with special equipment. They are found to be related to other phenomena in the solar atmosphere.

The sun may be viewed visually with a telescope, if proper precautions are taken. **Never look directly at the sun.** It is nearly 500,000 times brighter than the full moon and emits dangerous rays that can ruin the observer's eyes very quickly. The best way is to project the sunlight through a small telescope onto a piece of white paper held a few inches from the eyepiece. The eyepiece is left in place, but the focus must be adjusted until the image is sharp at the edges. The edge of the sun is called the *limb*, the same term that is used for the moon and for the planets and stars.

Even with such simple apparatus several things about the sun's disk may be noticed. First, if there are sunspots, they will be easily seen. Next, the sun looks darker at the edge than at the center. There are spots most of the time, although there are periods when no spots are visible. Near the spot areas, particularly when they are close to the limb, there are bright areas called *faculae*, meaning "little torches." They are most easily seen near the limb because at the center they are about as bright as the solar background. Near the limb the faculae are brighter than the background and are visible even in small telescopes. In Fig. 11-2 sunspots are shown as dark areas and faculae as bright areas.

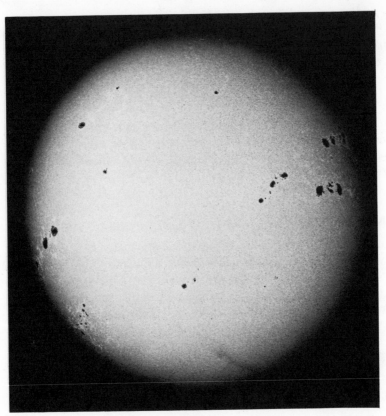

FIG. 11-2   The sun on May 17, 1951, showing sunspots and faculae. Notice that the sun is darker at the limb than at the center. (*Photograph from the Hale Observatories*)

Outside the spot and facula areas the sun is not uniformly bright, but has a mottled or spotted appearance. The small areas in Fig. 11-3 were originally called rice grains, but now are called *granules*. They have been photographed at high altitudes where the air is steady and the seeing good. But more recently they have been photographed with telescopes lifted above the earth to about 80,000 feet. The first photographs were made with a 12-inch telescope in an unmanned balloon in the northern part of the United States. The telescope was pointed at the sun by photoelectric guidance mechanisms. The photographs were taken with exposures of 0.001 second at the rate of one every second.

The granules are usually from 250 to 600 miles in diameter; but smaller and larger ones have been observed. Their lifetimes are from 1 to about 10 minutes.

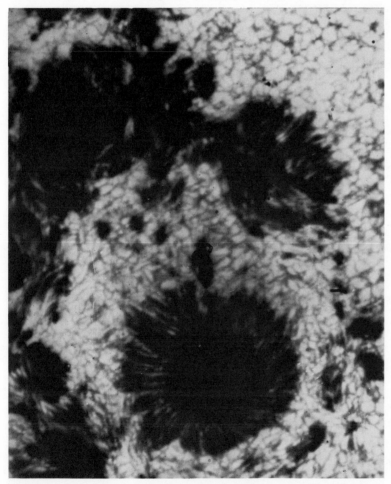

FIG. 11-3   Photograph of a portion of the sun taken from a balloon at about 80,000 feet. Granules or irregular white spots, last about 5 minutes; dark areas, lasting days or weeks, are sunspots. (*Project Stratoscope of Princeton University; sponsored by National Aeronautics and Space Administration: U. S. Office of Naval Research, and National Science Foundation*)

They are hot masses of gas from the solar interior that rise, cool, and then fall back. These motions were detected and measured in the balloon experiments. The temperature apparently varies between 120°K and 230°K above and below the average 6000°K of the solar disk. The granulations are visible almost to the limb.

11.6 THE PHOTOSPHERE

For convenience, the sun is divided into four layers (Fig. 11-4). They are, from the center outwards: the *interior*, the *photosphere*, the *chromosphere*, and the *corona*.

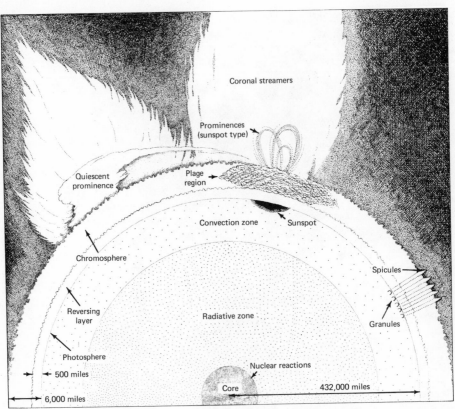

FIG. 11-4 The layers of the sun. The prominent features of each layer are also shown. *(From Mc-Laughlin; Introduction to Astronomy, Houghton Mifflin Company)*

The photosphere is that layer through which the solar energy escapes from the interior. As indicated by its name, it is the layer from which the light comes, and therefore it is the part we see. The temperature at the top of the photosphere (formerly called the reversing layer) can be determined by the radiation formulas and is about 6000°K, as derived in Section 11.4. An exact figure cannot be given, since the radiation escapes at different depths and the temperature varies from one level to another. This is also the reason for the darkening at the limb.

It has been estimated that the thickness of the photosphere is about that of the diameters of the granules, perhaps 600 miles. If it could be seen by itself, its spectrum would be continuous. But at the top the cooler gases absorb some of the energy, producing the dark-line spectrum discovered by Fraunhofer. The wavelengths of more than 25,000 lines have been determined, from which 60 different elements have been definitely identified, with seven more almost certain. There are still thousands of unmeasured lines, mostly in the ultraviolet and infrared.

With the use of space-borne instruments, the solar spectrum has been extended from 2900 Å, where the earth's atmosphere cuts off the ultraviolet, to 977 Å and down to the X-ray region. By infrared detectors, it has been extended to 23,700 Å, where the atmosphere cuts off again. One line in the ultraviolet spectrum of hydrogen, the Lyman-α line, has been of special interest. A few lines of carbon, oxygen, and silicon have been observed in the ultraviolet. Some lines in the X-ray region have been observed with photometers equipped with special filters.

Most of the solar lines are produced by atoms, but some molecules have now been identified. At 6000°K molecules are broken up into their fundamental atoms. But in sunspots and high levels where the temperature is lower, molecular spectra have been known for many years. Recent reports indicate that 14 molecules are identified with reasonable certainty, with nine others possible. Figure 11-5 shows a portion of the solar spectrum in which some identifications are shown.

11.7 THE CHROMOSPHERE

The chromosphere is located just above the photosphere. Its lower levels also produce some of the Fraunhofer lines. The temperature at the bottom is 6000°K and increases toward the top. It seems to be about 7000°K at the 2000-mile level, but there are differences in the computation of the temperature, depending on which spectrum lines are used. Hydrogen and helium give temperatures of 10,000°K to 20,000°K at 2500 miles, which may be in the corona.

The division between the chromosphere and the corona is visually very sharp, but physically the changes occur gradually. Hence it is very difficult to locate the boundary exactly. One conclusion is that it is at 3000 miles.

The density at the bottom of the chromosphere is 10^{-8} g/cm^3, and at the top it has dropped to between 10^{-11} and 10^{-12} g/cm^3. If the chromosphere could be seen by itself, it would show a bright-line spectrum, indicating that it is a hot gas under lower pressure than in the photosphere. This spectrum is visible during total eclipses of the sun. When the upper photosphere and lower chromosphere are the only parts of the sun not covered by the moon, the spectrum flashes out, with bright lines occupying the places of the dark lines of the absorption spectrum. This lasts only one or two seconds just before and just after totality. It is called

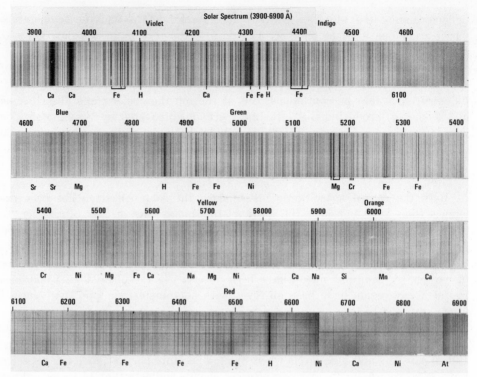

FIG. 11-5. A portion of the sun's spectrum. Wavelengths are given in angstroms and the color regions are indicated. *(Photograph from the Hale Observatories)*

the *flash spectrum*. Clouds of hydrogen give the chromosphere (the word means color sphere) its characteristic red color, which is quite prominent during eclipses.

About 3500 miles above the photosphere, bright *spicules* (spikes) rise into the hot gas of the corona. They are masses of hot gas rising to about 4500 miles at the equator and 6500 miles at the poles. They last about 5 minutes; their diameters are between 300 and 1200 miles; and their velocities reach 12 to 25 miles/sec. These phenomena should be accompanied by very loud noise, which of course is not transmitted through the vacuum of space. Fifty or more spicules are visible at one time on the solar limb.

11.8 THE CORONA

The most beautiful and most publicized layer of the sun is the corona, ordinarily seen only during a total eclipse of the sun. It is very large, having been seen to 7 million miles; it is thought to reach the earth and beyond. It is quite faint, about as bright as the full moon, but is 500,000 times fainter than the photosphere, which explains why it cannot be seen at all times. Surprisingly, it is very hot, with temperature estimated at 1 to 2 million degrees Kelvin. But radiation from the corona is weak because of its low density.

It is difficult to explain the high temperature of the corona. Probably the best explanation is that it is heated partly by the flow of hot material from the chromosphere and partly by supersonic waves from below. Kinetic energy from these waves heats the low-density gas to such a high level that it is given very high velocity. The density of the corona at 6000 miles is 10^8 to 10^9 atoms/cm³. At this low density and high speeds, collisions strip the electrons from the nuclei of the atoms. The speed of the electrons is 5000 miles/sec and of the protons 125 miles/sec. These particles also collide with atoms and assist in the process of removing electrons from them. It is possible that at least some of the coronal material comes from interplanetary dust and meteoroids.

The corona has a weak continuous spectrum crossed by bright lines, not absorption lines as in the Fraunhofer spectrum. It was once thought that, since the bright lines do not correspond with the positions of the absorption lines, a new element, called coronium, had been discovered, as helium had been discovered in the solar spectrum in 1868. But in 1941 the Swedish physicist, B. Edlén, showed that the strong green line of coronium was produced by iron with 13 electrons missing, and that the red line came from iron with nine electrons stripped from the atom. Other lines are due to similarly highly ionized calcium, nickel, and possibly argon. Thirty lines have now been observed in the spectrum of the corona.

The shape of the corona varies with the sunspot cycle. When there are many spots, the corona is almost spherical with streamers running out radially in all directions. At spot minimum, the corona is unsymmetrical with large lateral streamers and small polar ones.

Until all the correlations among coronal streamers, sunspots, and magnetic fields in and around the sun are made, the corona will not be completely understood. Figure 11-6 shows the changes in shape of the corona with sunspot activity.

11.9 THE INTERIOR

Since the interior of the sun is completely inaccessible, it can be studied only by theoretical computations. They must be based on the known facts about the sun: its mass, density, temperature as observed, and the structure and motions of its atmosphere. Then the known laws of physics must be applied to investigate the interior.

Because of the great mass of the sun, the pressure at the center is estimated to be 1 billion atmospheres (15×10^9 lb/in²). To support this enormous pressure, the temperature at the center is computed to be 15,000,000°K. It is assumed that the interior is entirely gaseous; so the gas laws hold. Its density increases to between 100 and 150 g/cm³. This is about 10 times that of the most dense solid known on the earth, but 1000 times less dense than some of the white dwarf stars. The spectrum is continuous, and the maximum radiation, from Wien's law, is mostly X rays.

It was formerly thought that the sun produces its energy by burning some combustible substance like coal. But it has been calculated that if the sun were made up entirely of coal, there would be a supply sufficient for only 8000 years.

FIG. 11-6 The changing shape of the solar corona, near sunspot minimum (above) and maximum (page 225). (Yerkes Observatory photograph)

Later, it was thought that the sun may have been converting its kinetic energy, due to gravitational contraction, to heat energy. If this were so and if the sun had contracted from the limits of the solar system to its present size, there would have been enough energy to keep it shining for about 25 million years. There is every reason to believe that the sun is much older and has been radiating at its present rate for several billion years.

It was realized after the discovery of radium and the emission of energy from radioactive material that the solar energy probably comes from the interiors of atoms. That is, the sun is now thought of as a huge atomic reactor. Two possible chains of events were proposed in 1938 by Hans Bethe (1906–   ), a German-American physicist, who received the Nobel prize in physics for this work. Bethe outlined the carbon cycle and the proton-proton reaction as possible reactions inside the sun.

The *carbon cycle* (Fig. 11-7) can be represented by the following nuclear reactions[1]:

(1) $^{12}_{6}C + ^{1}_{1}H \rightarrow ^{13}_{7}N + \text{gamma}$
(2) $\qquad ^{13}_{7}N \rightarrow ^{13}_{6}C + ^{0}_{1}e$
(3) $^{13}_{6}C + ^{1}_{1}H \rightarrow ^{14}_{7}N + \text{gamma}$

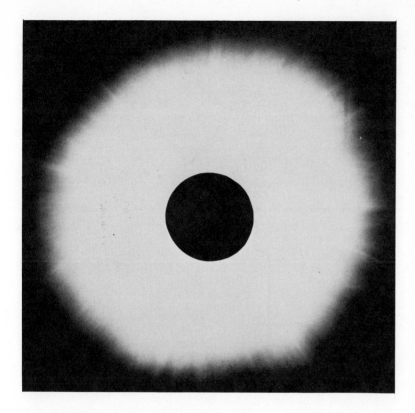

(4) $^{14}_{7}\text{N} + ^{1}_{1}\text{H} \rightarrow ^{15}_{8}\text{O} + \text{gamma}$
(5) $^{15}_{8}\text{O} \rightarrow ^{15}_{7}\text{N} + ^{0}_{1}\text{e}$
(6) $^{15}_{7}\text{N} + ^{1}_{1}\text{H} \rightarrow ^{12}_{6}\text{C} + ^{4}_{2}\text{He}$

where   arrow = becomes
                           gamma = gamma ray
                              $^{0}_{1}\text{e}$ = positron

The carbon cycle begins when a carbon nucleus captures a proton deep in the interior of the sun. The carbon is supposed to have been captured by the sun after having been formed in the interior of a star that exploded, sending its heavier-than-helium atoms into space. This cycle can take place only where the temperature is around 15,000,000°K.

The combination of the carbon nucleus and the proton forms an unstable nitrogen-13 nucleus, plus a small amount of energy in the form of a gamma ray. This nucleus immediately emits a positron and returns to a stable isotope of carbon, which combines with another proton, forming ordinary nitrogen plus a gamma ray. The nitrogen nucleus combines with a third proton, forming an

[1] In a nuclear equation, the charge on the nucleus (atomic number) is written as a subscript to the left of the symbol, and the mass is written as a superscript to the left. For example, the helium nucleus is $^{4}_{2}\text{He}$, indicating an atomic mass of 4 and atomic number of 2.

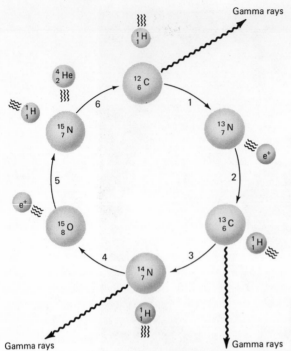

Gamma rays

Gamma rays

Gamma rays

Gamma rays

FIG. 11-7 The carbon cycle, a chain reaction inside the sun that releases energy when the nuclei of hydrogen atoms combine to form helium nuclei.

unstable oxygen-15 isotope. This particle decays by emitting a positron, returning to nitrogen-15, which in turn combines with a fourth proton.

Now this particle, instead of forming a stable oxygen nucleus, splits to a helium nucleus and the original carbon which started the reaction. Thus four protons (hydrogen nuclei) combine to form one helium nucleus, energy having been released in the process.

The *proton-proton reaction* takes place in three steps:

(1) $^1_1H + ^1_1H \rightarrow ^2_1H + ^0_1e$
(2) $^2_1H + ^1_1H \rightarrow ^3_2He + gamma$
(3) $^3_2He + ^3_2He \rightarrow ^4_2He + 2^1_1H$

In this series two protons first combine directly into a heavy hydrogen nucleus (deuteron), with the emission of a positron. The deuteron captures another proton, forming a helium isotope and some gamma radiation. This combining process is called *fusion*. Finally two of these helium isotopes combine. Instead of forming a still heavier nucleus, however, the particle splits into an ordinary helium nucleus and two protons. The splitting process is called *fission*. This reaction can occur at a temperature lower than that required for the carbon cycle and takes place above the solar center. It is estimated that the carbon cycle furnishes only 4 percent of the total energy and that the proton-proton reaction gives the other 96 percent of the solar energy. In the latter reaction, four protons combine to form one helium nucleus, as in the carbon cycle. Six protons are involved in the process, but two are returned in the final step.

The atomic weight of four protons is $4 \times 1.008 = 4.032$. That of the helium nucleus is 4.004. Hence, 0.028 unit of mass has been converted in the reaction. According to the theory of relativity, mass and energy are equivalent and are related by the formula

$$E = mc^2 \qquad (11\text{-}3)$$

where E is the energy in joules, m is the mass in kilograms, and c is the velocity of light in meters per second.

The amount of energy radiated by the sun is 3.85×10^{26} joules per second. Solving equation (11-3) for m and substituting:

$$m = \frac{E}{c^2} = \frac{3.85 \times 10^{26} \, \text{j}}{(3.00 \times 10^8 \, \text{m/sec})^2} = 4.28 \times 10^9 \, \text{kg}$$

or 4,280,000 metric tons. This is the amount of solar matter that is being converted into energy *each second!*

The amount of mass lost is 0.7 percent of the mass of the protons that enter the reactions. To produce this energy, it can be calculated that about 616 million tons of hydrogen are being changed into 612 million tons of helium each second. Suppose that the sun were composed entirely of hydrogen when these reactions started and that it has been radiating and will continue to radiate at the same rate as at present. The mass of the sun is now 2×10^{27} tons. At this rate of mass loss, it will continue to shine for 100 billion years. However, it is thought that the sun will not continue to radiate at its present rate in the future.

11.10 SUNSPOTS

Sunspots move across the disk of the sun from the east limb to the west limb. If a large group lasts more than a month, it will disappear from view on the west limb and reappear two weeks later on the east limb. This apparent motion is caused by the rotation of the sun. It will also be noticed that the sun does not rotate as a solid, but that spots away from the equator take longer to go across than those nearer the equator. The results of studies of the period of rotation are given in Table 11-1.

Spots appear in zones, usually from 5° to 40° latitude (both north and south), although an occasional spot is seen outside these limits. The rotation can be

TABLE 11-1

solar latitude	period of rotation
0°	24.65 days
20	25.19
30	25.85
35	26.62
75	33.00

checked by observing the Doppler effect at both limbs from the equator nearly to the poles, including latitudes where no spots are seen. The check agrees exactly with the method of spot rotation.

The center of a spot is called the *umbra*. Here the temperature measures about 4000°K, which is the reason for its much darker appearance than the photosphere, where the temperature is 6000°K. The spectrum shows bands typical of the spectra of compounds, which can exist only at temperatures lower than that of the photosphere. The umbra is surrounded by a *penumbra,* where the temperature and brightness are intermediate between those of the umbra and the photosphere.

A typical, large single spot has an umbra of diameter 10,900 miles and a penumbra of diameter 23,000 miles. It is surrounded by a bright ring 4000 miles wide. The intensities are 30, 80, and 103 percent, respectively, of the brightness of the photosphere. A spot is accompanied by a magnetic field more than 3000 times as strong as that of the earth. There is rapid motion of the material in the spot, with velocities up to about 2 miles/sec at the photosphere level. The magnetic field decreases in strength as the height increases. Many single spots are accompanied by an invisible companion detectable by the magnetic-field strength in the neighborhood. This companion lies completely below the visible layer of the sun. In other words, sunspots occur in pairs, with one frequently underneath, where it cannot be seen. If both spots are visible, they have opposite magnetic polarities; that is, one spot has a positive polarity like that of a north-seeking magnet on the earth and the other has a negative polarity. In any one sunspot cycle, if the leading spot of a pair or group—leading in the sense that it is ahead because of the solar rotation—has a positive polarity, all leading spots in that hemisphere will have positive polarity. In the opposite hemisphere, the polarity will be reversed; that is, the leading spots will have negative polarity. The polarity of spot pairs is shown for the years 1942 to 1971 in Fig. 11-8. As the figure shows, at spot minimum the old-cycle spots are last seen near the solar equator and the new-cycle spots are first seen at higher latitudes. During the next cycle the polarity is reversed.

The number of sunspots and spot groups is counted daily and averaged by months and years. Records have been kept and plotted since at least 1759. Rudolph Wolf of Zurich set up a formula to show the Wolf sunspot numbers. The formula is as follows:

$$\text{Wolf sunspot number } r = k(10g + f) \tag{11-4}$$

where g is the number of groups, f is the number of individual spots visible on a given date, and k is a number that depends on the size of the telescope used and conditions at the observing site. This formula gives a uniform number that can be plotted, as in Fig. 11-9.

There is an approximate regularity in the times of maximum and minimum of the Wolf numbers. As the figures show, the heights of the maxima vary, whereas the minima are fairly uniform, with few, if any, visible spots. The period is 11.1 years on the average, but the intervals vary from about 7 to 15 years. This

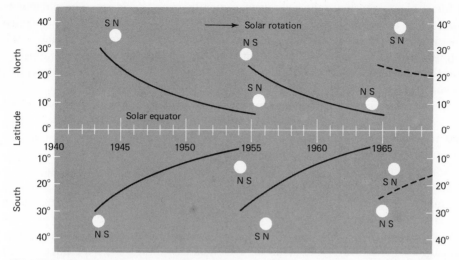

FIG. 11-8 The law of sunspot polarity. The average change in latitude and corresponding magnetic polarities of spot pairs from 1942 to 1966.

period is called the *sunspot cycle*. If the change of polarity is taken into account, the period should be doubled. The spot maximum in 1958 was one of the highest on record. This was the maximum selected for the IGY, which was followed by the IQSY at a time when the sun was quiet.

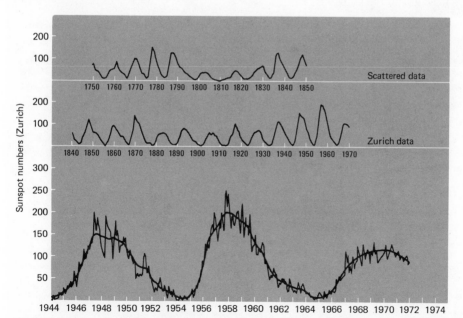

FIG. 11-9 The sunspot cycle from 1749 to 1970. Short-period fluctuations are shown below for 1944 to 1972. The upper curves have been smoothed and the short-period fluctuations eliminated. (Courtesy Robert Howard, Hale Observatories)

Ideas about sunspots have changed radically over the years. The old books drew spots as depressions in the photosphere. Herschel thought the surface underneath was cool enough to support life. But when a spot approaches the limb, it can be seen as a very turbulent region, with streams of gas extending up into the chromosphere. When seen on the limb, such turbulences are called *prominences.*

Spots change continuously in many respects. At first a spot is seen as a small pore between the granulations in the photosphere, but with a diameter of about 1000 to 2000 miles. This small spot may soon disappear, but may last a day or longer. It may develop into a much larger spot or group of spots.

A region in which sunspots and their accompanying phenomena appear is called a *center of activity* (Fig. 11-10). This center is considered to be a result of a magnetic field deep inside the sun and may last as long as 1000 years. Inside this field the material moves very slowly. The field itself is composed of magnetic tubes twisting into the upper layer of the sun, the convective zone. The diameter of the tubes is the same as the thickness of the zone, perhaps 60,000 miles. When the tubes are brought to the sun's surface by a mechanism not yet understood, the result is a spot group. This group starts as a small group of faculae, followed by a visible spot 1000 miles or so in diameter. It then grows

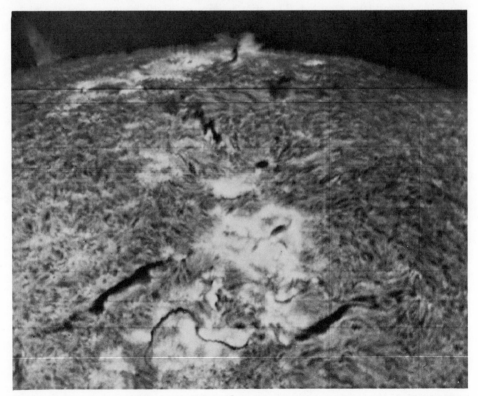

FIG. 11-10 An enlarged portion of a center of activity taken in hydrogen light, showing bright and dark flocculi near the sun's limb. *(Photograph from the Hale Observatories)*

and develops into a group that may be more than 100,000 miles in length. In 1946 and 1947 three of the largest groups ever seen appeared at intervals of about six months. They were visible for about two months each. The first of these groups is shown in Fig. 11-11.

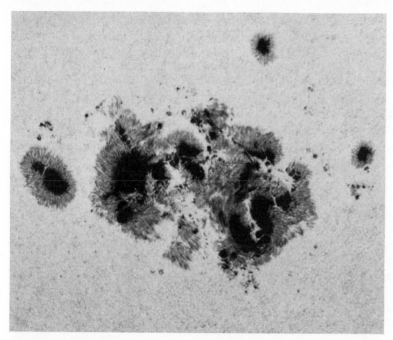

FIG. 11-11   A large sunspot group, photographed with the 60-foot tower telescope on Mount Wilson on February 2, 1946. (*Photograph from the Hale Observatories*)

A maximum occurs about 11 days after the group is first seen, at which time small filaments may appear. These are long, stringlike dark lines of hydrogen, and may reach a length of 200,000 miles. As the spot centers disappear, the filament continues to be visible, but soon stretches out in a direction parallel to the solar equator. During this time the faculae are still visible; they are eventually cut in two by the filament, but are still present after four rotations of the sun, or about 130 days. Finally, the filament begins to dissolve; after nearly a year, the activity center completely disappears from the surface. It is believed that the magnetic field is still present in the interior.

11.11   THE SPECTROHELIOGRAPH

The study of the chromosphere and solar prominences was greatly assisted by the invention of the *spectroheliograph,* an instrument that made possible solar photographs in a single line of the spectrum. The bright red line of hydrogen and the two violet lines of calcium are the strongest lines in the spectrum of the chromosphere. The photograph (Fig. 11-10) of a center of activity was taken in red hydrogen light.

Filters have been made that pass light from only a very narrow portion of the spectrum and are now used to photograph a large part of the sun at one time with exposures of a fraction of a second. If the resulting photographs are run rapidly through a motion picture projector, the motions of the solar material can be seen. The instrument for taking such photographs is called a *spectrohelio-kinematograph*.

The best type of telescope now in use for solar studies is the *coronagraph*, which was invented in 1930 by Bernard Lyot (1897–1952), a French astronomer. This telescope uses glass for the objective as free as possible from bubbles or scratches, which might scatter the sunlight. The tube is made carefully to exclude dust particles. The solar image is blocked off in such a way that only light from the solar atmosphere passes through. When used with filters and placed at a high-altitude station, especially good studies can be made.

11.12 PROMINENCES

When the foregoing techniques are used, photographs of the sun show large clouds of hydrogen or calcium gas distributed over the disk, particularly in the regions of centers of activity. Hale called them *flocculi*. When the centers of activity reach the limb, prominences can be seen, although prominences are also seen near the poles where there are no visible sunspots.

Prominences have been grouped by types, some of which are:

1. *Quiescent*. These appear to be motionless for long periods, but are fairly active. When projected onto the solar disk, they are seen as hydrogen filaments.
2. *Active*. This type shows much motion, rising or falling. The prominent direction is downward. Apparently the hydrogen or calcium rises in a state that does not emit light in the line being observed, particularly the Hα line. As the gas cools, it falls back onto the sun. Its motion is shown in spectrohelio-kinematograms.
3. *Eruptive*. These are seen to rise at very high velocities, sometimes exceeding 300 miles/sec, and are blown completely away from the sun. One of the most spectacular eruptive prominences is shown in Fig. 11-12. This type plays a part in the production of auroras.
4. *Sunspot type*. These are mostly arches that lie above centers of activity.
5. *Coronal type*. Some prominences are visible in the corona and appear to rain down onto the chromosphere.

Figure 11-13 shows a sunspot group near the limb and the same area taken

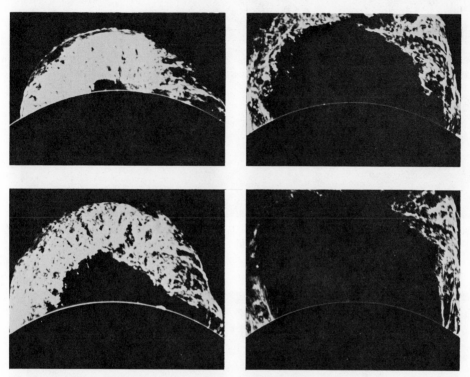

FIG. 11-12   Four successive (top left to bottom right) spectroheliograms of the great explosive prominence on June 4, 1946, taken with a coronagraph. Total elapsed time is one hour. (*Harvard College Observatory*)

in red hydrogen light. The sunspots in Fig. 11-14, near the sun's center, were also photographed in the red H-alpha light.

Occasionally a center of activity will produce a sudden and short-lived increase in brightness, called a *flare*. There may be as many as 10 centers on the sun at one time, any one of which might produce a flare. One large flare and 100 or more small ones per day may be seen near large spots. They brighten in from 5 to 10 minutes and die out more slowly. In general it is possible to see them only in a spectrohelioscope, but the brightest flares can often be seen without special equipment. They are being looked for at several observing stations, such as the McMath-Hulbert Observatory, where the photograph in Fig. 11-15 was made.

Flares radiate strongly in the ultraviolet and X-ray regions of the spectrum. They also send out subatomic particles with velocities of several hundred miles per second, and waves that are picked up by radio telescopes. Cosmic rays apparently are emitted by the large flares. The result of all this activity is that space is filled with high-energy particles and waves, which affect the magnetic field and radiation belts around the earth.

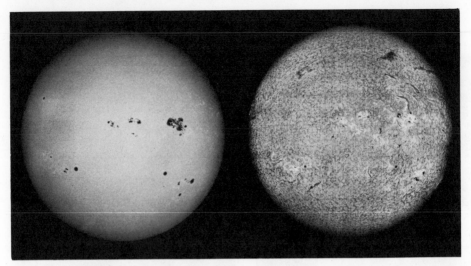

FIG. 11-13   Photograph of the sun (left) taken in ordinary light, shows sunspots and faculae. The sun on the right, taken in red hydrogen light, shows clouds of hydrogen at higher levels. *(Photograph from the Hale Observatories)*

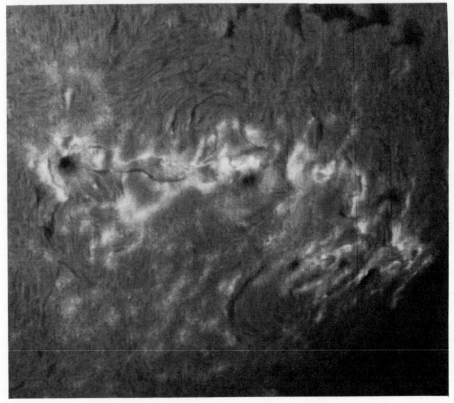

FIG. 11-14   H-alpha spectroheliogram of a pair of sunspot groups taken on March 17, 1969, near the central meridian of the sun. *(Courtesy Robert Howard, Hale Observatories)*

It is well known that there is a correlation between solar activity and terrestrial affairs, such as auroras, radio fadeouts, and (some people believe) even the weather. Displays of auroras are more frequent in times of sunspot activity, although they have also been seen when no spots are visible. They are produced by the particles from the sun, which have finite velocities, so there is a delay of a day or two between the ejection of the particles and the observation of an auroral display. However, some auroras are also thought to be the result of X-ray emission from flares. The resulting displays occur only minutes after a flare is visible. At the same time, the transmission of radio waves is interrupted, and strong changes in the earth's magnetic field sometimes result. The enormous spot groups in 1946 had very strong magnetic disturbances associated with them, but those of 1947 produced only small disturbances. So there is no direct correlation between the size of spots and the strength of the magnetic field of the earth.

The sun emits a high-velocity plasma, a mixture of about equal numbers of electrically charged subatomic ions and electrons. Because of its high velocity, this plasma flows out into space, reaching at least as far as the earth as the

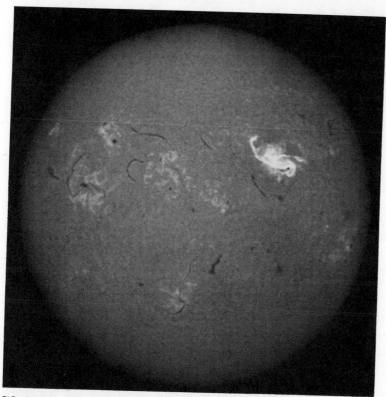

FIG. 11-15 Spectroheliogram of a solar flare taken during a time of considerable sunspot activity. Flares are usually associated with sunspots. (McMath-Hulbert Observatory of the University of Michigan)

solar wind. At the earth's distance, the solar wind has a velocity of about 280 miles/sec and a density of between one and ten particles per cubic centimeter. There is little doubt that it is responsible for the Van Allen radiation belts, which change in intensity with the variation in solar activity. The sun's corona, which is thought to reach as far as the earth, mingles with the interplanetary medium—plasma and dust particles—and is thought to extend throughout the solar system.

11.13 A MODERN SUMMARY OF THE STRUCTURE OF THE SUN

There have been several proposed models of the sun. The following is probably the most accurate of the models to date (see Fig. 11-16):

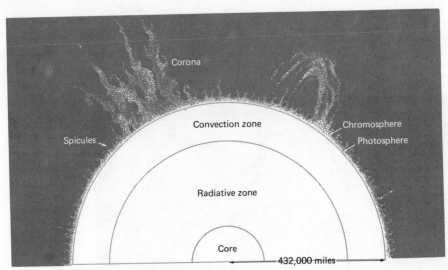

FIG. 11-16   A model of the sun's layers.

1. *Interior.*
   Radius: 432,000 miles (680,000 km).
   Mass: Practically 100 percent of the total mass of the sun, 1.99×10^{30} kg.
   Temperature: At center, about 15×10^6 degrees K.
   Density: Increases inward to between 100 and 150 g/cm^3.
   Structure: *Central core*, diameter about 150,000 miles (240,000 km), about 15 percent of the sun's diameter, composed of hydrogen and helium; the former is decreasing and the latter increasing because of the transformation of hydrogen to helium. *Radiative zone*, above the central core; thickness, 275,000 miles (442,000 km). *Convection zone*, above radiative zone; thickness, 93,500 miles (150,000 km).
   The central core produces energy by the carbon cycle and the remaining zones by the proton-proton reaction of transformation of hydrogen to helium.

The energy rises to the surface by radiation, except in the convection zone, where it is transported by currents (convection) due to the turbulence of the upper layer. It is possible that the transfer of energy is by sound waves. Because of the transfer of energy, the temperature decreases from the central core to about 6000°K at the photosphere and 10,000°K in the chromosphere.

2. The *photosphere* was described in Section 11.6. Its thickness is thought to be about equal to the diameters of the granules, perhaps 600 miles (960 km). It has been suggested that there are supergranules that may be about 50 times larger than the smaller ones. They are said to be related to the magnetic fields inside the sun.

3. The *chromosphere* is believed to be only 1300 to 3300 miles (2100 to 5300 km) thick. The spicules rise from the top of the chromosphere at a speed of about 20 miles/sec (30 km/sec) into the corona and soon fall back into the chromosphere. Their origin is still not understood, but they may be associated with magnetic cores which give them an upward motion.

4. The *corona* is the outer layer of the solar atmosphere and is thought to reach as far as the earth and probably beyond. Its temperature is more than 1,000,000°K, and it is said to be in a state of continual expansion, which gives rise to the solar wind.

11.14 THE FUTURE OF THE SUN

It is interesting to speculate on the sun's probable future. The following prognosis is based on present theories of stellar evolution.

Helium produced by the proton-proton and carbon-nitrogen cycles discussed earlier in this chapter is being deposited in the center of the sun. In other words, the sun is building up a helium core. At the same time, its temperature will rise and the sun will expand to compensate for the resulting increase in internal pressure. The sun is believed to have been in existence for about 6 billion years and to have consumed 6 percent of its original hydrogen. After another 6 billion years, when about 12 percent of the hydrogen is gone, the helium core will be too hot and the pressure will be too great for the sun to remain at its present size and it will expand.

The expansion stage will last perhaps 1 billion years, during which the sun will expand to 30 or 40 times its present diameter and will then be a little cooler than it is now. But the increased size, even at a lower external temperature, will mean that there will be an increase in the total radiation and the earth will become too hot to support life. The oceans will boil and the earth will be surrounded by a hot, steamy atmosphere. After its size has become stable, the sun will begin to contract and heat up again, but at some stage it may explode, blowing off perhaps 0.1 percent of its mass from the upper layers of the photosphere into space. It will then condense rapidly—in a few hundred million years —and become a hot, dense star of the white dwarf class. The earth will re-form its oceans, which will then freeze, and end its existence as a cold, dead planet still in orbit around a small, dying star.

QUESTIONS AND PROBLEMS

Group A

1. Describe two methods of measuring the distance to the sun.
2. The solar constant, 1.96 cal/min/cm² is equivalent to about 130 watts/ft². If there are no losses, how much energy would be incident on a 1500-ft² roof perpendicular to the sun's rays?
3. Would a body emit as much energy in the infrared region of the spectrum when blue-hot as when red-hot? Explain.
4. If the temperature of the sun were 4000°K, (a) at what wavelength would the radiation be most intense, and (b) what would be the sun's color?
5. What properties of the chromosphere account for its emitting so much less light than the photosphere?
6. Although telescopes are often located on mountains, many are used successfully at lower elevations. It is much more important, however, that coronagraphs be located at high altitudes. Why?
7. Calculate the percentage of the mass of hydrogen that is converted to energy during the formation of helium.
8. What is produced from the six protons that are consumed during the proton-proton cycle?
9. List the types of solar activity related to sunspots.
10. How could you distinguish between a small, round sunspot and Venus in transit across the disk of the sun?
11. Near the end of a sunspot cycle, how could it be determined whether a sunspot pair belonged to the old or the new cycle?
12. The true period of rotation of the sun with respect to the stars is less than that obtained by watching a sunspot; that is, it is less than twice the time required for a spot to travel across the visible portion of the sun. Explain.

Group B

13. What is the volume of the sun compared to that of the earth?
14. The ratio of the mass of the sun to that of the earth-moon system has been given as 328,912. In this chapter the mass of the sun is calculated to be 332,930 times the mass of the earth. From these figures, compute the ratio of the masses of the moon and earth. Compare your answer with the value given in Chapter 10.
15. The umbra of a sunspot has a temperature of about 4500°K. Using Stefan's law, compute the total radiation per square centimeter as a percentage of that for the average solar surface (6000°K).
16. Compute the solar constant if the temperature of the sun were 18,000°K instead of 6000°K.
17. Observe the sun by projecting the solar image through a telescope onto a piece of white paper. Plot the spots, observe their motions, and determine the rotation period.

12 eclipses

12.1 SOLAR AND LUNAR ECLIPSES

Eclipses, especially eclipses of the sun, are among the most beautiful and awesome phenomena of nature. In a solar eclipse the moon passes between the earth and the sun. An observer in a region reached by the moon's shadow sees the sun completely covered until only the eerie, pale corona is visible.

To see the sun almost suddenly extinguished must have been terrifying to primitive man. Imagine how our early ancestors felt when they witnessed a total eclipse of the sun, particularly if it took place on a bright, warm, sunny day. Without any warning it begins to get darker and the air feels cooler. If one looks at the sun, he sees a dark disk crawling across it. It gets darker. The birds and animals begin to go through their usual nighttime routines; bats fly about and birds go to roost in the trees. When the sun is completely blacked out, except for a small spot of light on its eastern edge, a shadow comes creeping over the earth from the west and suddenly the sun is gone. Then a faint light streams out from the edges of the dark disk. A few of the brighter stars appear in the darkened sky. The sun has disappeared. After a few minutes of darkness, suddenly a bright shaft of light breaks through on the western limb and the sun reappears. In about an hour the black disk is gone!

Not many people have seen a total eclipse of the sun, because the greatest width of the moon's shadow where it strikes the earth is only 165 miles and it is likely to pass over the ocean or uninhabited land (see Fig. 12-7).

Eclipses of the moon are not so dramatic, but more people are likely to see them, since they are visible over more than half the earth. In an eclipse of the moon, the earth passes between the sun and the moon, casting a shadow over the moon. The edge of the shadow is shaped like the arc of a circle or an ellipse. It seems to move across the face of the moon, beginning on the eastern side. When the moon is completely inside the shadow, it takes on a peculiar coppery color. The entire phenomenon may last as long as 4 hours. The moon moves out of the shadow, always toward the east.

It is not surprising that eclipses have fired the imagination of people from the earliest times. Some ancient civilizations kept records of them. The Babylonians had such an excellent reputation for making accurate observations of eclipses that the great astronomer Ptolemy (about A.D. 150) consulted their lists, going back to the 8th century B.C.

The Greeks had admired Babylonian astronomy records long before the time of Ptolemy. Five centuries earlier, when the Greek historian Callisthenes was in Babylon campaigning with Alexander the Great, he sent Socrates a complete list of eclipses from the Babylonian records, which extended over 1900 years!

The Chinese also kept records of eclipses, but there is no evidence that the Egyptians ever made reference to them.

Before a detailed study of eclipses is undertaken, perform the following experiments to find out for yourself some of the mechanics of eclipses. Keep in mind that the distance between the earth and the moon varies.

You will be looking for different types of eclipses. The following definitions are sufficient to identify them for these preliminary experiments:

A *total eclipse* occurs when the sun or moon is completely obscured. A *partial eclipse* occurs when the sun or moon is only partly obscured. An *annular eclipse* occurs when the moon is too far away to completely cover the sun, and a ring of sunlight surrounds the moon.

For your experiments you need the following equipment: (1) a bright, narrow-beam light, such as a strong flashlight, to represent the sun; (2) a ball, 2½ to 3 inches in diameter, preferably with a smooth, shiny surface, to represent the earth; and (3) a smaller ball or coin with a diameter about one fourth of that of the larger ball, to represent the moon. A dime or penny would be easier to handle than a ball. These experiments should be done, if possible, in a darkened room so that the shadows can be seen more easily.

EXPERIMENT I.   Place the earth in the path of the light and about 2 feet away. Move the moon into a position such that the earth is between it and the sun.
QUESTIONS: (a) In this position, what kind of eclipse is possible—solar, lunar, or both? (b) As seen from the earth, what is the phase of the moon—full, quarter, or new? (c) From what part of the earth will the eclipse be visible? (d) What types of eclipse are possible— total, partial, or annular? (e) Is it possible for the moon to be in this relative position and still not be eclipsed?

EXPERIMENT II.   Place the moon between the sun and the

earth and adjust the distance of the moon from the earth so that a sharp shadow is formed on the earth.

QUESTIONS: (a) In this position, what kind of eclipse is possible—solar, lunar, or both? (b) What is the phase of the moon—full, quarter, or new? (c) Where must you be located to see this kind of eclipse? (d) What can you say about the distances involved in order to obtain a sharp shadow? (e) What types of eclipses are possible—total, partial, or annular? (f) What kind of eclipse would be easier to predict—lunar or solar? (g) Why? (h) Under what conditions would you expect to see an annular eclipse of the sun? (i) Why is it not possible to have an annular eclipse of the moon?

Lunar eclipses, as you have seen from these experiments, are visible over half the earth, and therefore many people at widely separated places can view an eclipse at the same time. This fact revealed to ancient people that time is local and changes as one travels east or west.

In 331 B.C. there was an eclipse of the moon at the time of the battle of Arbela, a city east of Ninevah in Asia Minor. Pliny says that the moon was eclipsed at the second hour of the night at Arbela, while in Sicily it was eclipsed just as it was rising. Since the full moon rises at sunset, there must have been a difference of two hours, according to their time, between Arbela and Sicily. Astronomers—among them Eratosthenes, Hipparchus, and Ptolemy—knew it was possible to determine longitude by observing the times of eclipses from widely separated stations.

Since the solar corona is visible from the earth only during times of total eclipse, eclipses of the sun are of importance for the study of the corona. Astronomers have gone on expensive eclipse expeditions to faraway places, such as the high Andes in Peru and islands in the Pacific Ocean, in order to study the extent and shape of the corona. Spectra of the corona and the flash spectrum—the bright-line spectrum visible just before and just after totality, lasting about 2 seconds—were photographed to learn about the composition and nature of the regions of the sun where they are produced. High-flying jets follow the moon's shadow for observations above the clouds, and rockets are used to lift instruments to still higher levels. All levels of the solar atmosphere can be observed at any time, without waiting for a total eclipse, with the coronagraph, the orbiting solar observatories, and the manned missions to the moon.

12.2 PREDICTING ECLIPSES

Some ancient astronomers thought that the moon is closer to the earth than the sun is and that solar eclipses occur when the moon passes in front of the sun. This is the correct interpretation. They also believed that the moon passes through the shadow of the earth, causing lunar eclipses, which is also correct.

Since the Babylonians, Chinese, and Greeks thought eclipses influence human affairs, they attempted to predict them. There are certain similarities in the situations under which eclipses are possible. Both solar and lunar eclipses occur near the two nodes—that is, the points where the moon's path crosses the ecliptic. Lunar eclipses occur when the moon is full and runs through the earth's shadow and is near the nodes. This was known to the early astronomers, and lunar eclipses were easy to predict. Solar eclipses occur at new moon (see Fig. 12-1), but other factors, such as the varying distance from the moon to the earth and the relative angular diameters of the sun and the moon, were not accurately known to the early astronomers. These factors made solar eclipses difficult to predict.

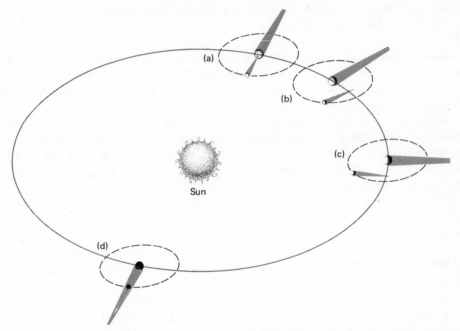

FIG. 12-1 Diagram showing that eclipses occur only when the moon is near a node while in new or full phase. A solar eclipse occurs when the moon is at new phase (between earth and sun); a lunar eclipse occurs when the moon is at full phase (on opposite side of earth from sun). Eclipses occur only at (a) or (d) where the moon is near a node and thus in the plane of the earth's orbit. At (b) and (c) the moon is not near a node and thus above or below the place of the earth's orbit.

A particular eclipse will recur about 18 years, 11⅓ days later in the same latitude but 120° west. This cycle is designated by the Greek word *saros*. Although some scholars believe that the astronomers of the Mesopotamian region predicted eclipses by the use of the saros, there is no evidence to support this belief.

The saros is an interval of 223 synodic months (the intervals between successive new or full moons), after which the sun or the earth's shadow will be in the same place on the ecliptic and at the same node of the moon's orbit. The time between successive passages of the sun through the same node is less than one year, because of the regression of the node as described in Chapter 10. This interval, the *eclipse year,* is 346.6 mean solar days. Two hundred twenty-three synodic months contain 6585.321 solar days or 18 years, 11⅓ days; 19 eclipse years contain 6585.4 days. That is,

$$223 \text{ synodic months} = 6585.321 \text{ solar days}$$
$$19 \text{ eclipse years} = 6585.4 \text{ days} = 18^{y} 11\tfrac{1}{3}^{d}$$

Therefore, if an eclipse occurs at a node at some point on the ecliptic, it will occur again at almost the same point after 18 years, 11 days.

The extra one-third day means that the next eclipse of the saros is visible one third of the circumference of the earth farther west. But after three saroses, an eclipse is visible in approximately the same longitude as the first one. There is a slight shift in latitude. As a result of these correlations, solar eclipses follow each other in a series, beginning at a pole and ending at the other pole. The entire series lasts about 1200 years, during which there are 29 lunar eclipses and 41 solar eclipses, of which 10 of the latter are total. Several series of saroses are in progress at the same time.

Today it is not necessary to depend on the saros for the prediction of eclipses, because with accurate data for the orbits of the earth and the moon, the computation can be done on a more scientific and more accurate basis. This depends on a knowledge of the nature and motions of the shadows of earth and moon.

12.3 LUNAR ECLIPSES

In Fig. 12-2, three circles represent the sun, the earth, and the moon. They cannot be drawn to the correct scale because of the difference in size and distance. The sun's diameter is 100 times larger than the earth's and 400 times larger than the moon's. If it were drawn with a diameter of 1 inch, the earth and moon would be too small to show satisfactorily.

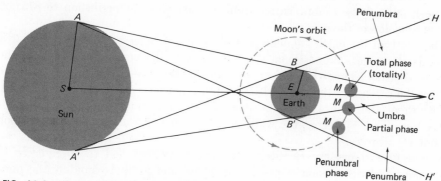

FIG. 12-2 The production of lunar eclipses.

Since the earth is an opaque body, it cuts off sunlight completely in a cone formed by straight lines drawn tangent to both bodies. The projection of this cone is shown in the diagram as lines tangent to the circles representing the sun and the earth. These tangents, ABC and $A'B'C$, are external tangents. The geometrical cone of the earth's shadow inside of which no part of the sun can be seen is called the *umbra*. If internal tangents, $A'BH$ and $AB'H'$, are drawn as shown, they form another cone from which part of the sun could be seen. This is the *penumbra*.

The length of the umbra from the center of the earth, E, to the vertex, C, can be calculated by plane geometry.

The radii of the two circles, SA and EB, are perpendicular to the tangents at A and B. Since the triangles, ASC and BEC, are right triangles that have a common angle ACS, they are similar, and their sides are therefore proportional. From this we can write:

$$\frac{SA}{EB} = \frac{SC}{EC} \tag{12-1}$$

Let $SE = D$ be the distance from the center of the sun to the center of the earth and $EC = L$ be the length of the earth's shadow. Substituting in equation (12-1)

$$\frac{SC}{EC} = \frac{D+L}{L} = \frac{SA}{EB} = \frac{432{,}300 \text{ miles}}{3{,}963 \text{ miles}} = 109.1 \tag{12-2}$$

whence,

$$D + L = 109.1\ L \tag{12-3}$$
$$D = 108.1\ L \tag{12-4}$$

and

$$L = \frac{D}{108.1} \tag{12-5}$$

The equation has been solved in algebraic form, since all the distances are variable. The average value of D is 92,900,000 miles, but it must be given precisely for the position of the earth in its orbit at the time of the eclipse.

The length of the umbra varies from 845,000 miles at perihelion to 875,000 miles at aphelion of the earth.

The width of the earth's shadow at the moon's distance can also be calculated by similar triangles. Proportional sides are used to derive an equation as in the above derivation. The width of the shadow varies depending on the distance between the earth and the sun and the distance between the moon and the earth. The width of the umbra averages about 5770 miles. Since the moon's angular diameter is about ½ degree, the width of the shadow at the distance of the moon is about 1⅓ degrees.

Figure 12-2 shows the earth's umbra and penumbra and the production of lunar eclipses. As the moon comes to full phase and in approximately the same plane as the earth's shadow, it passes through the penumbra, then through the umbra, and out in reverse order. As it touches the penumbra, where part of the sunlight is cut off by the earth, there is a slight amount of darkening of the moon's surface. This increases until, when the moon nears the umbra, the darkening is noticeable. This is the *penumbral phase* of the eclipse.

As the limb of the moon touches the umbra, the contrast between the two parts of the shadow is so great that there is a visible division between the bright and dark areas of the lunar surface. This is the beginning of the *partial phase* of the eclipse. In a telescope, this dividing line is not sharp. The limb of the moon crosses the umbra in about 1 hour, but the length of the partial phase depends on how near the moon is to the node, and how close the moon comes to the center of the shadow.

Finally, when the moon is completely inside the umbra, the eclipse is *total*. It might be expected that it would then be invisible, but this is not the case. Because of refraction by the earth's atmosphere, the light of the sun is bent enough for some of it to reach the moon, as shown in Fig. 12-3. The blue rays are filtered out and scattered, but the red rays get through producing a copper color, which may be quite dark near the center of the umbra. If there are high clouds above the earth in the region where the refraction occurs, the eclipse may be very dark. This was the case in the eclipse of December 30, 1963, at which time the moon was only dimly visible.

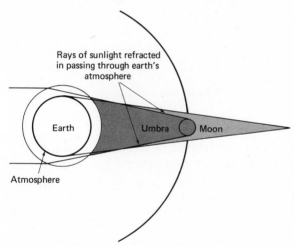

FIG. 12-3 The moon in the geometrical shadow (umbra) of the earth. Some of the light is refracted by the earth's atmosphere toward the moon. The shorter wavelengths (blue) are scattered and absorbed by the earth's atmosphere to a greater degree than are the longer wavelengths (red). The remaining light is predominantly red, which causes the moon to appear copper-colored during totality.

Figure 12-4 shows an end-on view of the umbra and penumbra, with the production of eclipses at different distances from the node. A is a total eclipse nearly central; B is a partial eclipse; C is a penumbral eclipse or *appulse;* and D shows the moon barely making contact with the penumbra (no darkening would be seen in this case).

The entire duration of a lunar eclipse may be as long as 5 or 6 hours, counting the penumbral phases. Totality can last up to $1^h 30^m$. The longest possible eclipse occurs when the moon passes through the center of the shadows. If it goes near the edge of the umbra, totality will be of short duration. In partial eclipses the moon is not entirely covered by the umbra. Occasionally the moon passes through the penumbra only.

Eclipses of the moon are not very important scientifically. It is possible to

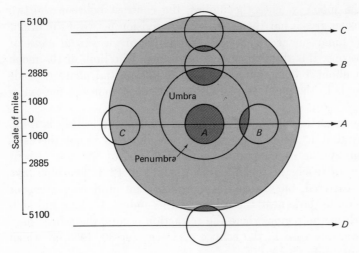

FIG. 12-4 End-on view of the umbra and penumbra of the earth's shadow, showing the path of the moon for total (A), partial (B), and penumbral phases (C) of lunar eclipses.

measure the drop of temperature as sunlight is blocked off. Such measures help slightly in studying lunar surface conditions and formerly helped show that the moon lacks appreciable atmosphere.

12.4 SOLAR ECLIPSES

The length of the moon's shadow can be calculated in a manner similar to that for the earth's shadow. As before, draw tangents, both interior and exterior, from the sun (not shown in Fig. 12-15) to the moon. Again there are similar triangles from which the length and width of the moon's shadow can be computed.

The length of the moon's umbra also varies with the distances involved. Its greatest length is 231,000 miles and its smallest length is 223,600 miles. Since the distance from the center of the moon to the earth's surface varies from about 221,800 miles to about 248,750 miles, the umbra is not always long enough to reach the earth. If it does reach the earth, the apparent diameter will be a little larger than the sun and the sun will be completely covered, except for the corona. This is a *total eclipse* of the sun. If the umbra does not reach the earth, the moon will be too small to cover the sun entirely, and, as seen from the center of the umbra extended to the earth, a perfect ring of the sun will be seen surrounding the moon. This is an *annular eclipse*. The production of annular eclipses is shown in Fig. 12-6.

The moon's shadow also has a penumbra, which is a little more than 4000 miles wide at the earth's distance. An observer inside the penumbra sees only a part of the sun's disk covered by the moon—a *partial eclipse*. The amount of the eclipse will decrease for observers near the umbra to zero for those at the edge of the penumbra, which may be more than 2000 miles distant.

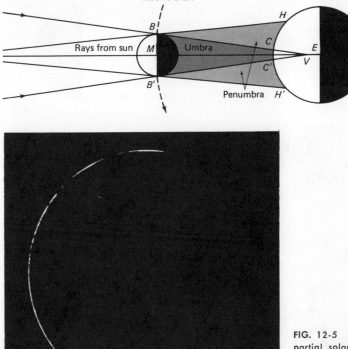

FIG. 12-5 The production of total and partial solar eclipses. *BC* and *B'C'* are externally tangent and *BH* and *B'H'* are internally tangent to the sun and moon. In the photograph, Baily's beads are visible just before the onset of totality. *(Yerkes Observatory photograph)*

The earth moves in its orbit around the sun and the moon moves around the earth in the same direction, west to east. The shadow also moves across the earth from west to east, and the earth rotates in the same direction. So the moon's shadow overtakes a point on the earth's equator with a speed of 0.3 mile/sec. At the poles the shadow moves at nearly 0.59 mile/sec. At intermediate stations, the speed is intermediate between the two extremes.

The greatest width of the moon's umbra where it strikes the earth is 165 miles and the maximum duration is a little more than 7 minutes. This is always near the equator. The duration of totality depends on the latitude and on the width of the umbra. If the vertex of the umbra barely touches the earth, there will be an instantaneous total eclipse. There is a possible situation wherein the eclipse will be total in the center of the eclipse track and annular at each end.

An eclipse track is a narrow strip of the earth that is swept by the moon's umbra. It is curved, since it is the intersection of a straight line and a sphere. Because an eclipse track is at most 165 miles wide, total eclipses are seen in very restricted regions. This is the reason eclipse expeditions are sent to parts of the earth where the eclipse is predicted to be total and where weather conditions are as favorable as possible.

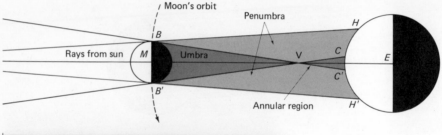

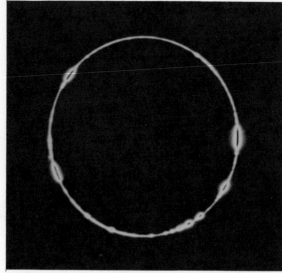

FIG. 12-6 The production of annular and partial eclipses of the sun. *BC'* and *B'C* are externally tangent and *BH* and *B'H'* internally tangent to sun and moon. The photo shows an annular eclipse. The bright spots are caused by sunlight shining through valleys on the moon.

Many times the umbra misses the earth entirely. At those times the eclipses are partial only, varying in magnitude from near zero to nearly total. The paths of seven total eclipses between 1972 and 1983 are shown in Fig. 12-7. The eclipse of June 20, 1974, in Australia, is off the map.

The calculation of the exact time, place, and duration of an eclipse depends on the following:

1. The distance of the moon from the earth and the earth from the sun.
2. The position and motion of the umbra on the earth.
3. The latitude and longitude of the observing station and its elevation above sea level.

Sixteen total and annular eclipses of the sun are listed in Table 12-1. Total eclipses of the moon visible in the western hemisphere are listed in Table 12-2. To see a total eclipse of the sun it is necessary to be inside the eclipse track, but a total eclipse of the moon may be seen from any part of the earth where the moon is above the horizon.

For an observer inside the eclipse track, a total eclipse of the sun is very spectacular. *First contact* is the instant when the moon begins to cover the sun.

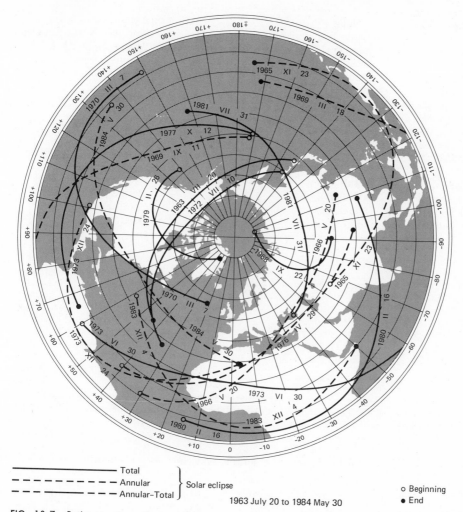

FIG. 12-7 Paths of solar eclipses visible in the northern hemisphere between 1963 and 1984. (Reprinted with permission from J. Meeus, K. Grossjean, W. Vanderleen; Canon of Solar Eclipses, 1966; Pergamon Press)

By projecting the sun on a white paper through a telescope, it is possible to estimate the time of first contact within a few seconds. The limb of the moon can be seen sharply defined at the limb of the sun. This is the beginning of the partial phase of the eclipse.

The partial phase ends in about 1 hour at *second contact*, when the sun is completely hidden by the moon. Shortly before the second contact, the sky becomes noticeably darker and the colors are quite different from those at sunrise and sunset, because the light comes from the limb of the sun, which is darker

TABLE 12-1   Central Eclipses of the Sun

date	type	duration	region
1972 Jan. 16	annular		Antarctica
1972 Jul. 10	total	3 min	Northern North America
1973 Jan. 4	annular		Argentina, South Atlantic
1973 June 30	total	7 min	Atlantic Ocean, Africa
1973 Dec. 24	annular		South America, Atlantic
1974 June 20	total	5 min	Australia
1976 Apr. 29	annular		Mediterranean Sea, Asia
1976 Oct. 23	total	5 min	Pacific Ocean
1977 Apr. 18	annular		Africa
1977 Oct. 12	total	3 min	Pacific Ocean
1979 Feb. 26	total	3 min	Northern U.S.A., Canada
1979 Aug. 22	annular		Antarctica
1980 Feb. 16	total	4 min	Africa
1980 Aug. 10	annular		Pacific Ocean
1981 Feb. 4	annular		South Pacific Ocean
1981 Jul. 31	total	2 min	Asia

TABLE 12-2   Total Eclipses of the Moon Visible in the Western Hemisphere, Central Standard Time[a]

date	begins	middle	ends	totality
1972 Jan. 30	3:11 A.M.	4:53 A.M.	6:35 A.M.	$0^h 42^m$
1975 May 24, 25	10:03 P.M.	11:46 P.M.	1:29 A.M.	1 30
1979 Sept. 6		4:55 P.M.	6:31 P.M.	0 46
1982 Dec. 30		5:26 P.M.		0 33

[a] For Eastern Standard Time, add 1 hour; for Mountain Standard Time, subtract 1 hour; for Pacific Standard Time, subtract 2 hours.

and more reddish than light from the center, and there is no blue and yellow light scattered by the atmosphere.

In the instant before totality the only visible parts of the sun are those that shine through valleys on the limb of the moon. They look like bright spots superposed on the inner corona, which is then visible. This phenomenon is called *Baily's beads*. They are shown in the bottom photograph of Fig. 12-6.

The corona suddenly comes into view as Baily's beads disappear. It can usually be seen to a considerable distance from the sun, and its shape changes from one eclipse to another. Totality lasts from an instant to about 7 minutes, during which time bright stars and planets are visible. At *third contact* the sun reappears, again with Baily's beads and at *fourth contact* the entire sun is visible again.

The kinds of observations made during a total eclipse depend on the field of interest of the observatory and the astronomers making up the expedition. Among the topics of interest are:

1. Photography of the phases of the eclipse, especially of the corona. The invention of the coronagraph has made photographs of the inner corona less important than before.
2. Observations of the flash spectrum and the limb of the sun with photography.
3. Motions of the material in prominences.
4. Polarization of the solar atmosphere.
5. Radio waves from the sun when the interior is completely hidden.

12.5 ECLIPSE SEASONS

We have shown that for an eclipse to occur, the moon must be new or full and at or near a node. The question to be answered now is: How near the node must the moon be and when will the eclipses be total, partial, or annular?

First, consider eclipses of the moon. Figure 12-8 shows the positions of the earth's shadow and the moon near a node. The descending node is shown, but eclipses will also occur under similar circumstances near the ascending node. It must be remembered that the earth's shadow and the sun move eastward along the ecliptic at about 1° per day, while the moon moves 13° per day, or about 1° in 2 hours. It has already been noted that the apparent diameter of the moon is ½° and that the diameter of the earth's shadow is about 1⅓° or 2⅔ times as large.

Figure 12-9 shows the full moon at the limiting position, outside of which an eclipse is impossible. The moon is just tangent to the earth's shadow. The angular distance between the center of the moon and the center of the shadow

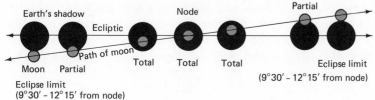

LUNAR ECLIPSES

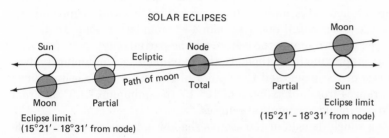

SOLAR ECLIPSES

FIG. 12-8   The production of eclipses showing the eclipse limits.

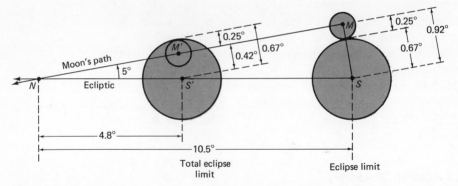

FIG. 12-9 The lunar eclipse limit, an enlargement of a portion of Fig. 12-8 showing the angular sizes of the moon M, and the umbra of the earth's shadow S. The moon's path makes a 5-degree angle with the ecliptic and crosses it at the node N. The farther the moon and the umbra are from the node, the greater the distance between their centers. Point S is the *eclipse limit*, the greatest distance the center of the earth's shadow can be from the node and still have it touch the moon; point S' is the *total eclipse limit*, the greatest distance at which the moon can be completely eclipsed.

at the extreme position is half the sum of the two diameters, or 0.92°. Figure 12-9 is an enlargement of the lunar eclipse portion of Fig. 12-8. Both apparent angular diameters are, of course, variable for reasons already discussed. Penumbral eclipses are not considered here, since they are relatively unimportant.

From Fig. 12-8 it is obvious that an eclipse of the moon will happen only inside the limits shown, when the moon runs through the shadow. If full moon is outside the limits, there will be no eclipse. At the extreme right and left, the moon is just tangent to the shadow, the limiting case. The distance of this point from the node in degrees is called the *eclipse limit*. The total distance on both sides of the node is the double limit.

The angle between the moon's path and the ecliptic is slightly variable, but is about 5°, as shown in Fig. 12-9. Assuming the angular distance from the center of the moon M to the center of the shadow S at tangency to be 0.92°, the angular distance, $R = MS$, can be found from the slender triangle formula (10-3):

$$R = 57.3° \times \frac{d}{p} = 57.3° \times \frac{0.92°}{5°} = 10.5°$$

The limiting distance from the node for total lunar eclipses can also be calculated by this method. This limit is about 4.8°. That is, total eclipses of the moon occur when the moon is full and less than 4.8° from either node. If the full moon occurs between the two limits, the eclipse will be partial only.

This is only an approximate method. If the maximum and minimum values are used for the size of the moon and the umbra of the shadow, the eclipse limit for lunar eclipses ranges from a maximum of 12° 15′ to a minimum of 9° 30′ from the node. They do not include penumbral eclipses.

The time during which the sun or the earth's shadow is inside an entire double eclipse limit is called an *eclipse season*. The earth's shadow and the sun move around the ecliptic in 365.24 days or at a rate of about 1° per day. Therefore, in 18.7 days the shadow would move completely across the minimum double lunar eclipse limit or in 24.2 days across the maximum double limit. In other words, the lunar eclipse season lasts at most about 24 days. The interval between full moons is 29½ days, the synodic month. It is therefore possible for the moon to be full before it reaches the limit and again after it has passed the limit. Hence, there may be a lunar eclipse season without an eclipse.

The solar eclipse limit is about 50 percent longer than the lunar limit. It lasts between 30.5 and 36.5 days, depending on the length of the limit. If a new moon occurs just before the limit is reached, there must be another new moon while the sun is inside the limit. In other words, there will be an eclipse of the sun at every eclipse season. In fact, there may be two eclipses during one season. For, if there is a small partial eclipse just inside the limit, there may be another 29½ days later, before the end of the same season. If there are two solar eclipses in successive months, there will be a total lunar eclipse at the full moon between the two, since the earth's shadow will be near the middle of the lunar eclipse limit at the other node.

There are two eclipse seasons in a single calendar year, and there may be part of a third because of the regression of the nodes. If the first season begins in early January, the second will begin in late May or early June, and the third in December. If there are two solar eclipses during each of the first two seasons, there is the possibility of another at the beginning of the third. That is, there may be five solar eclipses in a single calendar year. If so, there will be two lunar eclipses, for a total of seven. However, if the first eclipse occurs in December of the preceding year instead of January, there may be one solar and one lunar eclipse before the end of the year, for a total of four solar and three lunar eclipses. If penumbral eclipses are counted, there will be at least two lunar eclipses each year. The minimum number of solar eclipses in a single calendar year is two.

Since the penumbra is much wider than the umbra and since there are at least two solar eclipses each year, partial eclipses of the sun are fairly common and can be seen by many people.

12.6 OCCULTATIONS AND TRANSITS

The passage of the moon in front of a star or a planet is called an *occultation*. This phenomenon happens frequently, since the moon is ½° wide. However, occultations of stars bright enough to be seen near the moon are rather rare.

Since the moon moves eastward among the stars, when it occults a star the star disappears behind the east limb, which is the dark limb between new and full moon. To the unaided eye the star is seen when still some distance from the lighted part of the moon and seems to disappear instantly. But in a telescope the dark part is seen faintly illuminated, so the illusion is not quite as startling as it is with the unaided eye.

With modern photoelectric instruments and oscilloscopes, the light of the star is seen to vary rapidly because of diffraction by the limb. The entire phenomenon of disappearance lasts about 0.02 second. This method can be used to determine the angular diameter of a large star, which cannot be measured otherwise.

Occultations have been used to determine the exact position of the moon's limb and hence the center. Since the moon's motions are very complicated because of perturbations, it is impossible to calculate its position exactly. By the use of occultations, the computed position of the center of the moon is corrected occasionally.

The approximate times of occultation are listed annually in advance in the almanacs; but if accurate times are desired, it is necessary to substitute the latitude and longitude of the observing station. Methods for making these computations are also published in advance.

The occultation of a bright star or planet is of considerable interest (see Fig. 12-10). Binoculars are recommended if no telescope is available. Both disappearance and reappearance of stars occur so fast that the exact times should be computed beforehand. It is usually difficult to find the exact point of the moon's limb where the star will reappear. So it may be a second or two after reappearance that the star will be seen and it may then be well separated from the limb. A planet has a visible disk, which will gradually disappear or reappear. Occultations of Venus, Jupiter, and Saturn are well worth watching. Figure 12-10 shows the emergence of Jupiter and three of its four Galilean satellites. During occultations of Saturn, the moon can be seen crossing the ring system. And the occultations of the brilliant Venus are beautiful, particularly if both bodies exhibit their crescent phases.

FIG. 12-10   The emergence of Jupiter and three of its satellites after their occultation by the moon. (Griffith Observatory)

An occultation of a star by a planet is very rare, but such occurrences are calculated and listed in advance. Such an event is shown in Fig. 12-11.

The planets Mercury and Venus sometimes pass in front of the sun. This phenomenon is called a *transit*. Transits of Mercury occur about 13 times per century. They always occur in May or November when the planet is near one of its nodes. The planet moves diagonally across the sun's disk fairly near the

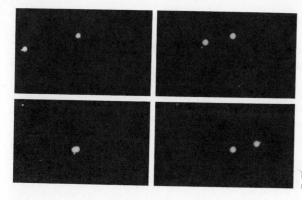

FIG. 12-11 Occultation of a star by the planet Neptune. The star is near the center of each photo. Neptune is moving from left to right. One of the planet's satellites is also visible right next to Neptune. *(Tersch Enterprises)*

limb. The entire transit lasts about 4 hours and can be seen only in a telescope. During this time, the planet is seen as a small, very dark, round spot moving slowly from east to west. The next transits of Venus will occur on June 8, 2004, and on June 6, 2012.

QUESTIONS AND PROBLEMS

Group A

1. What percent of the earth's surface can witness the following: (a) moon entering totality, (b) some part of a lunar eclipse, assuming it lasts 3 hours?
2. It is possible for a solar eclipse to be annular over part of its track and total over another part. Under these circumstances, is the total eclipse of long or short duration? Explain.
3. Give several reasons why the duration of totality of a solar eclipse varies from one locality to another.
4. Explain why the flash spectrum lasts only a second or two just before and just after totality.
5. In a coronagraph the visible portion of the sun is occulted by a disk. During a total eclipse this portion of the sun is occulted by the moon. What observations can be made during a total eclipse that cannot be made with the coronagraph? Why?
6. What would be seen from the moon while solar and lunar eclipses are observed on the earth?
7. If the moon revolved around the earth at the same rate but in the opposite direction, how fast would the moon's shadow overtake a point on the equator?
8. If there are five solar and two lunar eclipses during a year, what would be the nature of the lunar eclipses? Explain.

Group B

9. Compute the average length of the shadows of Venus and Jupiter (see Table 13-1). *Answer:* Venus, 597,000 miles.
10. The tip of the shadow of a spherical satellite just touches the earth's surface, 100 miles below, when the satellite is in line

with the centers of the sun and the earth. What is the diameter of the satellite?

11. By using similar triangles, compute the width of the earth's shadow at the moon's average distance.

12. Show that during a solar eclipse the shadow of the moon travels at about 2120 mph. Should the synodic or sidereal period be used in the computation?

13. Venus is 0.72 a.u. from the sun. Would there be solar and lunar eclipses if the earth-moon system were as close to the sun as Venus?

14. If the moon's nodes remained stationary instead of regressing, what changes would there be in the occurrence of solar and lunar eclipses?

13   the planets and their families

The solar system consists of the sun and all the smaller bodies under its gravitational influence. Among these are the nine planets, most of which control a secondary family of satellites (moons). There are also thousands of minor planets, usually called asteroids, and an unknown number of comets, which are collections of particles loosely held together. Everywhere in between there are myriads of smaller particles, called meteoroids. Still more finely divided particles are called interplanetary dust.

13.1 DISCOVERY AND CLASSIFICATION OF PLANETS

The first six planets in order of distance from the sun have been known since ancient times, although the earth was not recognized as a planet until the time of Copernicus. The seventh planet, Uranus, was discovered by visual observations made with a telescope by William Herschel in 1781. Its irregular motions led to a mathematical study ending in the discovery of the eighth planet, Neptune, in 1846. More computations led to a program of search that culminated in the discovery of Pluto, the ninth and last planet, in 1930. It is now believed that the system of planets in the solar system is completely known.

This system is isolated in space. Pluto is less than 40 a.u. from the sun. The nearest star is about 270,000 times the distance of the earth from the sun. The space between the stars and the sun is almost empty, although some astronomers believe there are billions of comets wandering around, some of which are deflected into the region of the sun. There are also interstellar dust particles so far apart that there is only about one atom per cubic centimeter in the space between the stars.

257

There are several ways of classifying the planets. One way calls Mercury and Venus inferior planets, meaning that they are inside the orbit of the earth. All the others are called superior planets. If they are classified by size and mass, Mercury, Venus, the earth, Mars, and Pluto may be called terrestrial planets, since they are much like the earth. Pluto, however, seems to be an almost unknown planet in many respects. Its size, mass, and density must be considered as uncertain. The larger planets are called major planets or sometimes Jovian planets. Jupiter is the largest, followed by Saturn and a pair of twins, Uranus and Neptune, of which Neptune seems to be slightly the larger. Table 13-1 lists the known data about all nine planets, plus Ceres, the first asteroid to be discovered and the largest. Its orbit is typical of the average asteroid. An asteroid looks like a star (hence the name), but is actually a small planet.

13.2 PLANETARY ORBITS

The sidereal periods, the times required for a single revolution around the sun, increase outwards from 88 days to almost 250 years. From the sidereal periods, the synodic periods can be calculated (see Table 4-1). The synodic periods are roughly in order. Mercury has the shortest because of its short sidereal period; Mars and Venus have the longest, since their sidereal periods are nearest that of the earth. The synodic periods then decrease to Pluto, which moves the slowest in its orbit. Pluto advances only about 1.5° per year, so the earth overtakes it in about one day more than a year.

The mean distances are correlated with the sidereal periods by Kepler's third law. They are given in Table 13-1 in both astronomical units and millions of miles. It is obvious from the table why the planets are placed in two groups on the basis of mass and diameter. The density of each planet can be computed from the mass and diameter as was done for the sun, moon, and the earth in previous chapters. The density is given in the table in grams per cubic centimeter; that is,

TABLE 13-1   Planetary Data

name	sidereal period	synodic period	mean distance from sun (a.u.)	mean distance from sun (millions of miles)	equatorial diameter (miles)
Mercury	$88^d.0$	116^d	0.387	36.0	3100
Venus	224.69	584	0.723	67.2	7524
Earth	365.26		1.000	93.0	7927
Mars	686.95	780	1.524	141.6	4220
Ceres	$4^y.604$	467	2.767	257.3	480
Jupiter	11.86	399	5.203	483.6	88,700
Saturn	29.56	378	9.561	888.8	75,100
Uranus	83.95	370	19.17	1782	29,000
Neptune	163.9	367	29.95	2784	31,000
Pluto	247.0	367	39.37	3661	3400?

it is compared to the density of water. It will be noticed that this figure, the average density, shows each planet, except Saturn, to be heavier than water.

The period of rotation of each planet is given, but is subject to change, as will be noted later in this chapter. The albedo, the reflecting power, of each planet is tabulated. Mercury's albedo is about the same as that of the moon. Both these bodies are without atmospheres. Mars' albedo is intermediate, as is expected for a planet with a moderate atmosphere. All the other planets reflect sunlight from their atmospheres, which accounts for their high albedos.

The determination of the position of a planet in its orbit and its right ascension and declination at any time is a problem belonging to the field of celestial mechanics and cannot be discussed here.

13.3 MERCURY

The planet nearest the sun and the smallest in the solar system is Mercury. Its mean distance is 36 million miles, with perihelion distance 28.6 million miles and aphelion 43.4 million miles. Since its synodic period is 116 days, Mercury comes into superior conjunction with the sun, and afterward into the evening sky, every four months. But because its maximum elongation is only 18° at perihelion and 28° at aphelion, it is too close to the sun to be seen easily. Since Mercury and Venus are inferior planets, they show all phases from new at inferior conjunction to full at superior conjunction, as shown for Venus in Fig. 13-1. The best times to see Mercury are at greatest elongation in the evening sky in March and April, when its orbit makes the greatest possible angle with the horizon, or in the morning sky in September and October. When brightest, Mercury is a little brighter than the brightest star, Sirius, but it is 10 times fainter than Venus at maximum.

Since the velocity of escape is low, only 2.6 miles/sec and the temperature is high, Mercury has no appreciable atmosphere of its own. It is now known to be in the sun's atmosphere. The temperature on the sunward side measures

mass (earth = 1)	average density (g/cm³)	period of rotation	number of satellites	albedo	velocity of escape	equatorial surface gravity (earth = 1)
0.054	5.2	59^d?	0	0.07	2.6 mi/sec	0.36
0.814	5.24	242.9	0	0.77	6.4	0.95
1.000	5.52	23^h 56^m	1	0.39	7.0	1.00
0.11 ?	4.0	24 37	2	0.20	3.2	0.39
317.8	1.33	9 51	12	0.51	37.1	2.54
95.2	0.68	10 11	10	0.50	22.3	1.06
14.5	1.56	10 49	5	0.66	13.9	1.08
17.2	1.58	15	2	0.62	14.6	1.12
0.11	4.85?	5 09	0	?	3.5?	0.60?

FIG. 13-1   Relative apparent size of Venus at five different phases. (Lowell Observatory photographs)

between 616°K and 690°K, depending on the distance from the sun. This is above the melting point of lead.

Because of the tidal forces from the sun, it was thought until recently that Mercury always keeps one side toward the sun. This was based on visual and photographic observations that seemed to show the same surface features at all times. However, radar observations at the Cornell University station in Arecibo, Puerto Rico, established a spin rate of 58.4 ± 0.4 days with respect to the stars, and its synodic period with respect to the sun is 172 days. Thus an area on Mercury faces the sun for 86 days at a time and faces away from the sun for the same length of time. The dark side was thought to have a temperature near absolute zero, but it has been found that the dark-side temperature may be as high as 150°K.

It will be remembered that the moon's orbit is rotated by perturbations caused by other members of the solar system. This orbital rotation is called the advance of the moon's perigee. There is a similar rotation of Mercury's orbit, called the *advance of the perihelion*, also caused by perturbations (see Fig. 13-2). In the case of Mercury, computations of all the known perturbations led to a rate of rotation of 531″ in 100 years. The measured rate is 574″, a difference of 43″ per

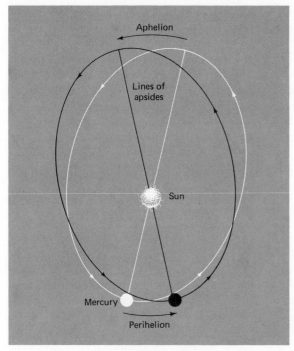

FIG. 13-2 Advance of the perihelion of the planet Mercury. The amount of rotation shown in the diagram requires more than 10,000 years.

century. This led to the theory of relativity, in which Einstein explained the difference.

In the theory of relativity, the mass and length of a moving body change with its velocity, according to the following equations:

$$l = l_0 \sqrt{1 - \frac{v^2}{c^2}} \quad \text{and} \quad m = \frac{m_0}{\sqrt{1 - \frac{v^2}{c^2}}} \tag{13-1}$$

In equations (13-1), l and m are the length and mass of the moving body; l_0 and m_0 are its length and mass when at rest; v is its velocity; and c is the velocity of light.

From Kepler's second law, the velocity of Mercury in its orbit increases from apogee to perigee by about 50 percent. Therefore, the mass of Mercury changes by a small amount. But this change of mass is enough to provide the extra perturbation and accounts for the $43''$ discrepancy in the advance of the perihelion. This was the first confirmation of the theory of relativity.

Mercury is much like the moon. It has the same albedo, no atmosphere, and its surface material probably has a low coefficient of heat conduction.

13.4 VENUS

The beautiful evening star, Venus, is also a conspicuous object in the morning sky. Before the two appearances were recognized as one planet only, it had different names: Phosphor, or Lucifer, in the morning and Hesper in the evening. After Venus passes superior conjunction, it remains in the evening sky for about nine months. Then it rapidly approaches the sun, passes between the earth and the sun at inferior conjunction and into the morning sky for another nine months.

In brightness, Venus surpasses all other objects in the sky, except the sun and the moon. There are several reasons for this. When closest to the earth, the planet is only about 26 million miles away and has an angular diameter slightly greater than 1'. At that time, inferior conjunction, it comes closer to the earth than any other major body, except the moon. Its phase is then new, since the dark side is turned toward the earth. Venus is invisible at inferior conjunction except at rare times when it passes the sun at a distance of a few degrees and can be seen with the telescope as a thin crescent, like a very new moon.

As Venus recedes from the earth after inferior conjunction, more and more of its illuminated side can be seen. After 36 days, it reaches its greatest brilliance at an elongation of about 39°. This is a compromise between phase and distance, then about 38.3 million miles. It then recedes to superior conjunction at a distance of some 161 million miles and is behind the sun, and therefore invisible again. Its brightness has faded by a factor of about two when last seen in the bright sky just before sunrise. After about two months of invisibility, it reappears in the evening sky, going through its changes of phase and brightness in reverse order.

Another reason for its great brilliance is that Venus is almost as large as the earth. Since it is about the same size and mass, it has nearly the same velocity of escape, 6.4 miles/sec. Its albedo, 77 per cent, indicates a very dense atmosphere, which is one of several factors that interfere with satisfactory observations of the planet. Another is that it is never more than 48° from the sun, and is observable at night only at low altitudes. Venus' albedo is greater than that of any other planet.

Spectrograms taken with terrestrial telescopes have shown only bands of carbon dioxide, although more recent observations show faint lines that were tentatively identified as carbon monoxide and nitrogen. Care must be taken to distinguish between lines that have their origins in the atmosphere of Venus and those originating in the earth's atmosphere. Observations in 1959 from a balloon high enough above the earth to be free of interference by water vapor in our atmosphere showed faint lines of oxygen and water vapor.

On December 14, 1962, the space probe Mariner 2 passed within 21,000 miles of the planet's surface and helped settle some questions about the planet that had been unanswered for many years. Some of the findings are listed below:

1. The temperature above the cloud layer is about 220°K (−60°F), but it is 700°K (800°F) on the surface. It seems to be uniform over the entire planet, except for one cold spot, which may be a mountain.
2. The clouds are very dense, thick, and cold, perhaps −30°F. They lie from about 45 to 62 miles above the surface. The sun

cannot penetrate the clouds, which are thought to be composed of hydrocarbons—familiarly known as smog.

3. The atmosphere below the clouds is mostly carbon dioxide. If there is oxygen, it is at least 1000 times less dense than terrestrial oxygen. Water is also about 1000 times less than on the earth.

4. Venus' surface seems to be composed of dust or sand.

5. Winds of solar plasma blow past Venus with speeds of 200 to 500 miles/sec, with temperatures up to 1 million degrees.

6. Venus has no magnetic field. This may be because the planet lacks a nickel-iron core or its magnetism may have been removed by the solar wind.

Since the first Venus probe there have been several close approaches by American and Russian spaceships. The data from these missions are not in agreement. The Russian ship Venera 4 crash-landed on the planet. It was assumed that Venera 4 landed on a peak or plateau some 15 miles high or stopped transmitting before reaching the solid surface. It may also be assumed that its altimeter was in error.

The American space probes are able to provide valuable information about planetary atmospheres, even though they merely pass by without entering the atmosphere. The flight paths of these vehicles are programmed to carry them behind a planet, so the radio signals to earth pass through the planet's atmosphere. Since radio waves are electromagnetic waves, they are refracted by the atmosphere. The amount of refraction is affected by the temperature of the gas and the average mass of its molecules. It is possible to measure the refraction with a high degree of accuracy and thus determine the temperature and molecular mass of the atmosphere.

The data on the size of Venus appear to agree. The radius is 6053.7 ± 2.2 km, giving a diameter of 7524 miles. Surface atmospheric pressure is estimated to be about 100 atmospheres and the temperature about 700°K, or 427°C. The Venera 4 results estimate the atmosphere to be composed of between 80 and 100 percent carbon dioxide. This agrees with the American estimate of at least 80 percent. The oxygen content is only about 1 percent and the nitrogen content is less than 7 percent. There is a trace of water vapor.

Assuming a surface pressure of 100 atmospheres, the atmosphere of Venus is superrefractive below about 20 miles. This means that light rays bend more sharply than the surface of the planet. As a result, if the atmosphere were transparent, a person could see over the horizon and would see the entire surface spread out around him as if he were in the bottom of a bowl.

Because of the opacity of Venus' atmosphere, it has been very difficult to determine its period of rotation. The rotation period is now given as 242.9 days with an uncertainty of 2 or 3 hours. It rotates in a retrograde direction (east to west). The day on Venus is 117 of our days. This, combined with the higher solar constant at Venus' distance from the sun (about twice the solar constant at the earth's distance) and the opacity of Venus' atmosphere that produces an intense greenhouse effect, yields the high temperature and would make Venus a very uncomfortable place to land. The Russian ships may have melted before landing.

In view of the uncertainty in the results of research in the vicinity of Venus, it is obvious that a surface landing with better equipment is highly desirable.

Neither Venus nor Mercury has a known satellite. Since the direct determination of the mass of a planet depends on a formula that includes the period and distance of a satellite, the mass of these two planets can be determined only by the effects of their perturbations on other bodies. This method is difficult, and for many years the results were uncertain. Values of the data obtained in 1969 for Mercury and Venus are included in Table 13-1.

13.5 MARS

Mars has probably attracted more attention than any other planet since the discovery of its so-called canals in 1877, the same year in which its two little moons were discovered. Much speculation about the possibility of life and the artificial nature of the canals has led to the publication of many books about the planet. Percival Lowell (1855–1916), an American astronomer, built his own observatory on "Mars Hill," on the edge of Flagstaff, Arizona, for the express purpose of observing the planet. While the Lowell Observatory has studied Mars almost continually since its founding, its work also led to the discovery of the ninth planet, Pluto, and to much other astronomical work of importance. Lowell's book entitled *Mars* is a classic. Earl C. Slipher (1895–1964) continued Lowell's work, photographed Mars from a location in South Africa where the planet was nearly overhead at favorable oppositions, and, just before his death, published a book about his latest findings.

Mars is usually not as easily recognized as Venus and Jupiter because it is not as bright, but its reddish color and its rapid motion among the constellations help in identifying it. Mars is absent from the evening sky for about a year near the times of conjunction with the sun. Then it brightens up as it nears the earth, stops its eastward motion, and makes a long loop toward the west that takes three months. It then resumes its eastward motion, fading slowly until conjunction again about a year later. The loops are traced every 26 months. They have slightly different sizes and shapes because of the eccentricity of the orbit and the position relative to the earth. Each loop is some 30° east of the preceding loop.

Because of the eccentricity of the orbit, 0.093, the opposition distances from the earth vary from about 35.5 million miles to 62.6 million miles as shown in Fig. 13-3. The closest oppositions unfortunately occur when the planet is nearly 30° south of the celestial equator, which was the reason for Slipher's expeditions to Africa to photograph it. The favorable oppositions in 1954 and 1956 were disappointing in that they failed to reveal anything new about the nature of the surface. In 1956 Mars developed a veil of haze or dust, which almost

FIG. 13-5 The surface of Mars as seen from Mariner 7. The photo was taken from an altitude of about 3300 miles, which was the closest approach of the spacecraft to the planet. An area of about 740 miles by 930 miles is shown. The Martian south pole is at the lower right. A cloudlike object is visible at the upper left and formations resembling snowdrifts at the upper right. (NASA)

around the planet, photographing the surface, measuring temperatures in the atmosphere, and studying its composition. By late April 1972, it had returned about 7000 photographs of the planet and was still sending back data as late as June 1972, but only at selected times.

Photographs taken from the earth have revealed a so-called "blue haze," a dense general haze that obscures the surface features taken in blue and violet light. This haze disappears at times, allowing the surface to be photographed even in these regions of the spectrum. This is called a "blue clearing." There is no trace of this blue haze in the Mariner photographs. There is, however, a scattering of light in layers in the Martian atmosphere at heights between 10 and 25 miles. This scattering was noticed at several latitudes.

Mariner 6 and 7 observations showed that the Martian craters are somewhat different from those on the moon, having fewer central peaks and less steep sides. They are thus intermediate between craters on the earth and those on the moon. There are a few signs of rills, but almost nothing of the system of canals. At least two classical "canals" were found to coincide with nearly straight alignment of dark-floored craters. Some of the classical "oases" have been identified as single, large, dark craters.

Three types of surface features were noted. In the *cratered terrains* some of the craters are large and flat-bottomed, with diameters up to 100 miles and more. They differ somewhat from lunar and terrestrial craters, where there is respectively less and more erosion. Other craters are smaller, bowl-shaped, and resemble lunar impact craters. In some areas relatively smooth, cratered terrain gives way abruptly to *chaotic terrain,* consisting of short ridges and depressions that are practically uncratered. A bright "desert" was found to be an area of *featureless terrain,* down to the resolution of 300 yards. No area of comparable size and smoothness is known on the moon. It is evident that surface processes must have been operating almost up to the present in the areas of featureless and chaotic terrains to account for the lack of craters.

The atmospheric pressure was determined from its effect on the Mariner signals passing through it when the vehicles passed behind the planet. The pressure at the surface is of the order of 1 percent of that on the earth's surface, and is thus comparable to the pressure in the earth's atmosphere at an altitude of 20 miles. (It is interesting to note that the pressure at the surface of Venus is 100 times that on the earth's surface.)

The atmosphere is between 60 and 100 percent carbon dioxide. Water vapor, ionized carbon dioxide, and carbon monoxide have been detected, along with atomic hydrogen and oxygen, but the amounts are very small. No trace of nitrogen has been found by the spacecraft.

The temperature on the Martian equator rises above $+60°F$ at noon, but, because of the thin atmosphere, it drops below $-100°F$ during the night. Apparently the surface material is a good insulator. The dark areas are warmer than the bright areas. At the south polar cap the temperature is near $-190°F$, which is close to the freezing point of carbon dioxide (dry ice) under Martian pressure. The polar cap seems to be composed of solid carbon dioxide with a small amount of water ice. Changes in the polar cap from day to night suggest that the caps partially evaporate in the sunlight and form condensation clouds during the night. Based on the estimated loss per square centimeter, it appears that the total thickness is equivalent to a few feet of solid carbon dioxide with a trace of solid water.

There was another severe dust storm in 1971. These storms seem to occur when Mars is near perihelion. Photographs taken in 1971–1972 after the storm cleared showed the south polar cap to be separated into two parts, which are seamed by a complex system of cracks. A fourth class of surface features was added: near the poles are *sedimentary layers* up to 100 yards (meters) in thickness.

There is no direct evidence of life on Mars; the scarcity of water is a most serious limiting factor.

Mars rotates in $24^h 37^m$ in direct motion (west to east). Except for the greater distance from the sun and the more extreme distances at perihelion and aphelion, the seasons are not very different from those on the earth. Its equator is inclined to the plane of the orbit by 25°, compared to 23.5° for the earth's axis. There are therefore 670 Martian days in a Martian year.

Mars has two small satellites, which were discovered by Asaph Hall (1820–1907) at the U.S. Naval Observatory in 1877. The inner satellite is named Phobos, meaning Fear. In the Mariner 7 photographs it measures about 11 by 14 miles, and in 1972 was estimated to be 21 by 25 km (13 × 15.5 mi). Its mean distance from the center of the planet is 5800 miles. Since its sidereal period of revolution is $7^h 39^m$, only about one third of the period of the planet and in the same direction, Phobos rises in the west and sets in the east.

The outer satellite—named Deimos, meaning "panic"—is only 7.5 by 8.5 miles in diameter according to the 1972 measures. Deimos is 14,600 miles from the center of the planet, or 12,490 miles from the surface. It revolves in $30^h 18^m$ and in the same direction as Phobos. It rises in the east and sets in the west, remaining above the planet's horizon for $2\frac{2}{3}$ earth days at a time. Phobos would be about 1000 times and Deimos five or six times brighter to an observer on the surface of Mars than Venus is to an observer on the earth.

It is interesting to note that Swift in *Gulliver's Travels*, written in 1726, reported observations of two satellites of Mars by the inhabitants of Laputa "whereof the innermost is distant from the center of the planet exactly three of his diameters, and the outermost five; the former revolves in the space of ten hours, and the latter in twenty-one and a half." The actual distances are 1.37 and 3.46 times the diameter of the planet.

The mass of a planet with a satellite can be calculated easily. Since Mars is a good example, the method is given here and the mass of Mars is calculated using Deimos as the satellite.

From Newton's modification of Kepler's third law of planetary motion (see Chapter 4 and Section 4.7),

$$P^2(M + m) = A^3 \tag{13-2}$$

Since this equation was derived by Newton as a result of gravitational attraction, it can be adapted to any pair of bodies moving in orbits about their common centers of gravity. The resulting equation is a proportion (stated as applying to Mars and its motion around the sun and Deimos in its revolution around Mars):

$$\frac{P^2_{Mars}}{P^2_{Deimos}} \times \frac{\text{mass of sun} + \text{mass of Mars}}{\text{mass of Mars} + \text{mass of Deimos}} = \frac{A^3_{Mars}}{A^3_{Deimos}} \tag{13-3}$$

Neglecting the mass of Mars as small compared to the mass of the sun and also that of Deimos compared to Mars, this proportion may be written:

$$\frac{P^2_{Mars}}{P^2_{Deimos}} = \frac{A^3_{Mars}}{A^3_{Deimos}} \times \frac{\text{mass of Mars}}{\text{mass of sun}} \tag{13-4}$$

provided the same units are used for corresponding terms.

Substituting for the periods and mean distances and solving:

$$\frac{\text{mass of Mars}}{\text{mass of sun}} = \frac{(686^d95)^2}{(1^d262)^2} \times \frac{(14,600 \text{ mi})^3}{(141,640,000 \text{ mi})^3}$$

$$= \frac{1}{3,081,000} \tag{13-5}$$

Multiplying by the ratio of the sun's mass to the earth's mass, the mass of Mars is 0.108 times the mass of the earth.

Gravitation on the surface of Mars is about 40 percent of that on the earth. Since the velocity of escape is low, 3.2 miles/sec, the amount of atmosphere retained is correspondingly low. This is confirmed by the low albedo, 20 percent, and the findings of the Mariners.

13.6 THE ASTEROIDS

During the 18th century a formula was developed that gives the approximate mean distances of the known planets from the sun. It is known as Bode's law, although he was not the first to state it; and it is not a law in the strict physical sense. If 4 is added to 0, 3, 6, 12 . . . , and each sum is divided by 10, a sequence of numbers results, each of which is the approximate distance of a planet in astronomical units. These numbers and the actual distances of the planets are given in Table 13-2.

TABLE 13-2   Bode's Law

planet	mean distance[a]		planet	mean distance[a]	
	Bode's law	actual		Bode's law	actual
Mercury	0.4	0.387	Jupiter	5.2	5.203
Venus	0.7	0.723	Saturn	10.0	9.539
Earth	1.0	1.000	Uranus	19.6	19.191
Mars	1.6	1.524	Neptune		30.071
Ceres	2.8	2.767	Pluto	38.8	39.518

[a] The mean distance is given in astronomical units.

When Bode's law was stated in 1766, there was an apparent gap in the series between the distances of Mars and Jupiter. It was thought that there was a missing planet in the gap. On the first night of the 19th century, January 1, 1801, an uncharted star was found by Giuseppe Piazzi (1746–1826), an Italian astronomer. This "star" was found to move and was observed for about a month before it became lost owing to the illness of its discoverer. By that time it had been recognized as a small planet from its motion. Also its orbit was computed by a new method developed by Karl Friedrich Gauss (1777–1855), a German mathematician, and fit almost exactly the mean distance computed by Bode's law. It was assumed to be the missing planet and was named Ceres.

Gauss' method depended on accurate positions of the body on three dates, preferably separated by a few weeks. It is still used in modified form. Ceres was found again at the end of 1801 in the location predicted by Gauss. It is the brightest of a system of small bodies now called *asteroids,* or *planetoids,* most of which are in orbits between those of Mars and Jupiter. Ceres sometimes reaches naked-eye brightness at favorable oppositions. Its diameter is 480 miles, much less than the diameter of the smallest of the nine major planets. The mass of Ceres has never been determined, but a reasonable guess is that it is about 1 percent of the mass of the moon, if it has the density of rock.

Much to the surprise of astronomers, another small planet was discovered in 1802; it was later named Pallas. The discovery of Juno followed in 1804, Vesta in 1807, and Astraea in 1845. At present nearly 2000 asteroids are known. Since they have orbits similar to those of the major planets, they are often called minor planets. Each one has been named by its discoverer. At first the asteroids were found by visual telescopic observations. After the invention of photography and its adaptation to astronomy, discovery was easy. If a photographic plate is exposed for an hour or more and the telescope is guided so the stars show sharp, round images, an asteroid shows a trail (Fig. 13-6). With the use of Schmidt cameras, asteroids are being photographed in great numbers. The orbits of about 1700 have been calculated, and it is estimated that 40,000 or more are within reach of modern telescopes. Their sizes vary, including "mere mountains broken loose," to use the expression of Henry Norris Russell (1877–1957), an American astronomer. The steady growth in numbers as magnitudes decrease indicates that the sequence continues down to bodies as small as micrometeoroids and dust grains.

The asteroids all revolve around the sun in the same direction, but the inclinations of their orbits vary considerably, up to 52°. The eccentricities of the

FIG. 13-6 A field of stars with trails of two asteroids (marked by arrows). (*Yerkes Observatory photograph*)

orbits are mostly moderate, but one is as large as 0.83. A few come inside the orbit of Mars and one comes closer to the sun than Mercury. Icarus came within 4 million miles of the earth in 1968 and Geographos within 6 million miles in 1969. One of the most famous is Eros, which was used to determine the solar parallax (see Chapter 11) when it came to opposition in 1931.

In studying the orbits of asteroids, Daniel Kirkwood, an American astronomer, discovered gaps in the system (Kirkwood's Gaps). Certain asteroids, which may at one time have occupied positions with periods that were submultiples of Jupiter's sidereal period, were pulled out of orbit by perturbations.

One group of asteroids is of importance. They are the Trojans, all named for heroes of the Trojan war. From the study of the problem of three bodies in celestial mechanics, it was found that if a small body, such as an asteroid, gets into the vertex of an equilateral triangle with the sun and Jupiter at the other two vertices, it will stay there unless pulled out by some perturbations. Twelve asteroids are known near the triangular points. This discovery is evidence that the problem of three bodies has been solved correctly. A thirteenth asteroid lies at a mean distance about the same as Jupiter's, but at perihelion comes close to Mars and at aphelion nearly to the orbit of Saturn.

The origin of the asteroids is still very uncertain. Some theories say that a planet may have been formed between Mars and Jupiter and later broke up under unknown forces. There may have been explosions at different times, since the asteroids seem to be grouped into families. If a planet broke up, it must have been a small one, since the mass of the entire system is estimated at only 0.001 times the mass of the earth.

It is also possible that a mass of material failed to condense into a planet. Certainly the solar system is full of particles in the form of comets and meteoritic material, in addition to solar wind particles and cosmic dust. Collisions between particles have never been observed, but they must be going on all the time. Perhaps the future will produce evidence that will lead to a satisfactory theory about the formation and evolution of these pieces of matter.

13.7 JUPITER

The giant planet Jupiter orbits the sun in an ellipse that is five times larger and somewhat more eccentric ($e = 0.048$) than the orbit of the earth. It comes to perihelion at a distance of 460,260,000 miles and recedes to 506,900,000 miles at aphelion. Its sidereal period is nearly 12 years and, since its synodic period is 399 days, it comes to opposition every 13 months. In spite of its opposition distance of about 400 million miles, which is variable because of the eccentricity of the orbits of the earth and of the planet, it can be seen fairly well through a moderate-sized telescope. Its rapid rotation with a period of only $9^h 50^m$ has produced a noticeable flattening at the poles.

The equatorial diameter is 88,700 miles and the polar diameter is 83,200 miles. On the disk of the planet a series of markings, called *belts*, can be seen parallel to the equator. These bands appear to change shape and position from year to year. Another feature, the *great red spot,* first noticed in 1878, has been visible intermittently since then. If the spot itself is not visible, the indentation of its position in a belt can usually be seen, except when it is behind the planet during half of the 10-hour period of rotation. The red spot apparently also changes size and position on the surface of the planet. Its real nature is still unknown. The belts and the great red spot are plainly visible in Fig. 13-7.

FIG. 13-7 Photograph of Jupiter taken through a blue filter to emphasize the great red spot; the cloud belts are also conspicuous. *(Photograph from the Hale Observatories)*

The mass of Jupiter can be determined from the periods and mean distances of its satellites as in the case of Mars. The mass is 317.8 times the mass of the earth. The average density is 1.33 g/cm³. The amount of flattening, combined with the low average density, indicates that the planet must have a core of rather high density. A recent model suggests that the planet is (by mass) at least 78 percent hydrogen, and that the core has a density of 31 g/cm³, under a pressure of 100 million atmospheres (1 atmosphere = 14.7 lb/in²).

The rotation of Jupiter can be studied by watching the motion of the indentations in the cloud belts and confirmed by the Doppler effect in the lines of the spectrum at the two limbs. The speed of rotation is 10 miles/sec, about five times the speed of rotation of the sun. The Doppler shift is about 0.25 angstrom and can be measured easily. Like the sun, Jupiter rotates most rapidly at the equator.

The temperature at the surface of Jupiter can be measured by radiation instruments on the earth. They indicate a cold −200°F (144°K). According to infrared photometric observations, Jupiter is radiating more energy than it receives from the sun, indicating that it may have an internal source of heat.

The giant planet is one of the strongest apparent radio sources in the sky. Some of the radio waves appear to come from radiation belts surrounding the planet, similar to the Van Allen belts around the earth, but with an intensity that may be 1000 times greater. Others seem to come from Jupiter itself. They have been attributed to some sort of electrical disturbances, perhaps like giant lightning discharges on the earth. These waves are not radiated continuously, but come in bursts.

Jupiter has a very interesting family of satellites. Four of them were discovered by Galileo. They have been named Io, Europa, Ganymede, and Callisto. These four are about as large as our moon; two are smaller and two are larger. Those in the next group are small and have orbits with mean distances from the planet slightly larger than 7 million miles. The outermost group require about two years to complete their orbits at distances of 14 million miles. In order of discovery, Satellite V is inside the orbits of the four Galilean satellites. Numbers VI, X, and VII are next outside. All eight revolve around Jupiter in direct motion. The four outer satellites—XII, XI, VIII, and IX—are the most distant and revolve in retrograde motion. It is possible that all the satellites were once asteroids captured by Jupiter. There is also a theory that Jupiter once had 25 satellites and has lost 13: the 12 Trojan asteroids and one other at about the same distance from the sun.

The atmosphere of Jupiter shows the spectrum of molecules of ammonia (NH_3) and methane (CH_4). The latter is the more abundant. There is also a trace of hydrogen in the infrared, but hydrogen does not radiate in the visual region of the spectrum at the low temperature of Jupiter's atmosphere. An occultation of a star by Jupiter in 1952 showed that the average weight of the molecules in the atmosphere is 3.3. Since hydrogen has a molecular weight of 2.0, there must be a heavier gas present. However, it is thought that the atmosphere is mostly hydrogen and helium.

Other elements that have been detected in Jupiter's atmosphere seem to be present in about the same relative abundances as those on the sun, lending support to the theory that Jupiter represents the composition of the original nebula from which the solar system is assumed to have developed.

13.8 SATURN

The most beautiful planet is unquestionably Saturn, the second largest planet and the only one with a ring system (Fig. 13-8). The rings and a few surface details are visible with telescopes of moderate size. Galileo saw them in 1609 but did not recognize them as rings. He thought the planet was triple, or "eared," and the rings are still sometimes referred to as *ansae*, meaning handles. Christian Huygens in 1655 saw the ansae as a ring. That the rings are flat is proved when the earth moves into their plane twice every 29.5 years. They are then seen in large telescopes as a very straight, thin line on each side of the planet. They are not visible at those times in small instruments. This happened in 1936 and again in 1966, but in 1951 the planet was behind the sun at the most favorable time.

Most of the time the ring looks like a flat disk, tipped at an angle. In the 17th century Giovanni Cassini (1625–1712), an Italian astronomer who became director of the Paris Observatory, discovered a sharp, black division in the ring,

FIG. 13-8   A photograph of the ringed planet Saturn. Cassini's division is shown separating the inner and outer rings. The crape ring is shown only as a shadow on the planet. *(Photograph from the Hale Observatories)*

which is called Cassini's division. It separates the ring into two parts, an inner, brighter ring and an outer, fainter ring. Still a third ring lies inside the other two, called the *crape ring* (British spelling: crepe). It is very faint and is usually invisible, but its shadow can be seen on the ball of the planet.

The dimensions of Saturn and its ring system are:

1. Equatorial diameter, 75,100 miles.
2. Polar diameter, 67,800 miles. Oblateness, $\frac{1}{10}$.
3. Gap between the planet and the crape ring, 7000 miles.
4. Width of crape ring, 5500 miles.
5. Width of inner ring, 16,000 miles.
6. Width of Cassini's division, 3000 miles.
7. Width of outer ring, 10,000 miles.
8. Diameter of entire ring system, 171,000 miles.

The ring was originally thought to be a solid between 10 and 25 miles thick. Then it was found that there are gaps through which stars can be seen faintly. These gaps are similar to Kirkwood's Gaps in the asteroid system and are produced by perturbations by the planet's satellites. Cassini's division is the most obvious gap. It is caused by the pull of the nearest satellite, which has removed particles from that part of the ring system.

At the end of the 19th century, James Keeler (1857–1900), an American astronomer, placed the slit of a spectrograph across the rings and the ball of the planet. The result is shown in Fig. 13-9. The rotation of the planet causes the lines in the spectrum to shift toward the blue on the approaching side and toward

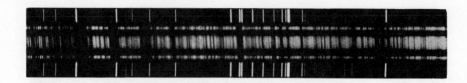

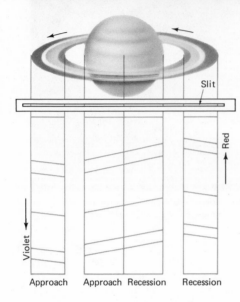

FIG. 13-9   James Keeler's spectrographic proof that the ring system of Saturn is composed of satellite-like particles. The lines in the center are tipped because of the rotation of the planet. The lines in the spectrum of the ring are tipped in the other direction because the ring particles act as small satellites and obey Kepler's third law. The bright lines at the top and bottom are comparison spectra. (*Lowell Observatory photograph*)

the red on the receding side, so the lines are tipped as shown in the spectrogram in the figure. But the surprising part of the study was that the inner edge of the ring is moving at a speed of 12 miles/sec, which is faster than the speed of the outer edge, 10 miles/sec. This proved immediately that the ring could not be solid, since in that case the outer edge would rotate faster than the inner edge. Furthermore, the speeds were exactly those predicted, if Kepler's laws were restated for the satellite system and applied to the particles at the distances of the inner and outer edges of the ring. Therefore, the ring must be composed of small particles; in other words, they are small satellites. Their sizes cannot be determined, but they are estimated to be like grains of sand with possibly a few larger pieces. They may be particles of frozen ammonia.

One final proof: French astronomer Albert Roche proved mathematically that if a satellite moved inside a certain limit, called *Roche's limit*, it would break up under the gravitational force of the bulge of the planet's equator. This limit is 2.44 times the radius of the planet and is between the outer edge of the ring and the first satellite.

The oblateness of the planet itself is a result of its rapid rotation of $10^h 02^m$ at the equator, and its low average density of 0.68 g/cm³. Like Jupiter and the sun, Saturn's rotation is fastest at the equator and slower toward each pole. The latest model shows a central density of 16 g/cm³ under a pressure of 50 million atmospheres. The mass is 95.2 and the surface gravity 1.06 times those of the earth. The velocity of escape is 22.3 miles/sec, so Saturn is able to retain atmospheric molecules of all types. Methane and ammonia have been detected. It is thought that 60% of the planet consists of hydrogen.

Saturn has 10 satellites, nine of which revolve in direct motion, the same as that of the planet's rotation and of the ring. But Phoebe, the outermost satellite, and one of the smallest, has retrograde motion. It may have been captured by the gravitational pull of the planet. Seven of the satellites are visible in telescopes of moderate size, especially when the ring is on edge.

Audouin Dollfus and his colleagues at the Meudon Observatory in France predicted the existence of a tenth satellite because of the perturbations it causes in the ring system. They established the existence of new gaps and also found brightness ripples that could not be accounted for. Other forces seemed to be acting that could be attributed to an unknown satellite very close to the inner ring. It had previously escaped detection because of the glare from the rings. It was finally found on three plates taken December 15, 1966, when the earth was passing through the plane of the rings, which were almost invisible. The new satellite, named Janus, has an orbital period of about 12 hours. Its diameter is estimated at about 200 miles.

Titan is the second largest satellite in the solar system. It is large enough to retain a trace of atmosphere and may be covered with snow.

13.9 URANUS AND NEPTUNE

In size, Uranus and Neptune are a pair of twins (Fig. 13-10), with diameters of 29,000 and 31,000 miles, respectively. Uranus was discovered by William Herschel in 1781, while "sweeping" the sky with his telescope for an estimate of the number of stars. It shows a small, yellow disk, which distinguishes it from the stars. Study of the motions of Uranus and the computation of its orbit, based on accidental observations made for many years before it was recognized as a planet, revealed that the orbit was strongly perturbed by some unknown object. Two astronomers, John Couch Adams (1818–1892) in England and Urbain Joseph Leverrier (1811–1877) in France, worked on the mathematical problem of determining the position of the perturbing body. Adams finished his calculations in October 1845 and sent the results to Sir George Airy (1801–1892), who was Astronomer Royal at the time. No search was made for the perturbing body.

Leverrier finished his computations in 1846 and published his results in June of that year. In July he suggested to Johann Gottfried Galle (1812–1910) in Germany that a search should be made. Galle had the proper star charts and in about an hour found and identified the body. It turned out to be a planet in almost exactly the position predicted by Leverrier and also by Adams. The planet was named Neptune. This was the first astronomical discovery made as a result of mathematical computation.

Uranus has a mass of 14.5 and surface gravity 1.08 times those of the earth.

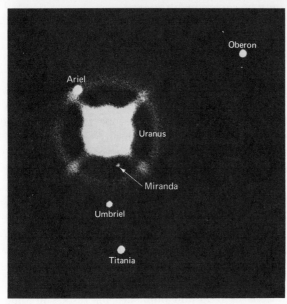

FIG. 13-10 Neptune and its two satellites (top) and Uranus and its five satellites (bottom). Miranda (just under the planet) was discovered in 1948. The rays on the four corners of the planets are caused by diffraction of light by the supports of the secondary mirrors in the telescopes used. (*Yerkes Observatory photograph*)

Its velocity of escape is **13.9** miles/sec. It has a dense atmosphere containing hydrogen and methane and an albedo of 66 percent.

One peculiar fact about Uranus is that it rotates about an axis with a period of $10^h 49^m$ in a plane nearly at right angles to the plane of its orbit. The exact inclination is 98°, so the planet actually rotates backwards. Its five satellites revolve in the plane of the planet's equator. Two satellites were discovered by Herschel, who thought he had discovered six, but four were stars. Two more were found by an astronomer named Lassell in 1851 and the fifth by Gerard Kuiper (1905–   ) in 1948 with the 82-inch telescope at the McDonald Observatory in Texas. The first four are 500 or 600 miles in diameter; the fifth, only 200 miles.

Photoelectric observations of the occultation of a star by Neptune were obtained on April 7, 1968. From these observations it is concluded that the diameter of Neptune is in the neighborhood of 31,000 miles. The average density is 1.58 g/cm³, closely matching the currently accepted figure of 1.56 for Uranus.

Neptune rotates in normal, direct fashion in about 15 hours. Its larger satellite was discovered by Lassell in 1846. It revolves backwards in 5^d 21^h. The smaller satellite's motion is direct with a sidereal period of about one year. Neptune has an atmosphere, its albedo is 62 percent, and it has a greenish color as seen in the telescope. Methane is present in the atmosphere, but apparently hydrogen has not yet been definitely detected. Both planets undoubtedly contain large amounts of hydrogen, but not as high a percentage as in Jupiter and Saturn, since their densities are higher. They are so far from the sun, 1.78 and 2.78 billion miles, that their temperatures are probably lower than −300°F (95°K).

13.10 PLUTO

The most distant planet of the solar system is Pluto. It is far from the other terrestrial planets and does not belong to the group of large planets. Little is known about it.

The computations of the perturbations of both Uranus and Neptune did not satisfy all astronomers. Some thought that another, more distant planet was also perturbing the two. A mathematical computation was made, but the disturbing body was not found at the predicted positions. Percival Lowell (1855–1916) (Fig. 13-11) started a search from his observatory on Mars Hill. After his death his brother gave a 13-inch photographic telescope to the observatory and the search continued. With this telescope, photographs of the sky were made along the ecliptic, where most planets are to be found. Any moving object could be found by viewing the plates in pairs with a special instrument called the *blink* microscope (Fig. 13-12).

In this method, two plates placed side by side can be seen separately in a single eyepiece. When a prism in the eyepiece is moved, first one and then the other plate can be seen. If the instrument is adjusted so the stars are superposed when the plates are seen rapidly in succession, a moving object appears to jump back and forth from one position to another. That is, in Fig. 13-13 the image of Pluto would be seen to move suddenly from its position in the top photograph to a lower position, as in the bottom photograph. In this way the new planet was discovered by an American astronomer, Clyde Tombaugh (1906–   ), at the Lowell Observatory.

Pluto was named for the god of the underworld, since it is certainly dark and cold at a distance of nearly 4 billion miles from the sun. The symbol for Pluto

FIG. 13-11   Percival Lowell (Yerkes Observatory photograph)

FIG. 13-12   The blink microscope used in the discovery of Pluto. (Lowell Observatory)

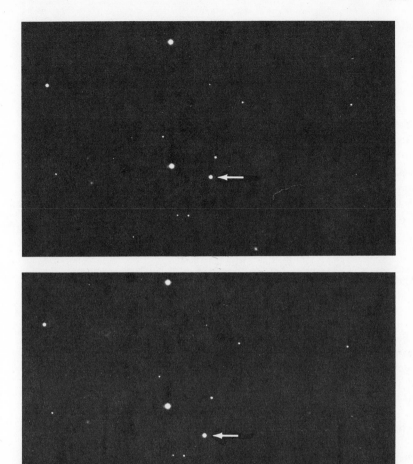

FIG. 13-13 Two photographs of Pluto, identified by the arrows. The motion of
the planet among the stars in the 24-hour interval between photographs is obvious.
(Photograph from the Hale Observatories)

is ♇ , the first two letters in the name of the planet and also the initials of
Percival Lowell, who started the search for the planet.

The planet's size is not definitely known. But in 1965 Pluto barely missed an
occultation of a faint star. This set an upper limit of about 4000 miles to the
diameter of the planet. Then in 1968 three astronomers at the U.S. Naval
Observatory recomputed its orbit using 158 observations made with the meridian
circle between 1960 and 1968. These observations, combined with all others
made before 1938, were used for a computation of Pluto's mass. They concluded
that the mass is not far from 0.18 that of the earth. The new mass and diameter
yield an average density for Pluto of more than 1.4 times the density of the
earth. They revised their data and in 1971 published their determination of the

mass, size, and density given in Table 13-1. The uncertainty of these figures is about 20 percent. This shows the difficulty of a study of Pluto.

Pluto's sidereal period is nearly 250 years. Its orbital eccentricity is so great, 0.246, that by the end of this century it will be closer to the sun than Neptune will be. However, because of the high inclination of the orbit, the two planets will never be closer together than several hundred million miles. One theory about the origin of Pluto is that it once was a satellite of Neptune. This is hard to accept because of the difference between the two orbits.

From Pluto the sun would have an angular diameter of less than 1' of arc. It would appear dazzlingly bright, 1600 times fainter than it does to us on the earth, yet some 250 times brighter than our full moon. The earth would always be less than 1.5° from the sun and would therefore be invisible from Pluto, as would Mercury, Venus, and probably Mars. All the other planets would show phases when seen from Pluto, as Mercury and Venus do to us.

QUESTIONS AND PROBLEMS

Group A

1. Observe the planets that are visible and note their positions in the constellations. Mark these positions on your star charts. Do this each night for Mercury, once a week for Venus and Mars, and once a month for Jupiter and Saturn. Continue for as long as possible and see what conclusions you can draw from them.

2. In what ways do the terrestrial planets differ from the Jovian planets?

3. How much would a 200-pound man weigh on (a) Mercury, (b) Venus, and (c) Mars? *Answer:* (a) 72 pounds.

4. How does the fact that the orbits of Mercury and Venus are inside the earth's orbit hinder observations of their surfaces?

5. Are any satellites of planets as large as a planet?

6. Why have few people ever seen Mercury? How can you find it for the observations asked for in Question 1?

7. Why does Mercury have the smallest albedo of the planets?

8. Make a drawing of the orbits of Venus and the earth and show why Venus is an evening star after superior conjunction.

9. Which is closer to the earth at maximum brightness, Mars or Venus?

10. If astronauts land on Mars during the opposition of 1973, how long will it take for radio signals to reach them from the earth?

11. Use the synodic period of Mars and the sidereal period of the earth to show that the center of each retrograde loop (at opposition) is about 26° east of the preceding one. How does the result agree with Fig. 13-2?

12. Phobos and Deimos both revolve around Mars in direct motion.

But as seen from Mars, Phobos rises in the west and Deimos in the east. Explain.

13. Give two reasons why Venus at maximum brightness appears brighter than Jupiter at its maximum brightness.

14. Why did Galileo discover only four satellites of Jupiter?

15. What conditions make the following planets unsuitable for life: (a) Mercury, (b) Venus, (c) Mars, and (d) the major (Jovian) planets?

16. If there is another planet beyond Pluto, what is its distance in astronomical units predicted by Bode's law?

Group B

17. Traveling at the speed of light, how long would it take to go from the earth to (a) Pluto, and (b) the nearest star? *Answer:* (a) $5^h 19^m$.

18. Calculate the length of a synodic day on Mercury in terms of earth days, assuming the sidereal period of rotation to be exactly two thirds of its sidereal period of revolution.

19. From the periods of revolution of Deimos and the rotation of Mars, show that Deimos remains above the horizon for $2\frac{2}{3}$ days.

20. Assuming that Phobos and Deimos have the same albedo, compute their ratio of brightness as observed from Mars. (*Hint:* Use the distances from Mars' surface and their surface areas.)

21. Compare the equatorial velocities of Jupiter and the earth due to their rotation.

22. Dione, a satellite of Saturn, is about the same distance from the planet as the moon is from the earth. Its period, however, is only about a tenth of the moon's period. Why? Using Newton's modification of Kepler's third law, calculate the ratio of Saturn's mass to the earth's mass. Neglect the mass of the moon and of Dione. Check your answer with the data in Table 13-1.

14 comets and meteoroids

Before the time of Tycho Brahe, comets were mysterious apparitions, thought to be in the earth's atmosphere and interpreted as omens of coming disasters. A bright comet was seen in A.D. 78 at the fall of Jerusalem, another in 1066 when England was invaded, and still another in 1453 at the fall of Constantinople.

14.1 ORBITS OF COMETS

In 1577 Tycho tried unsuccessfully to measure the parallax of a large comet. He reasoned that if the comet were in the earth's atmosphere it should have a parallax large enough to measure by observations from two stations a short distance apart. Since this experiment failed, he deduced that comets are celestial bodies at considerable distance from the earth, at least three times the moon's distance. He thought that they probably revolve around the sun like the planets.

Newton came to the conclusion that comets are attracted to the sun by gravitation. His work was extended by Halley, who calculated the orbits of 24 comets. Among them were several at intervals of about 75 years. Halley concluded that these comets were one and the same, and predicted a return about 1758. He did not live to see his prediction fulfilled, but the comet was sighted on Christmas Day of that year by an amateur astronomer. This comet has been called Halley's Comet since that time. Its last appearance was in 1910 and it is expected to return in 1986 (Fig. 14-1). However, comets are strongly perturbed by planets and the computation of the exact date of perihelion of Halley's Comet has not yet been made, but the opposition date will probably be April 15, 1986.

It is well known that comet orbits are very elongated. Methods of computing the orbits accurately date back to 1801, as has been noted in connection with

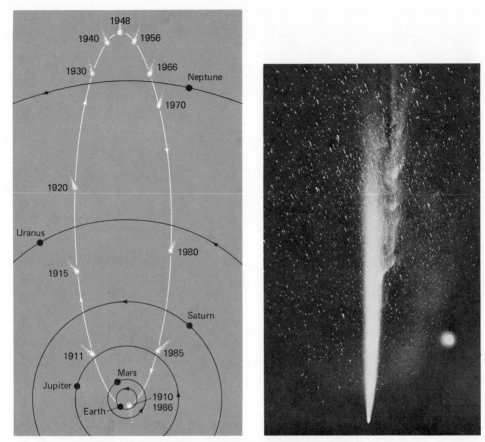

FIG. 14-1   The orbit of Halley's Comet and a photograph taken in 1910. The comet is now on its way back to perihelion, which is expected in 1986. (*Lowell Observatory photograph*)

the asteroid Ceres. All the elements, like those of the planets, can be calculated from three observations of the comet's position on three nearly equally spaced dates. Formerly, it was customary to assume the orbit to be a parabola, since the eccentricity was therefore equal to one, leaving only five more elements to be found. But with the invention of high-speed electronic computers, all six elements can be calculated at the same time in a few seconds. But the preliminary preparations still take several hours.

The parabolic orbit is the limiting case between elliptical (closed) and hyperbolic (open) orbits. Usually a comet is observed only during a short arc of its orbit. Near perihelion all three types of orbit nearly coincide, as Fig. 14-2 shows. But if observations can be made for several weeks or months before the comet comes to perihelion or after it has receded, it is possible to correct the elements and represent the motion of the comet more accurately.

Most comet orbits are nearly parabolic. Only a few have eccentricities greater than 1.0; that is, are hyperbolic. It is thought that comets that appear to be

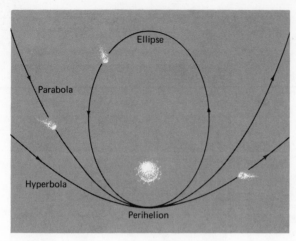

FIG. 14-2 In the neighborhood of perihelion, it is impossible to decide whether the orbit of a comet is an ellipse, a parabola, or a hyperboia. Earlier or later observations can distinguish among the three types.

moving in hyperbolas have been given a slightly higher velocity by the attraction of a planet and are only temporarily in large-eccentricity (hyperbolic) orbits. The rest are moving in ellipses, some of which are approximately parabolas.

If the determination of a comet's sidereal period shows that it returns every 200 years or less, it is considered to be a *short-period comet*. All comets that reappear at intervals greater than 200 years are called *long-period comets*. Comets moving in parabolas never return to the region of the sun. Almost two thirds of the long-period comets are listed in the tables as having parabolic orbits; that is, have eccentricities equal to 1.0. However, it is unlikely that a comet has an exactly parabolic orbit. Its orbit will most likely be an ellipse of eccentricity slightly less than 1.0. With present equipment it is not possible to follow a comet far enough from the sun to determine its period and eccentricity exactly. The periods of such objects range from about 250 years to perhaps several million years. A comet with a period of a million years would have a mean distance from the sun of 10,000 astronomical units. (Pluto is 40 a.u. from the sun.) If a comet went out to the distance of the nearest star, its sidereal period would be about 50 million years.

A long-period comet, because of the combined attraction of the sun and Jupiter, might appear to travel in a hyperbolic orbit when near the sun but in an ellipse beyond the orbit of Jupiter. In other words, if the orbit were nearly parabolic, it would actually travel in an ellipse of large eccentricity; but when near the sun it would travel in an arc of a hyperbola.

Six large comets have come in on the same large, nearly parabolic orbit. These comets are said to form a *comet group*. They came near the sun in 1668, 1880, 1882, 1887, and 1965. They are thought to have originated from one single, great comet that split into six or more pieces. They all passed exceptionally near the sun. For example, Comet Ikeya-Seki 1965 passed within 300,000 miles of the sun's surface. It was visible in daylight when only 1° from the sun and was estimated to have been 100 times brighter than Venus. The comet of 1882 broke into five pieces; Comet Ikeya-Seki broke into two pieces. Its period has been estimated as more than 600 years. Its velocity when nearest the sun was greater than 300 miles/sec. Of course it will be nearly stationary when at aphelion.

The short-period comets have orbits of eccentricity averaging about 0.5 and mean distances of about 5 a.u. Therefore their sidereal periods average about 11 or 12 years, the same as the period of Jupiter. It is quite certain that they have been captured by the gravitational attraction of Jupiter. If a long-period comet comes close to Jupiter, under certain conditions it will be slowed in its orbit and eventually be pulled into the planet's neighborhood.

About 45 comets have periods between three and nine years. In general they have aphelion distances near the orbit of Jupiter, are near the ecliptic, and their motion is direct. This group of comets is called *Jupiter's family*, since they were captured when the giant planet pulled them out of their former orbits into their present orbits.

From all the data on record, it is possible to conclude that nearly all comets go around the sun in elliptical orbits. Except for Jupiter's family, the directions of comet motions are almost completely at random, and their inclinations with the ecliptic vary from near zero to almost 90°. The final conclusion about the nature of comet orbits will depend to a large extent on an acceptable theory of their origin.

14.2 ORIGIN OF COMETS

A recent theory proposed by Jan Oort (1900–   ) of Holland suggests that there is a cometary cloud lying about halfway to the nearest stars. These comets are moving at random and are attracted gravitationally by the stars. If a comet is attracted in such a way that it is given a pull in the direction of the sun, it may move into the solar system. Its orbit would be a hyperbola or a parabola, depending on its velocity, with the sun at the focus. The comet would approach the sun only once and would then go back into space.

But if its velocity were slowed, as it might be if it passed a large planet, the perturbations might change the orbit into an ellipse and the comet would become a member of the solar system and stay in orbit around the sun, unless pulled out by some other force. In fact, Jupiter is known to have radically changed the orbits of several comets, either to make them smaller ellipses, or to send them back out into space at increased speeds. Oort has estimated that there are 100 billion comets in the interstellar group, but only about 800 have come close enough to the earth to have their orbits computed. About 20 percent are short-period comets.

Still more recently, Fred L. Whipple (1906–   ), director of the Smithsonian Astrophysical Observatory, has proposed a theory that there may be also a flattened belt of comets in orbit around the sun just beyond the orbit of Pluto. However, most comets come from the outer cloud proposed by Oort.

14.3 COMET DESIGNATIONS

Each comet is identified by the year of its discovery and a letter that indicates the order of discovery. For example, 1960c was the third comet discovered in 1960. Later, when all orbits have been computed, the comet is given a permanent number, the year it came to perihelion and a Roman numeral to indicate the order of perihelion passage. For example, Comet 1960c might have been the eighth comet to come to perihelion in 1959, but not discovered until 1960. Its designation would then be 1959 VIII. (This is a hypothetical case.)

Comets are also named for their discoverers, many of whom are amateur astronomers. In 1957 Comet Arend-Roland (Fig. 14-3) was named for two independent discoverers. Mrkos' Comet of 1957 was named for a Czech amateur, who first reported it. Halley's Comet and Encke's Comet were named for the men who first recognized them as periodic. The period of Encke's Comet is the shortest known, only 3.3 years. Its period is getting shorter at each return. This was originally assumed to be due to some resisting medium near perihelion. Now it is believed to be caused by jets erupting from the comet's nucleus, which act as retro-rockets to slow the motion. The comet seems to have lost most of its tail-making material. The nucleus may be protected by a layer of dust that

FIG. 14-3 Comet Arend-Roland of 1957 was unusual in that it seemed to have a short tail pointing toward the sun. (Photograph from the Hale Observatories)

slows evaporation. Encke's Comet has been seen at nearly every approach since 1786, the fifty-seventh in 1970.

14.4 COMPOSITION OF COMETS

All comets have approximately the same structure. When far from the sun a comet is small and loosely compacted. The most recent theory of the nature of a comet is that of Whipple, the leading American authority on comets. He has suggested that the cometary particles are composed of carbon, hydrogen, nitrogen, oxygen, and sodium, embedded in ice. The ice holds the solid particles together in the *nucleus,* which contains practically all the mass of the comet. It is being depleted in the periodic comets when they come near the sun. The discovery by the Orbiting Astronomical Observatory of hydrogen and hydroxyl halos for the two bright comets of 1970 strongly indicates the presence of water. This theory has been called the "dirty iceberg" theory.

When a comet approaches the sun, at a distance of 3 or 4 a.u., the heat of the sun melts part of the ice and releases the smallest particles to form a sort of atmosphere around the nucleus. This is called the *coma,* because of its hairlike appearance. The nucleus and the coma form the *head* of the comet. The head may be hundreds of thousands of miles in diameter. As the particles further divide, when they reach a size approximately equal to a wavelength of light, they are pushed away from the head to form the *tail* of the comet. In various stages the comet may consist of the nucleus only, or the nucleus may be hidden and the coma alone visible. This is the case in a good many periodic comets. Finally the tail forms, provided there is sufficient tail-forming material. Comets have been seen to lose their tails and grow new ones. Also several tails may be visible at the same time!

It is difficult to estimate the size of the nucleus. The diameters apparently range from a few miles for the larger comets to a half-mile for small ones. The coma changes size with distance from the sun. The head of a great comet in 1811 was 1 million miles in diameter. The coma is smaller near perihelion, expands until the comet is about 2 a.u. from the sun, and then decreases again. The nucleus reflects sunlight; the coma absorbs sunlight and reradiates it at a different wavelength—a process known as fluorescence. The spectrum of the nucleus is therefore the spectrum of sunlight; the spectrum of the coma is that of the molecules that compose it. The tail may be hundreds of millions of miles long, very diffuse, and usually curved. It shines by reflected light from the sun, with spectrum showing molecules of carbon and nitrogen and carbon monoxide ions.

In 1910 Halley's Comet came directly between the earth and the sun. The comet could not be seen then, confirming the belief that the coma is transparent, composed of very small particles and gas. The earth passed through the tail with no noticeable effects. It was expected to produce a shower of meteors, but the tail particles are much too small. The tail density was of the order of a billionth of a billionth (10^{-18}) of the density of the earth's atmosphere. Since the tail was known to have carbon monoxide molecules in it, some people were worried about possible asphyxiation, but the air prevented any harmful molecules from reaching the earth.

Some comets appear to have two kinds of tails. One is composed of gaseous particles and is straight, pointing directly away from the sun. The other is composed of finely divided dust particles and is curved. A notable example was the Arend-Roland Comet of 1957. There was an unusual spike pointing toward the sun (Fig. 14-3). This was the dusty tail seen on edge. Part of this tail was between the head of the comet and the sun. The angle from which it was seen gave it the appearance of a spike.

An example of a comet whose orbit was changed by Jupiter was Oterma's Comet, which was in the neighborhood of Jupiter between 1936 and 1939. Its period changed from 18 to 8 years. Between 1962 and 1964 it came near Jupiter again to a distance of about 10 million miles or less. It was strongly perturbed and went out so far from the sun that it may never be seen again.

Bennett's Comet, 1969ℓ, was called the Great Comet of 1970 when it was seen with the unaided eye for several weeks in April in the northeastern sky. It had two tails, Type I of ionized gas, and Type II, or dust tail, both of which showed considerable structure. On April 5 a bright nucleus could be seen inside the coma. Photographs in Lyman-α light showed a hydrogen cloud, 8 million miles across, surrounding the comet. This was the second comet known to have such a cloud. An unusual feature was seen and photographed: spiral jets apparently issuing from the head. It was agreed that Bennett's Comet was the finest comet seen in many years, perhaps since Halley's Comet in 1910.

14.5 METEOROIDS

If a comet is a loose collection of particles held together by ice, when the ice melts it is easy for the particles to become separated from the main mass of the comet. This happens when the comet is close to the sun. The particles, however, still have motion in the orbit and possibly a change of motion due to explosive ejections from the nucleus. They therefore become more or less stretched out along the orbit and are then referred to as a *stream,* or a *swarm,* if they are very numerous. The particles are small and not very close together, and are therefore invisible when seen from the earth's distance. Perturbations by a planet, such as Jupiter, help to pull them away from the comet and stretch them out in a path either in front of or behind the comet. Such particles are called *meteoroids.*

Occasionally a stream of meteoroids intersects the orbit of the earth. If the earth gets there at the same time, the meteoroids enter the earth's atmosphere at high speeds, that can vary from 7 miles/sec for those that overtake the earth to 44 miles/sec for those that collide head-on with the earth. Each particle, having mass and velocity, has kinetic energy. The friction between the particle and the molecules of air decreases the speed and the kinetic energy is transformed into heat and light, but mostly heat.

It has been estimated that only 0.0001 of the kinetic energy is changed into light for the slowest-moving meteoroids, but increases by a factor of seven for the fastest ones. If a meteoroid remains intact, it has a brightness that can be calculated from its velocity. But if it breaks up (this is known as fragmentation), the small pieces burn up more rapidly, because they are ejected into the air at lower altitudes than usual and the meteor is 100 times brighter than it would be if it remained whole. A 1-gram meteoroid with a velocity of 20 miles/sec seen at 60 miles is about as bright as the first-magnitude star Vega. The duration of its trail is about 1.6 seconds. In general, the heights vary from 70 miles for the fastest meteoroids to 40 miles for the slowest.

The term *meteor* is used to describe the entire phenomenon associated with the collision of a particle from space and the atmosphere of the earth. A *meteorite* is a piece of higher density, which has survived passage through the air and has hit the earth's surface. Extremely small meteorites are called micrometeorites. They are small enough to melt completely from the friction produced. Samples have been collected during space flights.

Meteors appear as moving streaks of light called *trails,* which are comparable in brightness to neighboring stars. They have been called shooting stars. The duration of a trail may be only a fraction of a second. Persistent light from a trail is called a *train.* The observed maximum duration of a train is 3 hours.

The meteoroid itself is too small to be seen. When it is about 70 miles above the earth, the heating is so great that atoms boil off from the surface of the particle at speeds of about 0.6 mile/sec relative to the particle. These atoms collide with air particles, causing the electrons in the air molecules to be put into violent motion or even pulled away from their nuclei. These processes are called excitation and ionization, respectively. A luminous cloud is formed around the meteoroid and may become visible. The particles lost by the meteoroid and the air molecules are heated to luminescence and form the meteor trail.

14.6 METEOR SWARMS

A group of meteoroids striking the earth's atmosphere was formerly called a *shower,* but is now called a *swarm.* A swarm is a group of meteoroids, moving in nearly parallel paths, that appear as moving, luminous streaks of light. They seem to radiate from a small point or area in the sky. This is called the *radiant point.* Many meteor swarms have been identified with the breakup of known comets. Only a few swarms produce meteors at a sufficiently high rate to be of interest to the casual observer. Some of them are listed below.

name	date of maximum	hourly rate	comet
Perseids	August 12	40–60	1862 III
Geminids	December 14	60	unknown
Leonids	November 17	variable	Tempel 1866 I

Swarms are named for the constellation in which the radiant point is located. If the paths of the meteors of a particular swarm are plotted on a star map, they all appear to come from the same area in a constellation, although they may appear to be going in any direction and may be seen in various parts of the sky.

The Perseids radiate from Perseus, a constellation that rises shortly after dark in August but is not on the meridian until about 5 A.M. This is the most dependable swarm and lasts for several weeks in July and August, although the rates are low except on August 12. More meteors may be seen when there is no interference by moonlight. Their speeds are about 35 miles/sec, so the trails are fast and long, somewhat reddish in color. Some 40 to 60 per hour may be expected at the time of maximum under good observing conditions, especially after midnight.

The Geminid radiant passes overhead at 2 A.M. on December 14 for locations in the southern United States. The hourly rate exceeds that of the Perseids. The velocity is intermediate at 21 miles/sec.

The Leonids are in the orbit of Tempel's Comet, which was discovered in 1866. This shower had been very spectacular in that year and previously in 1799 and 1833, when meteors were "as thick as snowflakes." But the swarm passed Jupiter in 1898, the particles were pulled away from the earth, and the expected shower in 1899 failed to appear. From an hourly rate of 10,000 in 1833, the numbers dropped sharply but rose to 240 in 1932. In 1962, 100 meteors per hour were seen; in 1966, 1000 per minute were reported (Fig. 14-4). Since the radiant point does not rise until almost midnight in November, these meteors are best observed after midnight. The radiant crosses the meridian after 6 A.M.

Two other notable showers were the Andromedids in 1872 and the Draconids of 1933 and 1946. Both were associated with short-period comets and both have been perturbed to such an extent that they probably will not be seen from the earth again.

Altitudes of meteor trails may be measured from ground stations. If a meteor is seen from two stations separated by a known distance, the trails plotted on the same map will be displaced by an angle that can be determined from the locations of stars on the map. This angle of displacement is the parallax of the meteor. As Fig. 14-5 shows, the parallax of the meteor is the angle p of the triangle formed by a point on the trail and the two stations.

Observations of meteors were formerly made visually, but they are now made by taking photographs simultaneously from two stations with wide-angle Schmidt cameras (Fig. 14-6). The altitudes are more accurately determined by this method. Still better determinations are made by radar, and can be made in the daytime as well as at night. A radar pulse is sent out and is reflected by the cloud of electrons in the trail. The time between the sending and receiving of the pulse is used for a determination of the altitude. The speed of the meteoroid may also be found by pulses from various parts of the trail. In the photographs the

FIG. 14-4 The Leonid meteor swarm on November 17, 1966. During this 3½-minute exposure, taken from Kitt Peak, Ariz., about 70 meteor trails were recorded. The two points of light near the radiant were made by meteors heading directly at the camera! (*Courtesy Dennis Milon, Sky & Telescope*)

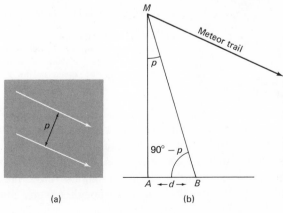

FIG. 14-5 The altitude of a meteor trail determined by the parallax method. In (a) the parallax p is the angular distance on a star map between the meteor trail as seen from two stations. In (b) p is shown as an angle of the triangle *ABM*, which can be solved for the altitude *AM* of a point of the trail.

FIG. 14-6 A Baker Super-Schmidt camera (bottom) used for photography of meteor trails. On top is a photograph showing three trails. Note the interruptions made 20 times per second for determining the speed of the meteoroid. (*Harvard College Observatory photograph*)

trail is interrupted about 20 times per second for accurate timing. These times combined with the altitudes permit a determination of the speed.

Spectrograms of the hot, excited vapor of meteor trails have been made. Bright emission lines of calcium, silicon, aluminum, and manganese have been found. These elements are similar to those in asteroids. Although oxygen and nitrogen lines would be expected from the atmospheric molecules, they have not been detected because these elements do not have lines in the visible part of the spectrum. The meteoroids themselves should also contain oxygen and nitrogen.

14.7 METEORITES

Meteorites can be grouped into three general classes. The irons, called *siderites,* contain about 10 percent nickel; the rest is almost all iron. Stony-irons are a combination of about equal parts of iron and stone. They are called *siderolites.* Since the stony material decomposes more rapidly than the irons, they have a tendency to separate into fragments. The stones, called *aerolites,* contain from 10 to 15 percent iron, the remainder being composed of silicates and associated material.

The iron meteorites are easy to identify because they are found frequently in regions where there is little native iron. The stones are more difficult to identify, but an expert can usually prove their meteoric nature by finding small pieces of iron embedded in the stone. When iron meteorites are cut and etched, their identification is made certain by the crystal structures that appear.

Another group of objects, called *tektites,* are thought to have their origins beyond the earth. It is not certain that this theory is correct, however. Tektites are rounded glassy bodies found in Australia, the Philippines, Indonesia, and other places. Some scientists even believe them to be pieces of the moon broken off by the impact of meteoroids and given a high enough velocity to escape from the moon and to be captured by the earth.

Figure 14-7 shows an aerolite and a siderite. The largest known meteorite in America is the large siderite in the American Museum of Natural History in New York City. It rests on a specially built scale and weighs a little over 34 tons (68,085 pounds). It was liberated from the Eskimos in Greenland by Admiral Peary in 1897. A still larger specimen in southwest Africa is estimated to weigh more than 50 tons. It has not been moved to a museum.

One theory of the origin of meteorites and their division into the various classes is that two asteroids about the sizes of Ceres collided. The collision broke the asteroid into pieces, the irons coming from the central core and the stones from the outer crust. These meteoroids are not found during showers, but come in contact with the earth at random. They are usually spoken of as sporadic and their falls are unpredictable. On the average, about 12 meteors per hour may be seen visually. About 80 percent are sporadic; the other 20 percent are swarm meteors.

Several scars on the earth's surface are believed to have been produced by the fall of large meteorites or collections of meteoric particles. The best-known crater of meteoric origin in the United States is the Barringer Crater near Winslow, Arizona (Fig. 14-8). This crater is nearly circular, 4200 feet wide and 600 feet deep, with mounded walls 150 feet high. There has been little erosion, but the age cannot be definitely determined; it is estimated at between 5000 and 75,000 years. Drillings in the floor of the crater have yielded no definite trace

FIG. 14-7   An aerolite (top) and siderite (bottom). The siderite is the largest known meteorite in America. It weighs 68,085 pounds, as measured by the specially built scales on which it is mounted. *(Nininger Collection; Courtesy American Museum of Natural History)*

FIG. 14-8 The Barringer Meteorite Crater near Winslow, Ariz. (American Meteorite Museum)

of the meteorite that produced it. It has been estimated that a mass of 1 million tons is located under the south rim, and there were plans to excavate it for the iron ore it is supposed to contain. Most astronomers believe that no large deposit will be found. Also, laws governing the disturbance of the water table in Arizona by deep drilling operations prohibit explorations of this kind.

Large pieces weighing hundreds of pounds have been picked up near the crater and many smaller ones have been found with a magnetic probe at a distance of several miles. It is likely that the small particles were from the outer skin of the meteorite, fragments that boiled off by the heat produced by friction with the air. The larger pieces may have been ejected when the entire mass broke up as a result of heat produced by the collision with the earth, and the resulting explosion.

The New Quebec Crater in northern Canada is the largest one known. It is about 2 miles in diameter and is now filled with water. Since no meteoric particles have been found, there is some doubt that it is actually of meteoric origin. In 1908 a fireball, which may have been the head of a small comet, was seen over Siberia, and pieces landed in a swampy, forested region. Trees were blown down for more than 20 miles, but the meteorite pieces were apparently buried in inaccessible locations. A second Siberian fall took place in 1947. This impact produced about 100 craters up to 90 feet in diameter and 30 feet deep. Iron fragments have been picked up near the craters. Isolated cases of meteorites falling near people have been reported, but there is no record of anyone having been killed by a meteorite.

14.8 INTERPLANETARY PARTICLES

In addition to meteoroids and micrometeorites, space in the region of the earth seems to be full of interplanetary dust. The *zodiacal light* is sunlight reflected by these minute particles. It can be seen as a faint band of light along the ecliptic (zodiac). It is best seen in the evening or morning sky when the light from the Milky Way does not interfere (Fig. 14-9). It is not visible in twilight or moonlight or from a city where the sky is bright from artificial lights. There is some

FIG. 14-9   Star trails and the zodiacal light are seen in this photo of the western sky. The observatory dome is in the right foreground. (Yerkes Observatory photograph)

indication that the zodiacal light is partly due to the reflection of sunlight from free electrons in addition to solid particles.

The morning and evening branches of the zodiacal light meet and spread out into what is known as the *gegenschein,* a German word meaning counterglow. This extremely faint light has been photographed and scanned with photoelectric cells, but is more difficult to see than the zodiacal light. It undoubtedly has the same composition. It appears as a circular patch some 10° in extent. One theory is that these particles are held temporarily in position by the attractions of the earth and the sun in locations predicted by the problem of three bodies. The solution of this problem states that if a small body comes near a certain point on a straight line with two major bodies, it will be captured and stay there unless pulled out by some outside force. There are three positions on this line with the earth and the sun, but only the gegenschein is observable.

It is believed that the earth is accompanied by a stream of particles much like the tail of a comet. No tail, however, has been observed on any other planet. It is also possible that the zodiacal light and the gegenschein are particles, ions, and electrons from the solar wind. This mixture is known as *plasma.* The plasma is dense enough to reflect sunlight in sufficient amounts to become visible. The particles would be trapped for a short time by gravitation at a distance of nearly 1 million miles from the earth, forming the gegenschein. (This word seems to be preferable to its translation.)

QUESTIONS AND PROBLEMS

Group A

1. The future positions of planets can be very accurately computed, but there is considerable uncertainty in predicting the time of return of comets. Why?

2. At first the orbit of the Ikeya-Seki comet of 1965 was computed as a parabola. Later it was found to be an ellipse of period about 4500 years, but uncertain to about 1000 years. How could this have happened?

3. How many names may a comet have? Give the meaning of each.

4. How would the earth be affected by encountering the (a) tail, (b) coma, or (c) nucleus of a comet?

5. At what part of a comet's orbit is it most easily observed? Why is a comet usually observable in a short arc of the entire orbit?

6. What type of spectrum is shown by (a) the nucleus, (b) the coma, and (c) the tail of a comet?

7. Assume the spherical nucleus of a comet to be at perihelion, traveling in direct motion. As the nucleus travels, it rotates slowly. Assume that each part facing the sun emits jets when heated by solar radiation, the jets gradually decreasing as that part rotates away from the sun. How will the comet's orbit be affected by the jets if the nucleus rotates direct or retrograde? Illustrate with a diagram.

8. If the meteoroids of a swarm enter the earth's atmosphere at a rate of 1200 per minute, how far, on the average, does the earth travel in its orbit between collisions? (The earth's velocity is 18.5 miles/sec.)

9. Do meteoroids, or meteorites from the asteroid belt, have the greatest velocity on the average when colliding with the earth's atmosphere?

10. Can a meteor shower be seen equally well from all parts of the earth? Explain.

11. Show by a diagram how it would be possible for the earth to meet the same meteoroid swarm twice in one year. Why is it unlikely?

12. Are meteors that are seen (a) before midnight and (b) after midnight overtaking the earth or colliding with it head-on?

13. About two thirds of the total number of recovered meteorites are siderites. About 90 percent recovered from an observed fall are aerolites. Explain.

14. Define the following: (a) shooting star; (b) meteor; (c) meteoroid; (e) micrometeorite.

Group B

15. One mechanism for dispersing the particles of a comet's head into the tail is light pressure from the sun, which is propor-

tional to the surface area of a particle. The opposite, attracting force of solar gravitation is proportional to the volume of a particle (assuming the particles all have the same density). Show that the ratio of the light pressure to gravitation is inversely proportional to the size of the particle.

16. Two stations 15 miles apart photograph the same meteor simultaneously. On comparsion of the photographs, an apparent displacement of 10° against the sky due to parallax is measured. If the meteor is directly above one station, what is its altitude?

17. If meteoroids traveled in nearly circular orbits, what would be the expected maximum and minimum velocities of impact with the earth's atmosphere?

15 the evolution of the solar system

Where did the solar system come from? This question has intrigued philosophers, theologians, geologists, astronomers, and others for hundreds of years. The beginning of what might be called a scientific approach to evolution dates back to the German philosopher Immanuel Kant (1724–1804) in 1755. Kant suggested that the sun and its family of planets with their satellites were formed from a single cloud of gas—in present terminology, a *nebula*. This cloud was in rotation and somehow condensed into discrete units, the smaller ones moving around the larger ones and the larger ones moving around the sun.

15.1 THE NEBULAR HYPOTHESIS

Kant's theory was based on the assumption that the entire space now occupied by the solar system was once filled with a thin gas—or, as we would say today, consisted mostly of hydrogen. The heavier elements began to separate from the lighter ones. He also suggested that both repelling and attracting forces set the gaseous mass into rotation. The condensations that gradually resulted grew by collisions, eventually becoming planets revolving in orbits around the central mass, which became the sun.

In 1796 Pierre Simon, Marquis de Laplace (1749–1827), wrote down his ideas on the possible origin of the solar system. Although he was one of the world's most famous mathematicians, he did not examine his proposed theory from a mathematical point of view, and he undoubtedly borrowed his basic ideas from Kant. Laplace's theory has been named the *nebular hypothesis*.

The structure of the solar system and the motions of all its parts as known at the time of Laplace, and requiring explanation by a satisfactory theory, were as follows:

1. The planets were known out to Uranus, which had been discovered 15 years before by William Herschel.
2. Some planets were known to have satellites: the earth, 1; Jupiter, 4; Saturn, 7; and Uranus, 2.
3. The planets were known to revolve around the sun in the same direction and in approximately the same plane.
4. The satellites were known to revolve around the planets in approximately the orbital planes in which the planets move around the sun. The exception is that Uranus' satellites were known to revolve in a plane nearly perpendicular to the plane of the planet's orbit. No small satellites had yet been discovered and none of the known satellites moved in retrograde motion.
5. The rotation of the sun was known from the motions of sunspots.

Laplace proposed the theory that the original cloud was a nebula, probably shaped like a spiral. It was composed of hot gas and was in slow rotation. As it cooled it began to shrink and rotate faster. This was in agreement with the theory of the conservation of angular momentum. There came a time when the inward forces of gravitation were balanced by the outward forces of rotation. The result was that a ring separated from the central mass, causing a further increase in the speed of rotation of the nebular material, but not of the ring. Other rings broke off, one for each planet.

There were condensations in each ring (Fig. 15-1). These condensations collided with the smaller particles and grew into the planets. The outer planets were thus formed first. In a similar way the growing planets split off rings, which formed into the satellites. Thus the planets revolved around the central mass that became the sun, and the satellites revolved around their planets. The outer bodies moved around their primaries at slower rates than those farther inside. The rotating nebula became flatter, and the rings split off in the equatorial plane. Therefore all the planets revolved around the sun in nearly the same plane and in the same direction. The satellites also revolved around the planets in about the same plane and in the same direction.

It is not certain that the rotations of the planets were known at the time of Laplace. Certainly, the rotation of the sun and probably Jupiter had been observed. It is known today that the rotation period of the sun is shorter than the periods of revolution of the planets. Also, the planets known to Laplace are now known to rotate in nearly the same plane and in the same direction as the satellites known in 1796.

There are at least three major errors in the nebular hypothesis. Granted that rings could split off, it is impossible for the ring material to form into

FIG. 15-1 The nebular hypothesis. One ring for each planet split off from a central nebular mass and condensed into a planet, which in turn, as it formed, split off rings to form satellites.

planets, since the speeds of the ring material would be so great that they would disperse into space rather than collect into solid bodies.

Then in 1877 the two little moons of Mars were found. The period of revolution of the inner moon, Phobos, was found to be shorter than the rotation period of the planet. This was contrary to the theory that a planet, having a smaller radius, should rotate faster than the revolution of the satellites.

Finally, about 1895 the American geologist Thomas C. Chamberlin (1843–1928) began to question the nebular hypothesis from a geological point of view. He and Forest R. Moulton (1872–1952), a colleague in the mathematics department at the University of Chicago, showed that the distribution of the angular momentum of the system is wrong. Angular momentum is the quantity of motion due to rotation or revolution. It is calculated by the formula:

$$\text{angular momentum} = mvr = mr^2\omega \tag{15-1}$$

where m is the mass, r is its distance from the center of motion, v is the linear velocity, and ω is the angular velocity (the number of turns in a unit of time).

The sun's angular momentum is difficult to compute because the sun does not rotate as a solid body. Moulton computed it and also found the angular momentum of all the planets and their satellites, using both the rotations and revolutions.

The law of the conservation of angular momentum states that the total linear momentum of a system of particles remains constant in amount and direction regardless of internal collisions or reactions. That is, equation (15-1) becomes

$$mr^2\omega = C \tag{15-2}$$

where C is a constant. So in a rotating nebula or any rotating system of particles, planets, or satellites, if the radius r decreases, then ω, the number of turns per unit of time, must increase, since the mass remains the same. Thus the system spins faster. Also, the nebula or other system keeps any angular momentum it had originally, unless acted on by an outside force.

The sun contains 99.8 percent of the total mass of the solar system and should therefore have nearly all of the system's angular momentum. But Moulton's computation showed that the sun's angular momentum is only 2 percent instead of 99.8 percent of the whole. Most of the rest is due to the motions of the largest planets.

15.2 THE PLANETESIMAL HYPOTHESIS

When all the objections are considered, it is impossible to accept the nebular hypothesis. Chamberlin and Moulton, about 1900, decided to try to put together a theory that would explain how angular momentum might have been added to that of the planets from outside. The resulting theory was named the *planetesimal hypothesis.*

They assumed that the sun was already in existence and in about the same state of activity (spots, prominences, and accompanying phenomena) as at present. They suggested that another star came close to the sun and raised two tides, one on each side, as the moon raises tides in the oceans on the earth. But since the sun is more explosive, the tides erupted, throwing great masses of matter out into space. This happened five times in rapid succession, an hour or so apart. The tides, called *bolts*, were given a motion in the same direction and in approximately the same plane as the orbit of the passing star. Thus motion was given to the material in the bolts by outside gravitational attraction.

If only the motion of the passing star relative to the sun is considered, its orbit must have been a hyperbola and the star's speed must have been several hundred miles per second. After the close approach, it went off into space and there would be no possibility of identifying it. The star came near the sun only once. (See Fig. 15-2.)

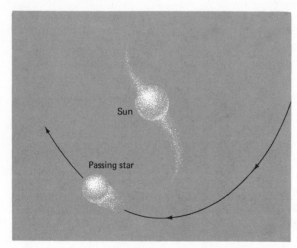

FIG. 15-2 The planetesimal hypothesis. A passing star pulled bolts of material from the sun, two at a time. A total of 10 bolts in rapid succession developed into the system of planets and satellites.

Some of the material in the bolts was assumed to follow the star or to disperse into space, particularly hydrogen and helium. The other, heavier particles fell back under the gravitational pull of the sun but were given forward motion by the attraction of the star. They went around the sun in elliptical orbits of high eccentricity. There were many collisions among the particles, which were called *planetesimals*. Moulton showed that, because of the collisions, the orbits became less eccentric and nearly circular. Also the collisions caused condensations to form, which grew by the infall of other planetesimals (accretion) and became the planets. Planetary atmospheres were formed after the planet became massive enough to retain gas molecules. The velocity of escape became larger, and the atmosphere was held to the planet by its attraction.

When the planetary condensations, called *protoplanets,* came to perihelion the first time, they were themselves disrupted by the tidal attraction of the sun and formed their systems of satellites in the same way the planets were formed from the solar bolts. Thus, the orbits of both planets and satellites and their direct motions were accounted for. The angular momentum was produced by the pull of the passing star, which provided the angular velocity ω for the planets. The 90° motion of the satellites of Uranus and its rotation were not explained. The few retrograde satellites were supposed to have been asteroids captured later by the planets.

Rotation was accounted for by the collisions of planetesimals with the limbs of the protoplanets and with the sun in such a way that they were given a forward spin. The sun's rotation is not exactly in the plane of the orbits of the planets, but is inclined about 7° to the ecliptic. This angle was thought to be a combination of the sun's original rotation and that given by the infall of planetesimals.

The sun lost only a small percentage of its original mass, according to the planetesimal hypothesis. Most of it condensed to form the planets and their satellites. Other particles did not escape from the solar system, but became comets and meteoroids, which occasionally came near or into contact with the planets. Collisions are still grinding the asteroids and small particles into still smaller pieces to form the interplanetary dust.

Many serious errors were pointed out, and thus the planetesimal hypothesis was discarded about 1935.

15.3 OTHER THEORIES

Sir James Jeans (1877–1946) and Harold Jeffreys (1891–   ) in England had been among the critics of the Chamberlin and Moulton theory. They proposed the theory that a passing star pulled only one bolt out of the sun. This bolt was thicker in the middle, and the major planets were formed there. Otherwise their theory is very similar to the planetesimal hypothesis and subject to the same criticisms. It was therefore not very widely accepted. Jeans and Jeffreys were apparently willing to modify their theory and proposed that the passing star actually hit the sun, knocking off the material necessary to form the planetary systems. This material would have been too hot to collect into smaller bodies.

One recent theory assumes that the sun was accompanied by a pair of small suns, which combined, became unstable, and broke up to form the planets.

Another theory suggested that the sun had only one companion, which exploded and completely disintegrated. It does not seem possible that such hot material could have formed the system of planets.

In Holland H. P. Berlage developed a theory combining many of the features proposed by Kant and Laplace, especially the nebular disk and rings. He assumed that the viscosity of the gaseous disk would cause it to develop into a system of concentric rings. The particles in the rings came together in just the right way to form the planets and their satellites. He developed equations that account for the spacings of the planets and most of their satellite systems. Some of Berlage's premises are questionable, but his theories have many good points, although they do not seem to have received much attention.

Another theory came from Germany near the end of World War II. C. F. von Weizsäcker assumed that the protosun was surrounded by a large disk-shaped envelope of extremely low density, as in the Laplace hypothesis, but with one important difference. He assigned a much larger mass to the nebular cloud and assumed that it acquired most of the angular momentum. This accounted for the fact that the sun has only about 2 percent of the angular momentum of the solar system. Also, he assumed the disk to be turbulent and developed vortices, five of them in a ring, one ring for each planet and at the correct distances to satisfy Bode's law (Fig. 15-3). Where the vortices came in contact, the material was moving in opposite directions, tending to cancel their motions. This caused material to condense in the regions between the rings, and from this the planets were formed. However, modern theory shows that the turbulent gases would not remain in such an arrangement long enough for the planets to grow in these regions.

FIG. 15-3   The Weizsäcker theory. The arrows show vortices formed in the equatorial plane of a nebula of gas and dust rotating about the sun. Accretion would take place along the heavy concentric circles to form planets and satellite systems with direct rotation and revolution. (*Yerkes Observatory photograph*)

15.4 THE DUST-CLOUD HYPOTHESIS

A discovery, made about 1945, has changed the thinking about the problem of evolution and gives support to the nebular hypothesis. It was found that small, dark areas on photographs of certain nebulae, as in Fig. 15-4, are dark clouds of opaque material in front of luminous material. They had been discovered by E. E. Barnard (1857–1923) of the Yerkes Observatory, but were not recognized as real objects. He thought them to be defects on the plates. Some of the dark areas are very small and almost perfectly round. They are called *globules*. Their diameters are calculated to be about 250 times the diameter of the entire solar system out to the orbit of Pluto. Their densities are extremely small, but they are dense enough to keep light from passing through from behind. The total mass of a globule must be at least as great as that of the sun. Gravitation is sufficient to hold it together, but in addition the particles are being pushed together by radiation pressure from stars outside. In other words, this mass of dust and gas is slowly condensing to form a star.

FIG. 15-4   A nebula in the constellation Scutum showing dark areas, called globules, which may be condensing into stars. *(Photograph from the Hale Observatories)*

Current theories now start with the assumption that the solar system originated from a globule, but they differ as to the way in which it condensed and the planets were formed. Kuiper has proposed a theory, the *protoplanet hypothesis,* that is a modification of the von Weizsäcker vortex theory. He assumed that in the case of the globule from which the solar system originated, the central condensation became the protosun. About 10 percent of the material of the cloud remained in the nebular cloud. Particles revolving in different planes collided when their paths crossed, causing them eventually to orbit in the same plane, and so the disk flattened out. The inner parts moved faster than the outer parts and the resultant friction produced turbulent eddies, as shown in Fig. 15-5. According to the modern theory of turbulent gases, these eddies would be more stable than those of von Weizsäcker. The friction caused most of the angular momentum to be transferred from the protostar to the nebular cloud.

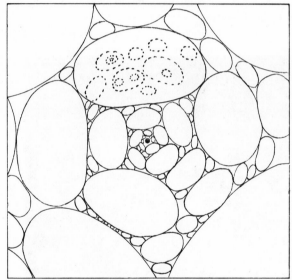

FIG. 15-5 The Kuiper theory. The turbulent pattern of vortices is shown that may have arisen in the primordial nebula around the sun, giving rise to the formation of planets and satellites. (*Yerkes Observatory photograph*)

The material in the disk condensed into smaller clouds, which became the protoplanets. These condensations existed at first as eddies moving in orbits around the central cendensation, the protosun. The protosun had not yet accumulated enough energy to begin the production of energy that would later become sufficient to start it shining as a star. The protoplanets accumulated all the loose material in their neighborhoods and grew to be larger than their present sizes. The estimate is that the earth's protoplanet had perhaps 1000 times its present mass and Jupiter's protoplanet was 10 to 20 times as massive as Jupiter is now. The heavier central mass of each became the nucleus of the planet surrounded by lighter material.

The common direction of revolution and rotation of the planets (except for the rotation of Mercury, Venus, and Uranus) is a direct consequence of the rotation of the original nebula. Their rotations speeded up as they became smaller, and each protoplanet developed a disk surrounded by a cloud of lighter particles, which was similar to, but much smaller than, the original large cloud. This planetary disk developed into the systems of satellites all in motion around the central planet and revolving in the same direction as the rotation of the planetary disk.

When the sun reached a sufficiently large size, it began to convert its hydrogen into helium with the emission of energy and the ejection of particles. These ejections drove the lighter gases away, especially from the inner planets. The earth, for example, lost its hydrogen and helium, but retained the solid metals, such as iron, and the rocky interior. Some of the gases, such as oxygen, were able to combine with the heavier elements. The larger planets from Jupiter to Neptune retained their lighter gases, and are probably composed mainly of hydrogen and helium. The planets lost most of their satellites. Jupiter kept its present 12, but Kuiper suggested that the 12 Trojan asteroids were originally satellites of Jupiter and that Pluto was a satellite of Neptune.

The asteroids came from either a protoplanet that broke up or one that did not form into a planet. In any case, its mass must have been smaller than any of the other planets. Pluto seems to be an exceptional case, since it is much smaller and less massive than its nearest neighbors. Most of the asteroids are assumed to have formed from the collision of two large asteroids and by later collisions between the remnants. Comets came from the outer regions of the nebular cloud. Their masses are too small to have developed into bodies with dense central condensations.

In view of the difficulties of explaining all the details of the evolution of the solar system and the fact that no single theory is adequate, the following suggestions have been made that combine many details of the old and new theories:

1. The solar system was formed from a cloud of dust and gas several hundred times larger than its present size.
2. The dust cloud may have been a double condensation with the larger cloud about as massive as the sun and the other about one tenth as massive.
3. The smaller dust cloud formed into the planetary system by some mechanism that disposed of 99 percent of the mass in the form of light gases and kept 1 percent to form the planets and their satellites.
4. Comets and meteoroids are debris left over.
5. The distribution of angular momentum is accounted for by the slow rotation of the larger cloud and the resulting slowly rotating sun. The planets obtained their angular momentum from the revolution of the small cloud around the larger one.

If this theory actually accounts for the solar system, there should be other similar systems, with perhaps one star out of 1000 having a system like ours. That would mean a total of about 100 million systems in our galaxy.

The stages in the formation of the sun would then be something like the following:

1. The interstellar cloud forms and is slowly condensed, partly by external radiation pressure and partly by internal gravitation.
2. The cloud becomes unstable and breaks up during the time of compression.
3. When the density reaches a point where the cloud is opaque to radiation, the temperature inside begins to rise and the breaking up stops.
4. The protosun collapses under its own gravitation.
5. When the temperature of the interior reaches about 100,000°K and the protosun is about 100 times the diameter of the present sun, or about the size of the earth's orbit, all the hydrogen and helium is ionized and further contraction takes place.
6. Production of energy begins when the internal temperature reaches about 1 million degrees Kelvin and hydrogen nuclei combine to form helium nuclei. In other words, the sun begins to shine as a star at this stage.

The future of the sun has been discussed in Chapter 11.

QUESTIONS AND PROBLEMS

1. What observational evidence is there that stars may be forming today?
2. Compare the angular momentum of the earth with that of (a) Mars and (b) Saturn. Assume that their periods are 2 and 30 years and their distances are 1.5 and 10 a.u., respectively. Neglect their rotations. *Answer:* (a) Earth's angular momentum is 8 times that of Mars.
3. During this century, astronomers have changed their thinking concerning the possible existence of life-bearing planets in the universe. How would this thinking be influenced by replacing the Chamberlin-Moulton planetesimal hypothesis with more modern theories of evolution?
4. What process do all the theories discussed in this chapter have in common?
*5. Look through reports based on studies of moon rocks and write in your own words a proposed theory of the evolution of the moon.
*6. Make up your own theory of the evolution of the solar system.

16 the stars

In discussions of theories of evolution of the solar system, the words sun and star were used interchangeably. The question is: How do we know that the sun is a star? Or, it might be rephrased: Are all stars suns? To answer these questions, the sun must be compared with the stars in a number of ways. How bright would the sun be, if it could be seen from stellar distances? How big are the stars? What is their composition? Do they have the same elements we know on the earth? What are their temperatures? In other words, the comparison of sun and stars depends on a knowledge of their physical nature. This study belongs to a branch of astronomy called astrophysics.

16.1 STELLAR DISTANCES

As has been pointed out, Tycho refused to accept the Copernican theory because he was unable to detect any parallax of the stars due to the motion of the earth around the sun. The reason Tycho failed to find this effect was that it is actually very minute and requires the use of a fairly large telescope to detect. Between 1833 and 1838 three astronomers, using the most accurate methods possible, measured this parallax and used the results to compute the distances of three stars.

Thomas Henderson (1798–1844) at the Cape of Good Hope used the meridian circle to measure the right ascension and declination of α Centauri several times in 1833. From these accurate measures he deduced the star's parallax, but he did not publish the results until 1838. Also in 1838 Wilhelm Struve (1793–1864) in Russia and Friedrich W. Bessel (1784–1846) in Prussia measured the parallaxes of several stars. Bessel observed 61 Cygni with his heliometer, an instrument that uses a telescope with a divided objective. This objective produces double images of all stars in its field. By moving the parts of the objective, it is possible to measure very accurately the distance between a parallax star and a comparison star nearby. The distances computed from the parallaxes were the first to show the vastness of the universe.

In determining distances in the solar system, it is customary to define parallax as an angle with the radius of the earth as the base line. If this definition were to be used for stars, the parallax would be less than 0.0001″, an angle much too small to be measured by any known method. Fortunately, another longer base line is available—the diameter of the earth's orbit, or 186 million miles. Stellar parallax is defined as the angle at a star subtended by one astronomical unit.

To determine stellar parallax, photographs of star fields are taken with long-focus telescopes. The shifts of the nearer stars are measured under a microscope with respect to the background stars and the parallax is computed in seconds of arc. The principle is shown in Fig. 16-1. The result is the relative parallax, unless that of the distant stars can also be computed. If so, a correction is made to find the absolute parallax. Since the parallax of the distant stars is so small, it can usually be neglected because it is smaller than the errors of measurement.

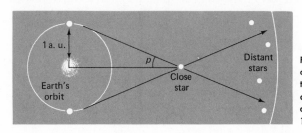

FIG. 16-1 Stellar parallax. The parallax, p, is determined by measuring the shift of a close star with respect to distant stars, as the earth moves in its orbit around the sun. The base line is 1 a.u.

The ideal procedure would be to take photographs of the star field frequently for six months or a year and measure the diameter of the small ellipse a near star would describe on the plates. But since this is not practical, observations at six-month intervals can be made when the star is at each side of its apparent orbit. In order to be sure that the star itself is not in motion, observations are continued at intervals of six months for about two years.

Referring again to Fig. 16-1, the earth, sun, and star form a slender triangle with p, the parallax, as the angle at the star. This triangle can then be solved by the radian method previously discussed, where the side opposite p is 1 a.u. in length, as shown. Here $p'' = 206{,}265''/d$, where d is the star's distance, also in astronomical units, and $d = 206{,}265''/p''$.

Since the distance in astronomical units is not convenient to use, a new unit of distance, the *parsec*, has been introduced. The word parsec is a combination of the first three letters of parallax with the first three of second. It is defined as the distance to an object whose parallax is 1″, or 206,265 a.u. Multiplying by the number of miles or kilometers in 1 astronomical unit, 1 parsec = 19.16×10^{12} (trillion) miles or 30.84×10^{12} (trillion) km.

The distance to a star is inversely proportional to its parallax; that is, the smaller the angle of parallax, the greater is the distance. The formula for distance is therefore simplified to

$$d = \frac{1}{p''} \text{ parsecs} \qquad (16\text{-}1)$$

where p'' is the parallax in seconds, and d is the distance in parsecs. However, the light-year is a more familiar unit of distance. Light travels 186,282 miles in one second. Mutiplying this by 31,557,000, the number of seconds in one year, the length of the light year is 5.88×10^{12} miles. From the preceding value of the parsec, 1 parsec = 3.26 light-years. The formula for distance can then be written as

$$d = \frac{3.26}{p''} \text{ light-years} \qquad (16\text{-}2)$$

EXAMPLE: The parallax of Sirius, the nearest bright star visible in most of continental United States, is 0.375″. Compute its distance in astronomical units, parsecs, and light-years.

SOLUTION: $d = \dfrac{1}{0.375''} = 2.67$ parsecs

$d = \dfrac{206{,}265''}{0.375''} = 550{,}000$ a.u.

$d = \dfrac{3.26}{0.375''} = 8.7$ light-years

Because of the difficulty of measurement of parallax, most determinations are accurate only to about 0.01″, and are independent of the size of the parallax. That is, a parallax of 0.10″ may be in error by 10 percent, whereas if the parallax is 0.01″, the error may be as great as 100 percent. This means that distances greater than 100 parsecs become very inaccurate by this method. Fortunately, other methods are available for more distant stars.

Data for the 20 nearest stars are given in Table 16-1. Only seven are brighter than magnitude 6.0 and visible to the naked eye. Also, seven are double stars. The nearest stars usually have the largest proper motions.

16.2 STELLAR SPECTRA

Spectra of most stars are dark-line spectra; that is, they have continuous bands of color crossed by dark lines. This type of spectrum indicates that the stars, like the sun, have incandescent interiors surrounded by cooler atmospheres under low pressure. It is assumed that the interiors are composed of gases under high pressure and temperature and that the gas laws hold throughout.

The gases in the atmospheres of stars are under low pressure and the atoms

TABLE 16-1   The Twenty Nearest Stars[a]

star[b]	right ascension (1950)	declination (1950)	parallax p″	distance (parsecs)	proper motion	radial velocity (km/sec)	tangential velocity (km/sec)	space velocity (km/sec)	visual magnitude of components (A)	(B)	absolute visual magnitudes of components (A)	(B)
1. α Centauri*	14ʰ 36.2ᵐ	−60° 38′	0.764	1.31	3.68″	− 23	23	33	− 0.01	+ 1.5	+ 4.5	+ 5.9
2. Barnard's Star	17 55.4	+ 4 24	0.546	1.83	10.30	−108	89	140	+ 9.54		+13.2	
3. Wolf 359	10 54.1	+ 7 19	0.430	2.32	4.84	+ 13	53	55	+13.66		+16.8	
4. Lalande 21185	11 00.6	+36 18	0.404	2.48	4.78	− 86	56	103	+ 7.47		+10.5	
5. Sirius*	6 42.9	−16 39	0.376	2.66	1.32	− 8	17	19	− 1.42	+ 8.7	+ 1.4	+11.5
6. Luyten 726-8*	1 36.4	−18 13	0.365	2.74	3.25	+ 29	42	51	+12.5	+12.9	+15.4	+15.8
7. Ross 154	18 46.8	−23 54	0.345	2.90	0.74	− 4	10	11	+10.6		+13.3	
8. Ross 248	23 39.4	+43 55	0.317	3.16	1.82	− 81	27	85	+12.24		+14.7	
9. ε Eridani	3 30.6	− 9 38	0.305	3.28	0.97	+ 15	15	21	+ 3.73		+ 6.1	
10. Luyten 789-6	22 35.8	−15 36	0.302	3.31	3.27	− 60	51	79	+12.2		+14.6	
11. Ross 128	11 45.2	+ 1 06	0.301	3.32	1.40	− 13	22	26	+11.13		+13.5	
12. 61 Cygni*	21 04.7	+38 30	0.292	3.42	5.22	− 64	85	106	+ 5.19	+ 6.02	+ 7.5	+ 8.3
13. ε Indi	21 59.6	−57 00	0.292	3.43	4.67	− 40	76	86	+ 4.73		+ 7.0	
14. Procyon*	7 36.7	+ 5 21	0.288	3.48	1.25	− 3	21	21	+ 0.38	+10.7	+ 2.7	+13.0
15. Σ 2398*	18 42.2	+59 33	0.284	3.52	2.29	+ 8	38	39	+ 8.90	+ 9.69	+11.1	+11.9
16. BD +43°44*	0 15.5	+43 44	0.282	3.54	2.91	+ 18	49	52	+ 8.07	+11.04	+10.3	+13.2
17. CD −36°15693	23 02.6	−36 08	0.280	3.58	6.87	+ 10	116	116	+ 7.39		+ 9.6	
18. τ Ceti	1 41.8	−16 12	0.273	3.66	1.92	− 16	33	37	+ 3.50		+ 5.7	
19. BD +5°1668	7 24.7	+ 5 23	0.266	3.76	3.73	+ 26	66	71	+ 9.82		+11.9	
20. CD −39°14192	21 14.3	−39 04	0.260	3.85	3.46	+ 21	63	66	+ 6.72		+ 8.7	

[a] Table is adapted from a list compiled by Peter van de Kamp (1969).
[b] An asterisk indicates a double star.

absorb energy from the continuous spectrum passing outward from the highly compressed interiors. Atoms of each element absorb specific wavelengths, producing a unique spectrum with lines in definite positions. However, the character of these lines is affected by conditions in the atmospheres: Stellar rotation produces widening by Doppler shifts, higher pressures in the lower levels also broaden the lines, and strong magnetic fields split the lines. These effects can be used to study atmospheric conditions in the stars and to distinguish those lines that are produced in a star's atmosphere from those that are produced by any intervening gas clouds in space.

In 1824 Fraunhofer, who had discovered the dark lines in the spectrum of the sun, also found dark lines in stellar spectra. In 1863 Pietro Angelo Secchi (1818–1878), a Jesuit astronomer at the Vatican Observatory, found important differences and divided the stellar spectra into four major classes, as follows Fig. 16-2):

I. Heavy dark lines of hydrogen.
II. Numerous lines of metals, less intense than the hydrogen
     lines in class I.
III. Bands, sharp toward the red, later identified as due to
     titanium oxide.
IV. Bands, sharp toward the violet, now identified as due to
     carbon and carbon compounds.

Father Secchi observed the spectra visually and was therefore unable to see the fainter lines that were not observed until the application of photography to this study.

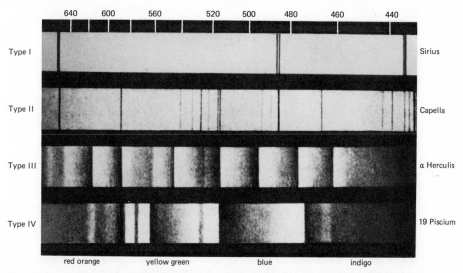

FIG. 16-2   Photos illustrating Secchi's spectral classification of stars. Secchi's classification was based on the visual examination of some 2000 stars. (From Graff's Grundriss der Astrophysik, with the permission of the B. G. Teubner Verlag, Stuttgart)

A new system of classification of spectra was devised at the Harvard College Observatory. The work of classifying the spectra was done by Miss Annie J. Cannon (1863–1941), beginning in 1911. The spectra were photographed using an objective prism (see Chapter 9). A catalog of stars giving their positions for 1900, their visual and photographic magnitudes, and their spectral classes, was published in 1918. It contained data for 225,300 stars in a 10-volume set. The catalog was called the Henry Draper (1837–1882) catalog in honor of the first astronomer to use photography in astronomy and the first American to study the spectra of stars. The system of classification has been called the Harvard system, but is officially the Henry Draper classification.

The original intention of the Draper system was to have the sequence of spectra arranged in order of the letters in the alphabet. This was changed later when it was found that certain classes really did not exist, but had been assumed from the poor photographs that had been used at the beginning of the studies. The order was changed again when it was found that the sequence was really dependent on temperature. For example, the order of B and A was interchanged because the B-stars are hotter than the A-stars. The O-stars are placed first because they are the hottest stars known.

An easy way to remember the order of this sequence is that the letters of the spectral classes are also the first letters of the sentence suggested by Russell: "Oh, be a fine girl, kiss me right now, sweetheart."

There are eight major classes and four minor ones, each designated by a letter. Each class is further subdivided into subclasses ranging from 0 to 9, in

TABLE 16-2   Draper Classification of Spectra

class	example	elements present	approximate temperature (°K)	color
O5	No bright star	H, He, O, N, with electrons removed	50,000	Blue-white
B0	Orion stars	He maximum intensity at B2	25,000	Blue
B5	Achernar	H, He strong	15,000	Blue
A0	Sirius	H maximum, Ca present	11,000	Blue-white
A5	Fomalhaut	Metals near maximum	9000	White
F0	Canopus	Ca increases; other metals	8000	White
F5	Procyon	Fe and other metals	7000	White
G0	Capella	Fe and others;	6000	Yellow
G5		Ca weakening, molecules increasing	5000	Yellow
K0	Arcturus	H weak; atomic lines, molecules strong	4500	Orange
K5	Aldebaran	Not much change	4000	Orange-red
M0	Antares	TiO present	3500	Red
M5	Faint stars	TiO stronger	3000	Red
R, N, S	Faint stars	Molecules strong	3000	Very red
I	ε Aurigae	Infrared spectrum, not yet observed	1000?	Not visible

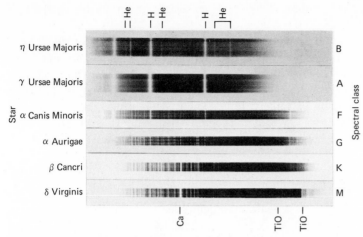

FIG. 16-3   The spectral classification of stars. The upper classes contain strong lines of hydrogen and helium. The lower classes show many lines of metallic elements and their compounds, such as calcium and titanium oxide. The lines are identified above and below. (*Yerkes Observatory photographs*)

order to take care of small changes in the appearance of lines from one class to the next. Table 16-2 shows the important features in the Draper sequence, including one subclass between the major classes. (See also Fig. 16-3.)

It is obvious from the table that the classes are arranged in order of temperature. It will also be seen that there is a continuous change in color. This is in agreement with the laws of Stefan and Wien discussed in Chapter 11. The latter predicts a change in maximum wavelength in exactly the order of the Draper classification.

Temperatures are known to be high in the O- and B-stars, because the removal of electrons from atoms requires a great deal of energy. In class O5 one or two electrons have been removed from the oxygen and nitrogen atoms. This is possible only at very high temperatures, estimated as high as 100,000°K by some astronomers. There is no known star hotter than an O5-star. Theoretically, an O0-star would have an infinite temperature!

At the lower end of the sequence the temperatures are low, as indicated by the red colors and also by the fact that compounds (molecules) are present. In the sun, for example, at a temperature of 6000°K in the photosphere there are no compounds, but compounds are present in sunspots, as shown by the molecular spectra, at lower temperatures. The classes M, N, and R differ from each other in the compounds present. All are red stars with temperatures of about 3000°K. In class M titanium oxide (TiO) is present, but in classes N and R it is absent and combinations of carbon with nitrogen and hydrogen are present. The R-stars are frequently called carbon stars, and most, if not all, at times have such heavy carbon compounds in their atmospheres that the light they emit is greatly weakened. They vary in light output as the compounds form and then dissolve at unpredictable times. There are no bright stars visible without a telescope in classes O, R, N, and S.

More recently, still another class has been added. There is evidence that a very few stars are so cool that they do not radiate visible light, but give out some energy in the infrared. These stars are I-stars, meaning infrared. One such star is a component of ϵ Aurigae, an eclipsing star system. The temperature has been estimated at 1000°K. It is extremely large, about 3000 times the diameter of the sun, and its density is so low that it is semitransparent. It is known to exist only because it eclipses its companion, a bright star, every 27 years.

16.3 STELLAR MAGNITUDES

Hipparchus and Ptolemy used a scale of magnitudes to indicate the brightness of stars on their charts. In this system the brightest stars were called first magnitude. Next in order were second, third, and so on to sixth for the faintest stars that could be seen without a telescope.

By 1856 stellar photometry had developed to such a degree that accurate magnitudes could be determined by visual methods. Photography was adapted to astronomy about the same time. Sir John Herschel (1792–1871) and Norman R. Pogson noticed that a first-magnitude star was about 100 times brighter than a star of sixth magnitude. It also appeared that each magnitude had about the same ratio of brightness to the next magnitude through the six classes.

When the magnitude scale between first and sixth magnitudes, an interval of five magnitudes, was studied carefully, it was found that each magnitude was 2.512 times brighter than the next fainter magnitude. This can be seen in Table 16-3, where a difference of magnitude 1.0 has a brightness ratio of 2.512. Since the brightness ratios multiply, the ratio for a difference of magnitude is the product of the ratios. For example, for a difference of two magnitudes the ratio of brightness is $2.512 \times 2.512 = 6.310$. It is helpful to remember that for each difference of five magnitudes, the ratio of brightness increases (or decreases) by a factor of 100. Thus, a difference of 15 magnitudes $(5 + 5 + 5)$ is equal to a brightness ratio of $100 \times 100 \times 100 = 1,000,000$.

We can apply this method to fractional magnitudes by using the first two columns of Table 16-3. For example, suppose we want to find the ratio of brightness for a difference of magnitude of 1.5. Use 2.512 for a difference of 1.0 magnitude and 1.585 for a difference of 0.5 magnitude. The product is $2.512 \times 1.585 = 4.00$. Sirius, mag -1.4, is 1.5 magnitudes brighter than Vega, mag $+0.1$. Sirius is therefore four times brighter than Vega.

The following equation has been developed and adapted to the magnitude scale:

Let l_1 and l_2 be the brightness of stars of magnitudes m_1 and m_2. Then

$$\frac{l_1}{l_2} = x^{m_2 - m_1} \tag{16-3}$$

If $m_1 = 1.0$ and $m_2 = 6.0$, from equation (16-3) we have

$$x^5 = 100 \quad \text{and} \quad x = \sqrt[5]{100} = 2.512$$

Putting this value of x in the equation, it becomes

$$\frac{l_1}{l_2} = 2.512^{m_2 - m_1} \tag{16-4}$$

For the student who has studied logarithms,

$$\log x = \frac{1}{5} (\log 100) = \frac{2}{5} = 0.4$$

and, from a table of logs, $x = 2.512$. The reduction from light ratios to magnitudes, or the reverse, can be done easily using logarithms. It can also be done with a log/log slide rule. Table 16-3 gives the ratio of brightness for given differences of magnitude.

TABLE 16-3   The Magnitude Scale

difference of magnitude	ratio of brightness	difference of magnitude	ratio of brightness
0.1	1.096	2.0	6.310
0.2	1.202	3.0	15.85
0.3	1.318	4.0	39.81
0.4	1.445	5.0	100.0
0.5	1.585	6.0	251.2
0.6	1.738	7.0	631.0
0.7	1.905	8.0	1585
0.8	2.089	9.0	3981
0.9	2.291	10.0	10,000
1.0	2.512	15.0	1,000,000

It soon became apparent that not all of the first-magnitude stars are of the same brightness. For the brighter stars, by the help of the equation, it is possible to use zero magnitude and even negative magnitudes. The scale can thus be extended to bright objects such as Jupiter, whose magnitude is -2; Venus, -4.4 at maximum brightness; full moon, -12.6; and the sun, -26.7. The magnitudes of the 20 nearest stars are given in Table 16-1 and of the 20 brightest stars in Table 16-4.

The equation makes it possible to compute magnitudes to decimals, provided the ratio of brightness can be determined accurately. For telescopic stars, the magnitude scale is extended beyond $+6$. A 1-inch telescope will show stars down to magnitude $+9$; 16-inch to $+15.0$; and the 200-inch telescope to magnitude $+20$ and even fainter by photography.

The eye is not a very sensitive photometer. A practiced observer using a telescope to compare stars close together can possibly distinguish differences to 0.1 magnitude. The photographic plate can be used to measure magnitudes

TABLE 16-4   The Twenty Brightest Stars[a]

star	right ascension (1967)	declination (1967)	visual magnitudes	spectral types of components[b]	proper motion (sec/year)	distance (light-years)	absolute magnitudes of components
Sirius	6^h 43.7^m	−16° 40.2′	−1.4 +8.7	A0 V wd	1.32	8.8	+1.4 +11.5
Canopus	6 23.2	−52 40.6	−0.9	F0 II	0.02	98	−3.1
α Centauri	14 37.3	−60 42.0	0.0 +1.4	G0 V K5 V	3.68	4.2	+4.4 +5.8
Vega	18 35.8	+38 45.1	0.1	A0 V	0.34	26	+0.5
Arcturus	14 14.2	+19 21.2	0.2	K0 III	2.28	36	−0.3
Capella	5 14.2	+45 58.0	0.2 +10	G0 II M1 V	0.44	46	−0.7 +9.5
Rigel	5 13.0	−8 14.3	0.3 +6.6	B8p Ia B9	0.00	815	−6.8 −0.4
Betelgeuse	5 53.3	+7 21.0	0.4v	M2 I	0.03	490	−5.5v
Procyon	7 37.6	+5 18.6	0.5 +11	F5 IV wd	1.25	11.4	+2.7 +13.0
Achernar	1 36.5	−57 24.2	0.6	B5 V	0.10	65	−1.0
Altair	14 49.2	+8 46.8	0.5	A5 IV	0.66	16.6	+2.2
β Centauri	14 01.5	−60 12.9	0.9	B1 III	0.04	293	−4.1
Aldebaran	4 34.0	+16 26.7	1.1 +13	K5 III M2 V	0.20	53	−0.2 +12
Spica	13 23.5	−10 59.4	1.2	B2 V	0.05	260	−3.6
α Crucis	12 24.7	−62 55.0	1.4 +1.9	B1 IV B3	0.04	391	−4.0 −3.5
Antares	16 27.4	−26 21.6	1.2 +5.1	M0 I A3 V	0.03	391	−4.5 −0.3
Pollux	7 43.3	+28 06.4	1.2	K0 III	0.62	39	+0.8
Fomalhaut	20 55.8	−29 47.8	1.3 +6	A3 V K4 V	0.37	23	+2.0 +7.3
Deneb	20 40.3	+45 09.7	1.3	A2p I	0.00	1400	−6.9
β Crucis	12 45.8	−59 30.5	1.5	B1 IV	0.05	490	−4.6

[a] "v" after a magnitude indicates that the star is a variable. "p" after the spectral type indicates that the spectrum is not exactly that listed, but is peculiar; that is, has some irregularity. "wd" means white dwarf.

[b] If two spectra are given, the star is a visual binary star.

accurately to 0.01 magnitude. The most sensitive instrument of all, the photo-electric photometer, under the best conditions and by averaging several observations, can make measures accurate to 0.001 magnitude or to an accuracy of 0.1 percent.

Magnitudes determined by eye estimates are called *visual* magnitudes. *Photographic* magnitudes are measures made on blue-sensitive plates. If a yellow filter is used, the magnitudes are called *photovisual*. *Photoelectric* magnitudes, made with photo cells, which are somewhat color-blind, are very similar to photographic magnitudes, although there are slight differences. All these magnitudes are for apparent brightness (brightness as seen by the photometer but reduced to the magnitude as it would be if the star were located at the zenith).

Because of absorption in the earth's atmosphere, corrections are made by calculating the angular distance of the star from the zenith at the time its brightness is measured and applying a correction factor, which has been computed from a series of changing positions of standard calibration stars. The corrections to photoelectric observations in particular are made very carefully; even the colors of the stars are taken into consideration.

The magnitudes of the moon and planets change with variations in distance and phase. The magnitude changes of the moon are very conspicuous. Mars changes by several magnitudes between conjunction and opposition. The changes in magnitude of all the other planets are noticeable to a practiced sky watcher.

Since the stars are not all at the same distance from the sun, it is desirable to calculate their magnitudes as if all were at the same distance in order to compare them properly with each other. The term *absolute magnitude* means the magnitude of a star calculated for a standard distance of 10 parsecs (32.6 light-years).

EXAMPLE: Compute the absolute magnitude of a star whose apparent magnitude is +10 and distance is 100 parsecs.
SOLUTION: The absolute magnitude is found by calculating the brightness seen from a distance of 10 parsecs. It is known that the intensity of light varies inversely with the square of the distance. Since the standard distance of 10 parsecs is 0.1 the actual distance, the brightness would increase by a factor of $(10)^2$, or 100, if the star were seen from a distance of 10 parsecs instead of 100.

From Table 16-3, we find that for a brightness ratio of 100 the change of magnitude is 5. Subtracting 5 from the apparent magnitude, because magnitudes decrease as brightness increases, we find the absolute magnitude to be $+10 - 5 = +5$.

The example above shows that, if the distance and apparent magnitude are known, it is possible to compute the absolute magnitude. By reversing the process it is possible to compute the distance if the apparent and absolute magnitudes are known. This is a very important statement, since it leads to a determination of the distances of galaxies by the use of Cepheid variable stars, which will be discussed in later chapters. This method has made it possible to estimate the extent of the known universe.

The sun's apparent magnitude m is -26.7. How bright would it be, if seen from a distance of 10 parsecs?

The brightness would decrease if its distance increased from one to 2,062,650 astronomical units (10 parsecs), by a factor of $(2,062,650)^2$. Substituting this figure into equation (16-4) and computing the magnitude change, we find the absolute magnitude of the sun to be about $+4.87$ (see the sample problem below). This figure may be slightly in error because of the difficulty of comparing the apparent brightness of the sun with the stars.

SAMPLE PROBLEM: Compute the absolute magnitude of the sun.

SOLUTION: By logarithms

$$2.512^{M-m} = 2,062,650^2$$
$$(M - m) \log 2.512 = 2 \log 2,062,650$$

or

$$(M - m) \times 0.4000 = 2 \times 6.31443$$
$$M - m = \frac{2 \times 6.31443}{0.4000} = 31.5722$$

and

$$M = -26.7 + 31.57 = +4.87$$

which is the absolute magnitude of the sun.

From the table of magnitudes (Table 16-3),

$$2,062,650^2 = (2.062 \times 10^6)^2 = 4.254 \times 10^{12}$$
$$= 4.254 \times 100^6 = 2.512 \times 1.694 \times 100^6$$

This large number has been broken down in order that the table of magnitudes may be used without further computation. Each part may be found in the table, or found by interpolation. Light ratios are multiplied together; magnitudes are merely added or subtracted. From the table, the magnitude corresponding to a ratio of brightness of 2.512 is 1.0; 1.694 is between 0.5 and 0.6; call it 0.57. A value of 100 is equivalent to 5.0 magnitudes, which must be taken 6 times. Therefore the difference of magnitude of the sun, the amount its brightness decreases for a change from its actual distance to the standard distance, is

$$M - m = 1.0 + 0.57 + 6 \times 5.0 = 31.57$$

which is exactly the result obtained by logarithms, neglecting the digits after the second decimal place.

It is possible to calculate all absolute magnitudes by using the inverse-square law, but computation is easier if reduced to a formula. Let l, m, and d be the amount of light, the apparent magnitude, and the distance of a star, and L and M be its brightness (luminosity) and absolute magnitude at the standard distance, $D = 10$ parsecs. Using the inverse-square law,

$$\frac{l}{L} = \frac{D^2}{d^2} = \frac{10^2}{d^2} = 2.512^{M-m} \tag{16-5}$$

Taking logarithms of both sides of this equation (the last two members),

$$2 \log 10 - 2 \log d = (M - m) \log 2.512 = 0.4(M - m)$$

and

$$M - m = \frac{2 \log 10 - 2 \log d}{0.4}$$

$$= \frac{2 - 2 \log d}{0.4} = 5 - 5 \log d \tag{16-6}$$

Thus,

$$M = m + 5 - 5 \log d \tag{16-7}$$

where d is the distance in parsecs. Since d is inversely proportional to the parallax p, the equation may be written

$$M = m + 5 + 5 \log p \tag{16-8}$$

where p is expressed in seconds of arc.

The *luminosity* of a star is its brightness compared to the brightness of the sun. To compute luminosity, it is necessary to know the absolute magnitudes of the sun and the star.

Inverting equation (16-5),

$$\frac{L}{l} = 2.512^{m-M}$$

Let $l = 1$ be the luminosity of the sun and $m = 4.87$ be its absolute magnitude. We then have

$$L = 2.512^{4.87-M} \tag{16-9}$$

where L and M are the luminosity and absolute magnitude, respectively, of any star. If M is less than 4.87, the star is brighter than the sun. If M is greater than 4.87, the star is fainter.

Magnitudes, both apparent and absolute, are called visual, photographic, or photoelectric, depending on how they are determined. These magnitudes differ because (see Chapter 9) an eye, a photographic plate, and a photoelectric cell are not equally sensitive to all wavelengths. The eye is most sensitive to yellow light, whereas photographic plates are more sensitive to blue light. As a result, yellow stars appear brighter to the eye than they do on photographic plates, whereas the reverse is true of blue stars. Another kind of magnitude, *bolometric* magnitude, is based on the total amount of energy emitted by a star. It can be observed only under special conditions by the use of a bolometer, an instrument sensitive to the infrared regions of the spectrum as well as to visible light.

If the temperature of the star is known, its bolometric magnitude can be computed from its photovisual magnitude by means of a correction that has been determined from theory. One definition of luminosity is that it is a measure of the total amount of radiation emitted by a star and therefore must be based on the bolometric magnitude.

The fact that the magnitude depends on the color in which it is measured provides a means of measuring color differences with a photoelectric photometer. To obtain the magnitude that would be measured visually, the light of a star is passed through a yellow filter and then is compared with the light of a standard star measured through the same filter. Similarly, the photographic magnitude is obtained by making the same comparison with the photoelectric photometer through a blue filter. The difference between the visual and photographic magnitudes is called the *color index*. Alternatively, the color index can be obtained by taking photographs of the star and the standard star through a yellow filter and then through a blue filter on the same plate. The magnitudes are obtained by measuring the size or amount of blackening of the stellar images. The photoelectric method is faster.

More recently, measures have been made with photometers equipped with three or more filters. One standard set of filters is the *UBV* system, meaning ultraviolet, blue, and visual. The color index of a star is then determined by comparing the magnitudes when the filters are used in various combinations. There is a definite relation between the spectral type, based on temperature, and color expressed by the color index, as shown in Table 16-5.

TABLE 16-5   Stellar Spectra and Color Index

spectral class	color index	spectral class	color index
B0	−0.32	G0	0.60
B5	−0.16	G5	0.68
A0	0.00	K0	0.82
A5	0.15	K5	1.18
F0	0.30	M0	1.45
F5	0.44	M5	1.69

16.4 THE SPECTRUM-LUMINOSITY DIAGRAM

As the data of absolute magnitudes and spectral types were being collected, Ejner Hertzsprung (1873–1967) in Holland in 1911 and Russell in the United States in 1913 noticed that, on a diagram correlating the two, the plotted points fell into distinct groups. They called the stars giants and dwarfs. This was a fortunate guess, since it was shown later that there are indeed groups of stars based on size. This diagram has been continued with the addition of data from thousands of stars. It is called the *spectrum-luminosity* diagram or the H-R diagram in honor of the two originators (Fig. 16-4).

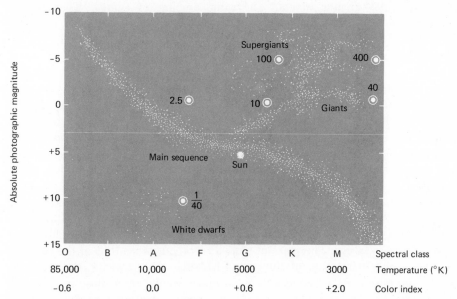

FIG. 16-4 The Hertzsprung-Russell diagram for stars of known distance. The numbers refer to star sizes compared to the diameter of the sun.

As the temperature scale on the diagram shows, the highest temperatures are plotted on the left, the lowest on the right. They vary from 85,000°K for the O-stars to 3000°K for the M-stars. Absolute magnitudes at the top are −10 for stars 1 million times brighter than the sun and +15 at the bottom for stars 10,000 times fainter than the sun. Others still fainter are not shown. The diagonal line is called the *main sequence*, since most stars are represented by points on that part of the diagram.

Another group with absolute magnitude about zero is composed of *giant* stars. *Supergiants* have magnitudes above −2 (an arbitrary division). Giant stars are therefore about 100 times brighter than the sun and supergiants are more than 600 times brighter. *White dwarfs*, at the lower left, have temperatures about 10,000°K and absolute magnitudes averaging +10. They are actually stars, but are more like planets in size.

Stellar diameters can also be calculated from the diagram. A star of class G with $M = 0$, a giant star, radiates 100 times as much energy as the sun. But, since its temperature is the same, the amount of radiation per unit area must be the same. Therefore, the surface area of a star of this class must be 100 times larger than the area of the sun and its diameter must be 10 solar diameters. A supergiant of class G with $M = -5$ would be another 10 times larger, or 100 times the diameter of the sun.

The red giant stars radiate at a lower rate than the sun and must therefore be still larger. To calculate the diameter, the rate of radiation can be found from Stefan's law (Section 11.4). If the temperature is 3000°K, half that of the sun, the radiation per unit area is only $(\frac{1}{2})^4$ or $\frac{1}{16}$ that of the sun and the area must be 16 times greater to emit the same amount of energy. That is, the diameter must be 4 times the solar diameter. Likewise, a star of $M = 0$ and spectral type M is 4 times larger than a star of $M = 0$ and spectral type G. So the cool, red giants and supergiants must be very large.

Similarly, the white dwarfs must be very small. A white dwarf whose temperature is 12,000°K and $M = +10$ has a diameter only $\frac{1}{40}$ that of the sun. Another kind of investigation shows that this white dwarf must have nearly the same mass as the sun and therefore its density must be $(40)^3 = 64,000$ times the density of the sun. This density is unlike any known on the earth and can be possible only if the electrons are stripped from the atoms and the nuclei packed closely together. This is the predicted state of the sun 7 or 8 billion years from now.

Fortunately, the absolute magnitudes of stars can also be determined by the appearance of their spectra. Consider the sun and the bright star Capella, which has the same type spectrum as the sun, G0. But it is 170 times brighter and therefore 13 times larger. The mass of Capella is about twice that of the sun and its density is only 0.001 that of the sun. With such a low density, the atoms in the star's chromosphere have more room to move around than those in the sun's chromosphere. The lines in the spectrum of Capella are therefore sharper than the solar lines, because the atoms do not interfere with each other as much. Also the temperature in the chromosphere is lower in the star.

The difference in temperature between main-sequence stars and giants was noticed in 1913 by Walter S. Adams (1876–1956) and A. Kohlschütter at the Mount Wilson Observatory. They also noticed that there are differences in the ionization of atoms, since ionization is easier at low pressures than at high pressures. Certain lines due to ionized gases are stronger in the spectra of giant stars than in the main-sequence stars. Also, in the latter the neutral lines are stronger than in the giants because there is a higher percentage of nonionized atoms, where the pressure is greater. This led to a new method of determining absolute magnitudes, by comparing the strength of certain lines in the spectra.

The method was improved by William W. Morgan (1906–   ) and his associates at the Yerkes Observatory. They divided the stars into luminosity classes, as follows:

Ia. Brightest supergiants.
Ib. Less luminous supergiants.
II. Bright giants.
III. Giants.
IV. Subgiants (between giants and main sequence).
V. Main-sequence stars.
VI. Subdwarfs (Sd).
VII. White dwarfs (Wd).

The sun in this classification is G0V, and Capella is G0III. The authors also worked out a graph showing the absolute magnitudes for each of the first six classes.

Since the absolute magnitudes are known with considerable accuracy, it is possible to substitute M and m in equation (16-7) or (16-8) and determine the distances and parallaxes of stars for which spectra can be obtained and classified. Parallaxes determined in this way are called *spectroscopic parallaxes*. (See Fig. 16-5 for the relation between the luminosity classes and the H-R diagram.)

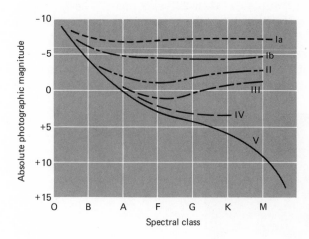

FIG. 16-5 The luminosity classes of stars on the Hertzsprung-Russell diagram. Class VI lies to the left of class V; for the location of class VII see Fig. 16-4.

16.5 STELLAR MOTIONS

The fact that stars are not "fixed" but are in motion was first discovered by Halley. He noticed that the positions of the bright stars Arcturus and Sirius had changed by a noticeable amount (1° and ½°, respectively) since the time of Ptolemy's *Almagest* (A.D. 150). This change of position was interpreted as due to the real motion of the stars. Although the amount of change per year was small, a substantial change had accumulated over 1500 years.

As shown in Chapter 6, a star's position on the celestial sphere is continually changing because of precession, nutation, and aberration; also, its measured position is changed by refraction. So when a star is observed with a meridian circle to determine its right ascension and declination, corrections must be made for these changes. It has also been noted that the celestial equator is not a fixed circle but moves with respect to the ecliptic.[1]

Therefore, it is customary to select some equator, such as the equator of 1900, 1950, or 2000, and refer all star positions to it as a standard. After this is done, the star may still show a change in right ascension and declination. This is the star's *proper motion,* defined as the rate of change of position in one year. It is measured with respect to the faint background stars, which are assumed to be too far away to show any measurable proper motion.

[1] The positions of stars in the tables in their chapter are for the equators of 1950 or 1967.

The corrections are practically the same for all stars in a given small region—for bright stars as well as faint ones. In general, the faint stars are farther away than the bright ones and do not show much, if any, proper motion. This is not always true, as can be seen from Table 16-1.

Proper motion is easily measured by comparing two photographs taken some years apart. The measuring instrument is the same as was used for the discovery of Pluto—the "blink microscope." The proper motion is computed from the amount of shift of the fast-moving star with respect to the background stars between the two dates.

Figure 16-6 is an example of two plates taken 22 years apart. The faint star shown by the arrow is Barnard's Star, which has the largest proper motion known, 10.30″ per year (see Table 16-1). It was discovered by Barnard at the Yerkes Observatory in 1916. It is the second nearest star to the sun, only six light-years away. Usually the stars with the largest proper motions are among the nearest stars and are put on observing programs for parallax determination.

FIG. 16-6 The proper motion of Barnard's star. The interval between the two photographs is 22 years. The arrows show the position of the star at the two dates. (Yerkes Observatory photograph)

If the earth did not move around the sun, the proper motion of a star would appear as a straight line between two points on a photograph. In Fig. 16-7 the proper motion is drawn as the straight line, AC, in the time interval of one year. If the star were stationary, its parallax would produce a small ellipse because

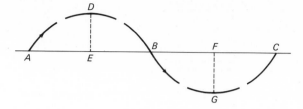

FIG. 16-7   The apparent path of a star in one year against the background of distant stars due to the parallax and proper motion of the star.

of the earth's motion. Combining the two, Fig. 16-7 shows the apparent motion in one year as a wave, the amplitude of the wave being DE or FG, the parallax of the star. If observations could be continued for several years, the parallax would remain the same. The length of the line would be proportional to the time interval, provided that the proper motion remained constant. For accuracy, the longer the line, the more accurate would be the determination of the proper motion, which is the length of the line divided by the interval in years. Hence, parallax and proper motion are both determined from the same series of observations.

As an illustration of the proper motion of stars, consider the nine stars in the Big Dipper, data for which are given in Table 16-6. It will be noted that Mizar, ζ Ursae Majoris, is a double star and is accompanied by a fourth-magnitude star 80 (g) named Alcor. These two stars are sometimes called the Horse and Rider. Mizar was the first telescopic double star to be discovered.

Several remarkable things can be seen from the table. First, the seven middle stars are all at approximately the same distance. Their parallaxes are about 0.040″, corresponding to distances of about 25 parsecs or 80 light-years. Second, these seven stars are all moving in approximately the same direction, as indicated by their proper motions. This is not quite so obvious from the table, because of the different positions of the stars in the sky, which partly conceal the fact that they are moving in parallel paths. Third, their space velocities average 18 km/sec, all being within 2 km/sec of the average. This group of stars is part of a larger cluster, all the stars of which are moving toward a common vanishing point. They thus form a *moving cluster* (Fig. 16-8).

The end stars, α and η, do not belong to the cluster, but are moving in the opposite direction. All the middle stars of spectral class A have approximately the same temperature. The conclusion might be drawn that they are members of the same family, are of the same age, and will continue to shine as stars for about the same length of time.

In computing stellar motion, the sun is considered to be stationary. This is known to be a false assumption, but historically all proper motions have been referred to the sun without consideration of whether or not it is also in motion. The *space velocity* of a star is its motion with respect to the sun. It is desirable to find both the speed and direction of this stellar motion. Space velocity can be

TABLE 16-6   The Stars of the Big Dipper

star	apparent visual magnitude	spectral class	parallax (sec)	proper motion (sec)	absolute visual magnitude	radial velocity (km/sec)	distance (light-years)	space velocity (km/sec)
Alpha, α	1.95	K0	0.030	0.136W	−0.65	−9	109	23
Beta, β	2.44	A0	0.043	0.089E	+0.63	−14	76	17
Gamma, γ	2.54	A0	0.041	0.095E	+0.64	−11	79	16
Delta, δ	3.44	A2	0.044	0.113E	+1.63	−16	74	20
Epsilon, ϵ	1.68	A0p	0.045	0.117E	−0.03	−10	72	16
Zeta 1, ζ_1	2.40	A2p	0.043	0.134E	+0.59	−10	76	18
Zeta 2, ζ_2	3.96	A2	0.043	0.134E	+2.15	−10	76	18
80 (g)	4.02	A5	0.040	0.125E	+2.13	−12	82	19
Eta, η	1.91	B3	0.013	0.115W	−2.52	−2	251	42

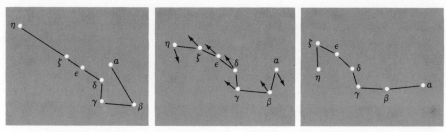

FIG. 16-8   The Big Dipper, a moving cluster, as it looked 100,000 years ago (left); as it looks today (center); and as it will look 100,000 years hence (right). In the center diagram, proper motions are shown by the lengths and directions of the arrows.

represented, therefore, as a vector—a straight line having both length and direction.

Let the space velocity of a star be represented by the white straight line S_1S_2 in Fig. 16-9. It can be resolved into two straight lines perpendicular to each other: S_1Y and YS_2, also shown in white. Let S_1Y be directed away from the sun. (Another case could be directed toward the sun.) The vector S_1Y is called the *radial velocity* of the star that is in motion. If a spectrogram of the star is taken, its lines are shifted according to the Doppler law. By measurement of the amount of shift, the component of the velocity in the line of sight, the radial velocity, can be found and is usually given in kilometers per second. Thus S_1Y can be determined by the Doppler effect.

The other component of the space velocity, YS_2, is the *tangential* or *transverse* velocity. It cannot be measured directly, but it may be computed. Figure 16-9 cannot be drawn to scale. The radius of the earth's orbit, SE, is 1 a.u. The distance to the nearest star, SS_1, is 270,000 a.u. It is obvious from the figure that the space velocity, S_1S_2, forms an angle with the sun that is the star's proper motion, μ. Also the star's parallax is the angle p at the star between two ends of the radius of the earth's orbit, SE.

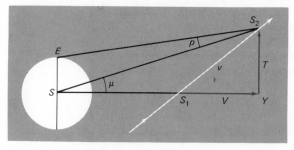

FIG. 16-9   The space velocity of a star, v, and its components, radial velocity V and tangential velocity T. It is also necessary to know the parallax p.

Now consider the two slender triangles, SS_2E and SS_2Y. From radian measure, $p = SE/ES_2$ and $\mu = S_2Y/SY$, both in radians. Dividing the second by the first,

$$\frac{\mu}{p} = \frac{S_2Y}{SY} \times \frac{ES_2}{SE} = \frac{S_2Y}{SE} \tag{16-10}$$

since ES_2 and SY are approximately equal and their ratio can be taken as unity without loss of accuracy. The equation above can now be rewritten

$$S_2Y = \frac{\mu \text{ sec/yr}}{p \text{ sec}} \times 1 \text{ a.u.} \tag{16-11}$$

Hence the tangential velocity,

$$T = S_2Y = \frac{\mu}{p} \text{ a.u./yr}$$

But in combining tangential velocity and radial velocity, the units must be the same, kilometers per second. Substituting in equation (16-11), 1 a.u. = 149,500,000 km and 1 year = 31,500,000 sec,

$$T = \frac{\mu}{p} \times \frac{149,500,000 \text{ km}}{31,500,000 \text{ sec}} = 4.74 \frac{\mu}{p} \text{ km/sec} \tag{16-12}$$

Finally, since S_1YS_2 is a right triangle,

$$v^2 = V^2 + T^2 \tag{16-13}$$

where v = space velocity, V = radial velocity, and T = tangential velocity, each expressed in kilometers per second. This equation gives the value of the space velocity. The direction of the angle between S_1Y and S_1S_2 can be found by trigonometry.

QUESTIONS AND PROBLEMS

Group A

1. How much is the parallax of a star increased by using the radius of the earth's orbit instead of the radius of the earth, as a base line?

2. Assuming observations could be made with the same equipment on Mars as on the earth, what would be the limit of accurate distance measures by the parallax method?

3. The parallax of the nearest star, α Centauri, is 0.76". Compute its distance in (a) astronomical units, (b) parsecs, and (c) light-years. *Answer:* (a) 272,000 a.u.

4. The distance of Barnard's star is 5.9 light-years. Compute its distance in parsecs and astrononomical units. What is its parallax? *Answer:* 1.81 parsecs = 374,000 a.u.

5. The wavelength of maximum radiation of class O5 stars is in the ultraviolet. Do these stars appear to be as bright as stars of equal size whose maximum radiation is in the visible spectrum? Explain.

6. What evidence is there that differences in stellar spectra are due to differences in temperature rather than in chemical composition?

7. Since differences of magnitude correspond to ratios of brightness, magnitudes are added and ratios are multiplied. (a) verify this by multiplying the brightness ratio for a magnitude

difference of 0.4. That is, for example, take a magnitude difference of 0.2. Multiply its brightness ratio, 1.202 by 1.202. (b) Repeat for magnitude differences of 3 and 6.

8. Brightness ratios not listed in Table 16-3 can be computed by using magnitude differences that add up to the desired total. The desired brightness ratio is the product of the corresponding ratios, as shown in Question 7. By using this method, find the ratio of brightness of (a) the sun and the moon; (b) Sirius and β Crucis. *Answer:* (a) magnitude difference is $14.1 = 10 + 4 + 0.1$; brightness ratio $= 436,000$.

9. The intensity of light decreases as the square of the distance. Assume that an observer is comparing the apparent brightness of identical street lights, A and B, at different distances from him. Fill in the missing numbers in the following table:

distance of light A (blocks)	distance of light B (blocks)	brightness ratio	difference of magnitude
1	10	100	5
1	100	10,000	
10	100		
10			10
	1000		15

10. The brightness ratio and difference of magnitude for stars can be found in the same way as for the lights in Question 9. Fill in the missing numbers in the following table; the first row is an example:

distance of star (parsecs)	distance ratio (to 10 parsecs)	brightness ratio	difference of magnitude	apparent magnitude	absolute magnitude
100	10	100	5	+8	+3
1000	100	10,000		+11	
	1,000,000				+2
				-2	+3

11. Show that a star of absolute magnitude $M = -10$ is 1,000,000 times more luminous than the sun, and a star of $M = +15$ is 10,000 times less luminous than the sun. Assume the sun is $M = +5$.

12. Distinguish between giant and dwarf stars as to size, density, temperature, color, and ionization of atoms.

13. Why are the motions of stars measured with respect to the sun?

14. How can proper motion be distinguished from motion due to parallax?

15. What factors produce a large proper motion?

16. Stars that appear rather far apart, such as those of the Big Dipper, may actually be much closer together than some that appear to be close together. Draw a diagram to show how this can occur.

17. What common properties would indicate that stars that may appear far apart are actually related?

Group B

18. An observer can see stars six magnitudes fainter with a 16-inch-diameter telescope than with a 1-inch telescope. Verify this by comparing the brightness ratio with the light-gathering power of the two telescopes.

19. By using Wien's law, determine in which spectral classes the maximum wavelengths of stars are in the visible spectrum. Assume limits of the visual spectrum to be 4000 to 7000 Å.

20. Two stars have the same surface temperature. One is five magnitudes brighter than the other. (a) Calculate the ratio of their surface areas and diameters. (b) Repeat for the case where one star is 10 magnitudes brighter than the other. *Answer:* (a) area ratio, 100; diameter ratio, 10.

21. Two stars have the same absolute magnitude. The temperatures are 6000°K and 3000°K. (a) Compute the ratios of their energy output per square mile, their areas, and their diameters. (b) Repeat for temperatures of 5000°K and 15,000°K. *Answer:* (a) Ratios: energy, 16; area, 16; diameter, 4.

22. Compute the ratio of the diameters of the following stars to the diameter of the sun: (a) $T = 12,000°K$; $M = 0$; (b) $T = 2000°K$; $M = +15$. Assume for the sun, $M = +5$. *Answer:* (a) 2.5.

23. Compute the ratio of the diameters of Sirius A and Sirius B from their temperatures and absolute magnitudes. Their distances are the same. (See Chapter 18 for data.)

24. Compute the tangential and space velocities of the following stars of the Big Dipper: (a) γ; (b) ϵ; (c) η. *Answer:* (a) $T = 11$ km/sec; $v = 15.6$ km/sec.

17   variable stars

In the year 134 B.C., Hipparchus observed a star in the constellation Scorpius where no star was visible before. This was contrary to the belief at that time that the stars were set in a crystal sphere where nothing could change. Hipparchus' star remained bright for a few months and then faded and disappeared. Another new star was seen in the constellation Cassiopeia from November 1572 to the spring of 1574. It was discovered in Germany and was observed by Tycho Brahe in Denmark. Tycho described it as a "nova stellis," meaning new star. The name *nova* is still used for similar stars, several of which are discovered every year. Kepler observed a nova in 1604.

Since the time of Hipparchus and the discovery of novas, other stars have been seen to vary in brightness. A *variable star* is usually discovered by observing its changes in brightness, but it may also vary in other ways, such as in its spectrum or even in its diameter. A plot of the magnitude of a variable star against time is called a *light curve*. A study of a light curve, combined with observations of the spectrum, gives a great deal of information and is fundamental to the study of the physical nature of this type of star.

17.1 CLASSIFICATION OF VARIABLE STARS

Variable stars have been grouped into many classes. We shall adopt a simple classification, since further subclasses would be confusing. They are:

  I. Novas, also called temporary or exploding stars.
 II. Cepheid variables, also called pulsating or eruptive stars.
III. Semiregular variables.

IV. Irregular variables.

V. Eclipsing binary stars.

The first four classes are also designated as *intrinsic variables,* since their variations are caused by changes inside the stars. The fifth class is variable only because of orientation, and the variations are caused by eclipses of one star by a companion.

The *General Catalog of Variable Stars* in 1958 listed the following numbers of variables: 959 novas, 9855 Cepheids, 1134 semiregular and irregular variables, and 2763 eclipsing binaries, for a total of 14,711 variables of all classes in our galaxy. This catalog has two supplements, bringing the total number to 18,791. It was further revised in 1970 and enlarged to include 20,437 stars.

All variable stars, including novas, are named for their constellations, with combinations of letters showing the order of discovery in the constellation. The first discovery in a constellation is given the capital letter R, followed by S, T, . . ., Z. Then RR, RS, RT, . . ., RZ; SS, ST, . . ., SZ. When ZZ has been reached, the series goes back to AA, AB, . . ., QZ (except that the letter J is omitted). This lettering series permits the listing of 334 variables in each constellation. When all 334 combinations are filled, the designations V335, V336, . . . are used as far as needed.

There are also three-letter abbreviations for the constellation names. For example, R Coronae Borealis is R CBr, the first variable star discovered in that constellation. RZ Cassiopeiae is RZ Cas, the eighteenth, AA Cygni is AA Cyg, the fifty-fifth discovered in their constellations. V335 Cygni is V335 Cyg, the 335th in that constellation. However, if a variable star already has a Greek letter or a name, that designation is retained. Examples: Algol = β Persei; and Betelgeuse = α Orionis.

17.2 OBSERVING TECHNIQUES

There are three ways to observe a variable star: visually, by photography, and by photoelectric photometry. The oldest way is by visual observations. It is sometimes possible to notice that a star does not always remain the same brightness compared to a star close by. This happened in 1934 when Nova Herculis was discovered by a British amateur astronomer, who noticed that "there was something wrong with the Head of the Dragon." A bright star appeared where none was found on his star map. It was then compared with neighboring stars both visually and photographically and its spectrum was photographed.

Visual photometry may be made with the naked eye, with a pair of binoculars, or with a telescope. Usually the amateur astronomers use small reflecting telescopes. The accuracy is only about 0.1 or 0.2 magnitude.

It is also possible to take photographs of the sky with a telescope. If the star images are in focus, the magnitudes are compared with the magnitudes of other stars on the plate by comparing the amount of blackening of the images. This method gives an accuracy of about 0.01 or 0.02 magnitude. In the left-hand photograph in Fig. 17-1 no star is visible at the end of the arrow. This means that the nova was fainter than the dimmest star on the plate. In the right-hand photograph, the nova is definitely much brighter than the brightest star, as shown by the size of the image.

Another technique by photography is to place the plate slightly behind the

FIG. 17-1 Two photographs of Nova Herculis 1934 (DQ Herculis), which increased in brightness from about fifteenth magnitude to second magnitude before it was discovered to be a nova. (Yerkes Observatory photograph)

focal plane of the telescope. This method is called *extrafocal photometry*. Here the star images are round disks and the magnitudes are determined by measuring the blackness of the images.

The most accurate way is by photoelectric photometry. Here it is necessary to observe one star at a time, by alternately exposing a photocell to light of a star and another close by. Since the brightness of a star is proportional to the amount of photocurrent released from the sensitive photosurface of the cell, the ratio of the deflections of an electric meter or recording device (see Chapter 9) is used as a basis of comparison and for computing the difference of magnitude. The photoelectric method is accurate to 0.001 or 0.002 magnitude.

Photography has a great advantage over photoelectric photometry because it is possible to photograph many stars, even millions, on a single plate and to record stars much fainter than those observable with the photocell. Furthermore, the photocell is somewhat color-blind, being most sensitive in the blue region of the spectrum. Red-sensitive cells have been developed and are being improved, but they do not approach the blue-sensitive cells for acceptable performance. Photoelectric photometry has been used since early in the 20th century and has become increasingly popular all over the world where accurate photometry is required.

To plot a light curve it is desirable to have a series of continuous observations. By photography it is sometimes the custom to move the plate a short distance in the telescope and take several exposures at different times on the same plate. With the photoelectric photometer it is customary to observe a variable star by first observing it, then shifting the telescope to a star close by and continue alternate observations as long as the stars are in reach of the telescope, but not too near the horizon, or until enough observations have been obtained to satisfy the needs of the observer.

17.3 NOVAS OR TEMPORARY STARS

A nova is given a designation of the word nova, the constellation in which it is located, and the year in which it was discovered; for example, Nova Puppis 1942.

It is now known that a "new" star is not new but has been in existence for billions for years. Because it flares up unexpectedly, very little is known about the prenova stage. Usually a nova is discovered when near its maximum brightness. Subsequent changes can be followed for years, but the chance of observing the rise to maximum is extremely small. The figure showing the light curve of Nova Puppis 1942 is typical (Fig. 17-2). Nova Herculis 1934 was discovered when it was near maximum. Between 1890 and 1960 ten novas have been visible to the unaided eye. It is estimated that about 25 appear in our galaxy every year, but most of them go unnoticed or increase in brightness to telescopic visibility only.

Changes in the spectral class of Nova Puppis 1942 are indicated on the light curve in Fig. 17-2. Before maximum brightness its spectrum was about class A

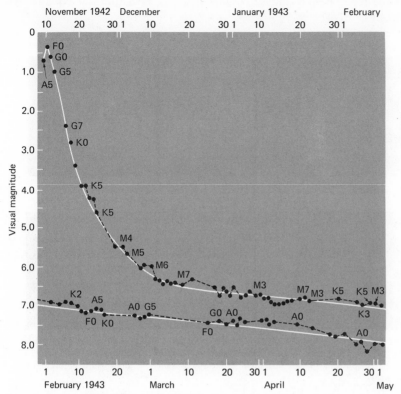

FIG. 17-2 The light curve of Nova Puppis 1942, indicating the changes in spectral type with change of brightness. The star was not observed during its rise to maximum before November 10, 1942. The top dates correspond with the upper curve.

or F (continuous spectrum crossed by a few dark lines). The dark lines are produced by that part of a star's atmosphere between the star and the earth. At maximum, the spectrum was class F. After maximum was reached, the spectrum changed to G and M. This is typical of all novas. Their spectra show large shifts of the lines toward the blue, indicating that the atmosphere is moving rapidly toward the earth on the near side with speeds reaching nearly 1000 miles/sec. It is assumed that the stars have exploded uniformly in all directions, the lines being broadened because the component of velocity from the observable part of the atmosphere varies from zero at each limb to a maximum approach in the center, the far side being hidden from view.

The usual nova slows its brightening process during the last two magnitudes of rise; then, as the brightness decreases, the spectrum shows a sudden change. The absorption spectrum is still visible, but now it has also strongly widened bright lines. These lines are from hydrogen and from ionized calcium and iron, widened about the zero-velocity position. The explanation is that the star's atmosphere is no longer opaque, but is so transparent that the light from the far side, which is receding from the earth, is seen and the lines are strongly shifted toward the red. These changes are shown in Fig. 17-3, and the explanation is diagrammed in Fig. 17-4.

FIG. 17-3   Spectra of Nova Herculis 1934 at maximum brightness (top) and after the atmosphere had formed. Note the widened lines in the lower spectrum. In each case, the narrow spectrum between the pairs of comparison spectra has been artificially widened to show more detail. (*Lick Observatory photographs*)

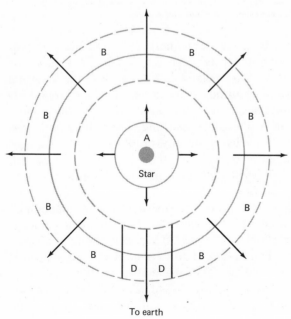

FIG. 17-4 Expanding atmospheric shells around a nova. *A* is a small shell around a star at its center. The larger circles represent the principal expanding shell. Material in regions *D* between the nova and the earth produce absorption lines. All other regions in this shell at *B* produce emission lines.

Meanwhile, the star's color has changed from white through yellow to red. During that time there are other changes in the spectrum, until the "forbidden" lines of doubly ionized oxygen and neon appear (somewhat like the ionized lines in the spectrum of the solar corona). The nova then has the appearance of a very small gaseous nebula and may take on a greenish color (Fig. 17-5). The nova finally fades away; the star loses its nebular spectrum and returns to its original white color and shows a continuous spectrum, sometimes without any visible lines.

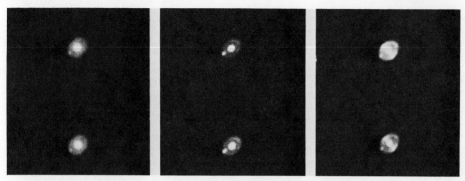

FIG. 17-5 Shells of gas around a nova. At the left are two photos of Nova Herculis in ultraviolet light, in the center photo in green light, and at the right in red light. The pictures were taken in 1934, 17 years after the outburst of the nova. (Photographs from the Hale Observatories)

During the nova stage of several stars, luminous clouds of atoms have been observed around them. The material is composed of gas and dust particles in the region, which reflect light from the nova. Since most novas appear in the Milky Way, it is not surprising to find these dust clouds, since they are located almost exclusively in the arms of the galaxy. A year or two later, the material ejected by the star at the high speeds measured by the Doppler effect has moved far enough from the star to be seen. After the star resumes its former brightness, it is at the center of a faintly glowing shell of hot gas called a *planetary nebula*. Figure 17-6 shows photographs of two planetary nebulae, each of which was probably produced by a nova several thousand years ago.

The distance of a nova can be calculated from the angular rate of expansion of the nebular shell in seconds of arc per year and the known speed of its material from the velocity measures. It is like the relation between the tangential velocity of a star and its proper motion and distance. Adapting equation (16-12) to this problem and solving for p or d, since $T = 4.74\mu/p''$ km/sec and $d = 1/p''$, we have

$$p'' = 4.74 \times \frac{\text{angular rate of increase}}{\text{velocity from Doppler effect}} \qquad (17\text{-}1)$$

and

$$d = \frac{1}{4.74} \times \frac{\text{velocity (km/sec)}}{\text{angular increase (sec/yr)}} \qquad (17\text{-}2)$$

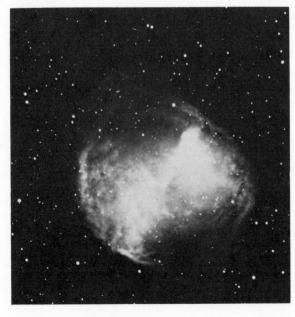

FIG. 17-6 Two planetary nebulae. The nebula in Monoceros (left) was photographed in red light. The small, black spots are dark globules—basic objects in the dust-cloud hypothesis of evolution. The nebula in Vulpecula (right) is known as the Dumbbell Nebula. (*Photographs from the Hale Observatories*)

EXAMPLE: If the measured velocity is 1000 km/sec and the angular rate of increase is 1″ per year, the distance to the nova is

$$d = \frac{1}{4.74} \times \frac{1000 \text{ km/sec}}{1 \text{ sec/yr}} = 211 \text{ parsecs} = 690 \text{ light-years}$$

After the distance has been calculated, it is possible to compute the absolute magnitude of the nova at any stage. At maximum the absolute magnitudes average about −7, or 12 magnitudes brighter than the sun. That is, the average luminosity of a nova is 60,000. At minimum the stars are thought to have about the same luminosity as the sun. That is, a nova brightens by about 60,000 times. If the star were hotter than the sun (class A or F), its diameter could be computed as was done from the H-R diagram. The average is about one fourth the solar diameter. Thus the prenova star is off the main sequence in the subdwarf class. At maximum it would be classed as a supergiant, about 60 to 200 times the diameter of the sun.

The novas seen by Tycho and Kepler were much brighter than the usual nova. Other similar novas have been found in distant galaxies, Andromeda and others. Their absolute magnitudes range from −12 to −18 and their luminosities from 6 million to 10 billion! These luminosities are greater than some entire galaxies composed of billions of stars. Such novas are called *supernovas*. Their velocities measure up to 3000 miles/sec, and it is probable that almost the entire mass of the nova explodes into space. It is estimated that a supernova should appear in a galaxy about once every 300 or 400 years. Since our galaxy has not produced a supernova since 1604, we are due for another one at any time. Let's hope it is not the sun! But it is a safe bet that the sun will not become a supernova for billions of years, if ever.

A well-known remnant of a supernova is the Crab Nebula, shown in Fig. 17-7. According to records from China, this supernova appeared in 1054. It is still expanding at a rate of about 0.21″ per year at a speed of 1300 km/sec. From formula (17-1) its distance can be computed as 1300 parsecs, or about 4250 light-years. The Crab Nebula is one of the strongest sources of radio waves from space. It contains a pulsar (see Section 1.2), a star that emits pulsating radio and optical waves with a period (in this case) of 0.033 second.

17.4 CEPHEID VARIABLES

The star δ Cephei was found to be a variable star in 1784 by the British astronomer John Goodricke (1765–1786). It is one of three stars in a small triangle in Cepheus and is circumpolar for northern latitudes. All three stars are visible to the unaided eye, but to study the variations of δ, a good pair of binoculars is recommended. The three stars can be compared in pairs. Two stars, ε and ζ, are constant in brightness, but δ varies in magnitude from 3.7 to 4.4 in a period of 5.37 days. (See map of circumpolar stars, Fig. 3-4.)

Other variable stars that have similar light curves have been found (almost 10,000 of them) and have been named Cepheids. The light-curve of a Cepheid

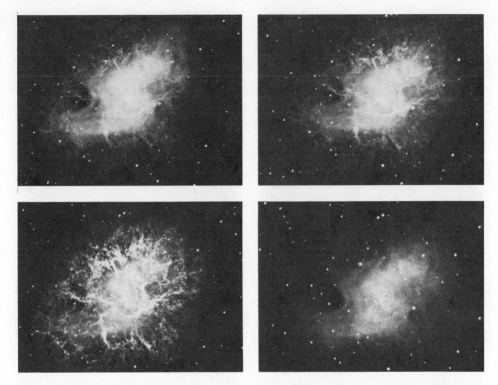

FIG. 17-7   The Crab Nebula photographed in blue, yellow, red, and infrared light. It is a remnant of a supernova and is still expanding. *(Photographs from the Hale Observatories)*

is not at all like that of a nova (see Fig. 17-8). The observations used for drawing the curve in the figure were made with a blue-sensitive photocell and are not quite the same as those made in yellow light, to which the eye is most sensitive. The variation in blue light is about 1.3 magnitudes and would be less if made in yellow light. The average variation of a nova is 12 magnitudes. Nova Puppis 1942 (Fig. 17-2) changed eight magnitudes between minimum and maximum light.

The light curve of a Cepheid is periodic; that is, the change in light repeats exactly (or almost exactly) in the period of a few hours or a few days. There is evidence that the periods of some Cepheids are getting shorter. These changes, of course, prevent the light curves from repeating exactly. To the eye, δ Cephei will be bright on one night, will fade to about half its maximum brightness during the next four nights, and will then regain its maximum brightness in about 1.5 days.

At the same time, there are variations in the spectrum caused by the Doppler effect. As shown in Fig. 17-9, the velocity curve has the same period as the light curve. The velocity changes by about 20 miles/sec in 5.37 days. The shapes of the two curves are almost mirror images of each other. This is typical of stars

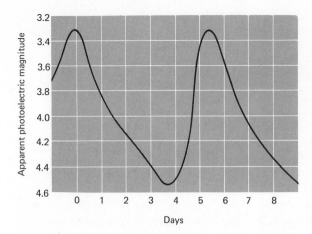

FIG. 17-8 A photoelectric light curve of Delta Cephei.

of this class. The variations in light are roughly one magnitude, although changes of less than 0.1 magnitude have been found for some Cepheids, including Polaris, the Pole Star. In most, but not all, cases the rise in brightness is faster than the decline, as in the case of δ Cephei. Some have distinct humps in the descending branch of the light curves (Fig. 17-10). The periods range from 3 to 50 days.

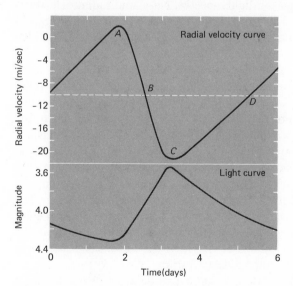

FIG. 17-9 Radial velocity and light curves of Delta Cephei, a well-known Cepheid variable star, interpreted as an expansion and contraction of the star. Maximum velocity inward is at A, maximum velocity outward is at C. The star is smallest at B and largest at D.

The spectral classes to which these stars belong are mostly F, G, and K, and their median (midway between maximum and minimum) absolute magnitudes vary from 0 to −5. In fact, they fit pretty well into the H-R diagram, the fainter stars being giants of class F with periods of about three days, ranging to super-giants of class K with periods of 45 days. But during the light variations, the spectral class changes by approximately one class. This indicates a change of temperature, amounting to about 1500°K in the case of δ Cephei. These variables

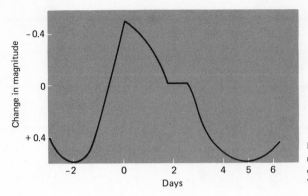

FIG. 17-10   Light curve of η Aquilae redrawn approximately from a photoelectric light curve by Stebbins and Wylie.

are hottest when they are at maximum light and coolest when at minimum. In addition, the lines in the spectrum oscillate about their normal positions during the light variations, as shown in the velocity curve. Near maximum, the lines are shifted the greatest amount toward the blue and near minimum toward the red. This indicates that the variable is brightest and hottest when expanding at the maximum rate and not when it is largest. It is coolest when contracting at the maximum rate.

The interpretation of the changes in light, spectral class, and radial velocity is as follows: The lines are produced in the part of the star's photosphere directed toward the earth, and are absorption lines. The Doppler shift toward the blue at maximum is now believed to be caused by a movement of the stellar material toward the earth, followed by a recession away from the earth as the star fades to minimum light. This indicates, on the assumption that the motion takes place in spherical shells, that the star is alternately expanding and contracting.

When the star contracts, the internal temperature and pressure increase to the point where they exceed a state of equilibrium. The star then expands to reduce these conditions. This expansion acts like a mild explosion, sending the material out to a distance where the balance is again upset, and the star contracts under its gravitational attraction. This contraction and expansion is a periodic adjustment of the pressure and temperature about conditions of equilibrium. The amount of pulsation depends on the size of the star. The smallest stars require the shortest times for their pulsations, and the periods increase in length as the size of the star and the amount of change of size increase. Since this theory was first accepted, these stars have been called *pulsating* stars.

Actually, this interpretation is a little too simple and does not fit all the facts. Martin Schwarzschild (1912–   ) at Princeton University has suggested that there is an internal pulsation but that the outer regions of the star do not expand and contract together with the inner regions. So there is a lag in the light curve. Also some lines in the spectrum are shifted toward the blue at the same time others are shifted toward the red. It appears that there may be two shells moving in opposite directions at the same time. In fact, the two shells seem to collide with a relative speed of some 100 km/sec. The humps in some light curves have never been satisfactorily explained.

The class of pulsating stars to which δ Cephei and η Aquilae belong is called *classical Cepheid*. The classical Cepheids that have periods between 2 and 10 days belong also to spectral classes F and G. Those with periods between 10 and 45 days are of classes G and K. They are larger and cooler than the fast-pulsating members of this class.

Besides the classical Cepheids just described, there are other variables with somewhat similar light curves. One group, the *RR Lyrae* or *cluster variables*, have periods of variation that are less than one day. Most of these stars are located in vast clusters of stars, globular in shape, that are very distant from the sun. The RR Lyrae stars, named for the variable star RR Lyrae, have absolute magnitudes near zero, belong to spectral class A or F, and are also assumed to be pulsating stars. Because of the difficulty of determining the parallaxes of these very distant stars by the trigonometric method, the scale of their absolute magnitudes is still uncertain. That is, their absolute magnitudes are near, but not necessarily exactly, zero. This scale will almost certainly be adjusted in the near future.

17.5 THE PERIOD-LUMINOSITY LAW

The relationship between period and luminosity of the classical and RR Lyrae variables is well established and has led to their use in determining distances where other methods fail. The relationship is shown in the curves of Fig. 17-11. It was discovered by Henrietta Leavitt (1868–1921) at the Harvard Observatory in 1912. She found that the fainter variable stars in the Small Magellanic Cloud have shorter periods than the brighter ones. At that time the distance to the cloud was not known, but later the absolute magnitude scale was fixed approximately by observations of stars of this type nearer the sun. All Cepheids were plotted on the same scale with absolute magnitudes between 0 and −4.

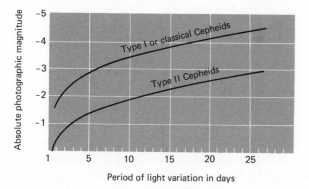

FIG. 17-11 The period-luminosity law for Cepheid variables.

The Magellanic Clouds are two large collections of stars about 20° from the south celestial pole. They look like detached pieces of the Milky Way, but are known to be collections of stars and masses of gas and dust. Another distant galaxy is faintly visible to the unaided eye in Andromeda (see Chapter 3 and Fig. 3-4). This is called the Great Galaxy in Andromeda. Our solar system also belongs to a galaxy comparable in size but slightly smaller. The stars in these two galaxies are strung mostly along spiral arms that come out of a central nucleus. There are also clouds of gas and dust in the spiral arms. (See Chapter 21 for a fuller description.)

Walter Baade (1893–1960) of the Mount Wilson-Palomar Observatories looked for cluster variables in the Andromeda Galaxy with the 200-inch telescope, but could not find any. He reasoned that they are too far away to be observed even with the largest telescope. He did, however, find stars of the classical Cepheid type. This led to the discovery that there are two distinct types of stars, which Baade called Population I and Population II. He estimated that the former are about 1.3 magnitudes brighter than the latter. This difference is shown in Fig. 17-11. Population II or Type II stars are found in globular clusters (which is the reason for calling them cluster variables) and are also found in the nucleus of our galaxy and in a halo around it. Some Type II Cepheids with periods between 10 and 30 days are also called W Virginis stars.

Type I or classical Cepheids are found in the spiral arms of our galaxy. Almost all the Cepheids in the neighborhood of the sun belong to this type. None of them have periods shorter than one day. They are supergiants from 400 to 5000 times brighter than the sun and their diameters are from about 20 to 300 solar diameters.

There is no difficulty in identifying RR Lyrae variables, since their light curves are unmistakable and their periods are less than one day. Hence, if the period-luminosity curve is right, their absolute magnitudes are zero. If the median apparent magnitude of a variable star of this type is known, equation (16-7) can be used for a determination of the distance.

EXAMPLE: Suppose a fifteenth-magnitude star in a globular cluster is found to be a variable star of this type. Rewriting equation (16-7) as:

$$5 \log d = m - M + 5$$
$$= 15 - 0 + 5 \text{ (in this case)}$$
$$\log d = 4$$
$$d = 10,000 \text{ parsecs} = 32,600 \text{ light-years}$$

This is the way the distances to the globular clusters have been determined. The value $(m - M)$ is called the *distance modulus*.

From the velocity curve it is possible to compute the change in diameter of a Cepheid variable. And from the absolute magnitude and temperature the maximum and minimum diameters can be found. In the case of δ Cephei, for example, the maximum diameter, which is 39.4 times the diameter of the sun, occurs at minimum light. This is after the star has expanded and the temperature has dropped to about 5500°K. Minimum diameter occurs near maximum light, where the temperature has risen to about 6700°K. Here the light is at maximum because the radiation is more effective at the higher temperature in accordance with Stefan's law. The diameter is then about 37.2 times the solar diameter. The change in diameter is about 6 percent and is about the same percentage as in other classical Cepheids.

17.6   IRREGULAR VARIABLES

Semiregular stars, as the name suggests, have periods that are not quite regular. Their light curves have irregular and unpredictable maxima and minima. Mira (o Ceti) was the first to be discovered and is the best known of this class. Mira means "wonderful." Its spectral class varies from M9 at minimum to M6 at maximum, and its temperature varies from 1900° to 2600°K. The period of Mira averages about 330 days, but the times of maximum or minimum cannot be predicted accurately in advance. The maxima vary from second to fifth magnitude and the minima from eighth to tenth. From Stefan's law, the change in temperature shows a variation of total energy by a factor of 3.5. However, most of this energy is in the infrared. The light in the visible region may vary as much as 15 times to as much as 1500 times, as indicated in Fig. 17-12. There are changes in the spectrum, but the real cause of the variation is not known. One explanation is that there is some irregular pulsation deep inside the star and a lag in the radiation due to the large atmosphere.

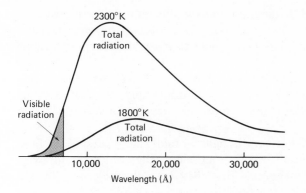

FIG. 17-12   The energy curves of a semiregular variable star at maximum and minimum temperature. The total radiation, the area under the curve, varies by a factor of about 2.5, while at the same time the visible radiation (the shaded area) shows a much greater amount of variation.

No periods can be found for the variation of the irregular variable stars. A study made at the Washburn Observatory of the University of Wisconsin showed that practically all stars of spectral class cooler than about M1 and brighter than about absolute magnitude −1 are variables of this type. The bright reddish star Betelgeuse in Orion belongs to this class, as do Antares, α Herculis, and μ Cephei. Betelgeuse varies irregularly by nearly one magnitude in the photoelectric (blue) region of the spectrum with a period of five or six years, with irregular, small variations in addition. During the light variations, the star also varies in diameter. The distance of Betelgeuse is very great and its parallax correspondingly small. Its angular diameter has been measured, and the linear diameter, which is very difficult to compute because of its uncertain parallax and absolute magnitude, may be nearly 1000 times that of the sun.

Other variables of this class have nearly constant brightness for days or even weeks, then flare suddenly by several magnitudes. These stars are known as *nova-like variables.* A star of this class that is well known to amateur observers is SS Cygni. R Coronae remains constant for long periods and then unexpectedly becomes fainter. After some irregular variations in brightness, it returns to normal brightness. The cause may be the formation of absorbing clouds of carbon in its atmosphere, since the spectrum shows strong bands of carbon. It belongs to spectral class R, which are yellow to reddish stars with spectra similar to G and K spectra with the addition of bands of carbon compounds.

T Tauri stars are red variables of rather low luminosity. They have rapid and irregular increases in brightness and are always found in dark nebular masses of dust and gas, such as the clouds in Orion. G. Haro and G. Herbig found variables of this class in a dark area in Orion where no star was found in photographs taken some years before. T Tauri stars have spectra with bright lines, mostly hydrogen and ionized calcium, and absorption spectra like the main-sequence stars from F8 to M. Those with type G spectra lie approximately on the main sequence, but those with spectral types K and M are two or three magnitudes brighter than typical main-sequence stars of those classes. It is thought that the T Tauri stars are young stars that may still be in the stage of gravitational contraction and have not yet come onto the main sequence. The T Tauri stars are located in young clusters and may be new stars in the process of formation.

The relations between the H-R diagram and the different classes of variable stars is shown in Fig. 17-13. The lines for Cepheids and the main sequence have been drawn straight; no attempt has been made to show the slight variations from linear relationships.

Some dwarf stars have been observed to flare. The flares last from a few minutes to perhaps half an hour. The cause of flaring is unknown, but it seems to take place in some part of the star's photosphere, much as the sun flares in areas of solar activity.

Some stars have been observed to vary in light, in total energy, and in the lines of their spectra. The light variations range from many magnitudes, as in novas, to such small fractions of one magnitude that they are detectable only by the sensitive photocell. The sun's total energy (the solar constant) varies by 1 or 2 percent. So the conclusion might be drawn that all stars are variable stars.

QUESTIONS AND PROBLEMS

Group A

1. What was the number in the order of discovery of the following variable stars in their respective constellations? (a) U Ophiuchi; (b) RZ Cassiopeiae; (c) ST Tauri. *Answer:* (b) eighteenth.

2. What are the advantages and disadvantages of photography, compared to photoelectric photometry, in the discovery and observation of variable stars?

3. Why is the cloud of gas called the Crab Nebula still expanding, 900 years after the explosion?

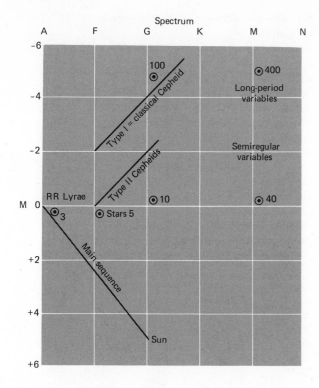

FIG. 17-13 The relation between the H-R diagram and the variable stars.

4. It is possible to determine the distance to a nova even though its absolute magnitude is not known, whereas for other stars the parallax or Cepheid-variable method must be used. Explain.

5. Can a better estimate be made of the number of novas, or supernovas, that will occur in our galaxy in a given year? Explain.

6. If a weight is hung on a spring, the position it occupies when the opposing forces of gravity and the tension in the spring are balanced and it is not moving is called the rest position. If the weight is pulled down and then released, it will oscillate about the rest position. (a) What kind of energy (kinetic or potential) does the weight have at the ends of its motion? (b) What kind of energy does it have when passing through the rest position? (c) Why does it not stop at the rest position? (d) The oscillation of the weight is somewhat analogous to the oscillations of the surface of a Cepheid-variable star. What are the opposing forces in the star that correspond to the opposing forces acting on the weight? (e) What condition would be necessary for the star to stop oscillating? (f) Answer questions (a), (b), and (c) in terms of the star's surface.

7. How could Miss Leavitt be sure that the variation in the average magnitudes of the Cepheids she observed in the

Magellanic Cloud was not due to differences in distance rather than related to the period?

8. In what ways does the light curve of a Cepheid variable differ from that of (a) a nova and (b) an RR Lyrae star?

9. The apparent magnitude of a Type I Cepheid variable is 1.7; its period is 10 days. (a) Find its absolute magnitude from the period-magnitude curve in Fig. 17-9. (b) What is the difference between its absolute and apparent magnitudes? (c) Compute its distance in parsecs. (d) Repeat for a Type I Cepheid of apparent magnitude +5.4 and a period of 30 days. *Answers:* (a) M = −3.3; (c) 100 parsecs.

10. From Fig. 17-9 find the absolute magnitudes of Type I and Type II Cepheids with periods of (a) five days and (b) 20 days. *Answer:* (a) Type I, −2.9; Type II, −1.4.

11. What error would be made in the computation of the distance of a Type II Cepheid, if its absolute magnitude were found from the curve under the mistaken assumption that the star was Type I?

12. Compute the distance to an RR Lyrae star, if its apparent magnitude is +20. *Answer:* 100,000 parsecs.

13. Is it possible that the variations in brightness of Cepheid variables are caused by giant sunspots? Discuss your answer.

14. What evidence indicates that T Tauri stars are young stars in the process of forming?

Group B

15. The nebular shell of a nova is observed to expand at a rate of 0.25″ per year. The velocity measured by the Doppler effect is 750 km/sec. What is the distance to the nova in light-years?

16. (a) Verify the quoted distance to the Crab Nebula by using the angular rate and velocity of expansion in the distance formula. (b) Compute the present diameter of the nebula in kilometers and astronomical units.

17. Refer to the magnitude and velocity curves of δ Cephei in Fig. 18-8. (a) What is the velocity of the star with respect to the sun? (b) What are the maximum velocities of expansion and contraction of its surface with respect to the center of the star? (c) Explain from the diagram why the star has its maximum and minimum diameters at the indicated points. (d) How can it be deduced that the surface temperature is lower at maximum size?

18. Show from Stefan's law that the energy radiated per unit area of Mira's surface changes by a factor of 3.5 during its cycle.

*19. Identify δ Cephei from a star map. Observe it every clear night with a good pair of binoculars. Try to estimate its magnitude by comparing its brightness with nearby stars and draw its light curve. Estimate the period of variation. Two stars near δ are ε Cephei, 4.2 F0 and ζ Cephei, 3.6 K0.

18 double stars

We have pointed out in describing the constellations that Mizar, the middle star in the Handle of the Big Dipper, is not single, but is really a double star. Gian Baptista Riccioli, an Italian astronomer, discovered in 1650 that it is double. This was about 40 years after the invention of the telescope. Since that time as larger and better telescopes were built, thousands of other double stars have been discovered.

William Herschel was one of the most famous double-star observers. Herschel, in his survey of the sky about the beginning of the 19th century, found that most double stars are unequal in brightness. He thought the brighter star was closer to the earth than the fainter one and proposed to determine its parallax by watching it describe a small ellipse around its companion in the course of one year. However, he found that in fact the two stars move around each other in periods of many years. He probably did not actually observe a complete period in any case, since all the stars he observed require years or even centuries to complete an orbit.

Several types of double stars are recognized. If two stars are accidentally lined up, one being at a much greater distance then the other, they are called *optical double stars*. But if they are actually close together and describe orbits around a common center of gravity, they are called *binary stars*. A *visual binary* is a binary star that is observed visually or photographically with a telescope.

If a binary system is composed of two stars too close together to be seen separately, or if the pair is so far away that they cannot be separated visually in any telescope, they may still be recognized as binary by spectroscopic observations. In that case the system is called a *spectroscopic binary*. Sometimes spectroscopic binary orbits are in line with the earth, so each star passes in front of its companion. They are then said to eclipse each other and the pair is called an *eclipsing binary* star. Thus the following classes of double stars are recognized:

1. Optical doubles.
2. Visual binaries.
3. Spectroscopic binaries.
4. Eclipsing binaries.

18.1 OPTICAL DOUBLES

Optical doubles are recognized by their common proper motions or by the fact that they have different proper motions. For example, the two pairs of stars in Lyra might be said to form an optical double star, since they have the same proper motion. It is possible that they are in orbit around each other; but this is improbable unless the orbital period is measured in thousands of years and is too long to produce even a short detectable arc of motion.

In another optical double star, the proper motions may be at right angles to each other and therefore easily recognized. The proper motion of each star would be along a great circle, appearing as a straight line. This motion would take the two stars away from each other in a relatively short time.

18.2 VISUAL BINARIES

Visual binary stars are recognized by their curved motions with respect to each other. In Fig. 18-1 the changing positions of the two stars in 12 years plainly show the binary nature of the pair. Three things must be known: the time of measurement, the angle between the two stars and a north-south line, and the distance between the two components in seconds of arc. A special instrument, the filar micrometer, is used on the telescope for these visual measures, or the angles and distances may be measured on photographs made for this purpose. A plot is then made, as in Fig. 18-2, showing the angle and distance for each date of measurement.

FIG. 18-1   Changing positions of a binary star in an interval of 12 years. The star is Kruger 60. *(Yerkes Observatory photographs)*

When the observations are plotted, usually as the fainter star around the brighter, the motion is in an ellipse. If the law of gravitation holds for binary stars, one star should lie at the focus of the ellipse. This does not always appear

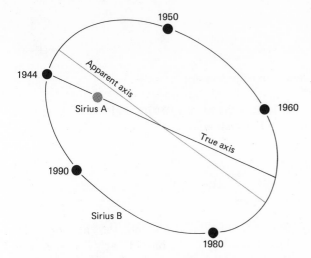

FIG. 18-2 The elliptical orbit of the companion of Sirius around the brighter star, Sirius A. The displacement of Sirius from the focus of the apparent ellipse is due to the inclination and orientation of the true orbit in space.

to be the case. But assuming that the law does hold and that the bright star does lie at the focus of the true ellipse, the displacement of the brighter star from the focus of the apparent ellipse can be explained as an effect of the orientation of the orbit in space.

From the apparent ellipse, the following data may be found:

1. The period P in years.
2. The semimajor axis a in seconds of arc.
3. The eccentricity e of the orbit.
4. The inclination of the orbit, i.
5. The orientation of the orbital plane in space.
6. The orientation of the ellipse in the plane.
7. The time of periastron, T; that is, the time when the two stars are closest together. (Compare this word with perihelion and perigee.)

Of particular interest in the study of double stars is the computation of the total mass of the system. In Chapter 4 Newton's modification of Kepler's third law was stated, and in Chapter 13 it was applied to the determination of the mass of a planet. It can be used in the same form to determine the combined mass of the two components in a binary star.

Let m_1 and m_2 be the masses of the two stars. Then

$$m_1 + m_2 = \frac{A^3}{P^2}$$ (18-1)

where A is the semimajor axis in astronomical units and P is the period in years. Since the semimajor axis is determined in seconds of arc, if the parallax of the stars is known,

$$A = \frac{a''}{p''} \text{ astronomical units}$$ (18-2)

or, rewriting equation (18-1),

$$m_1 + m_2 = \frac{a^3}{p^3 P^2} \qquad (18\text{-}3)$$

The total mass, $m_1 + m_2$, is expressed in solar masses.

EXAMPLE: The period of Sirius is 49.9 years, $a = 7.67''$, and $p = 0.374''$. Compute the total mass of the system.
SOLUTION: From equation (18-3),

$$m_1 + m_2 = \frac{(7.67)^3}{(0.374)^3 (49.9)^2} = \frac{8615}{2490}$$
$$= 3.46 \text{ solar masses}$$

Now if the center of mass of the two stars can be found, the ratio of their masses can be found from the relative sizes of the orbits of each star and the individual masses can be computed. Fortunately, this can be done in some cases, including Sirius. The binary nature of Sirius was first discovered when the bright star's proper motion was found to be a curve instead of a straight line. This is shown in Fig. 18-3, where the curves drawn are the proper motions of each component. The ratio of the heights of the two curves is inversely proportional to the ratio of the masses, which by measurement is found to be 2.5. Hence,

$$m_1 = 2.5 m_2$$

and $2.5 m_2 + m_2 = 3.46$ from the previous example. Hence

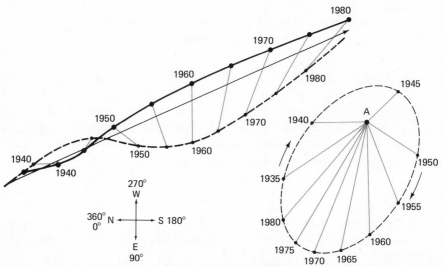

FIG. 18-3   The proper motions of Sirius A and B. The heavy curved line is for Sirius A, the dotted line for Sirius B, and the heavy, straight line represents the motion of the center of gravity of the pair.

$$3.5m_2 = 3.46$$

and

$$m_2 = 0.99$$
$$m_1 = 2.5 \times 0.99 = 2.47$$

both in terms of the sun's mass. This is one of the fundamental methods for determining the masses of stars.

Continuing the story of Sirius: The existence of the less massive star was predicted from the curved proper motion of the bright star. It was found visually in 1862 by Alvin Clark (1808–1887) while he was testing the objective for the 18-inch refracting telescope now at the Dearborn Observatory of Northwestern University in Evanston, Illinois. Its magnitude is +8.7, some 10,000 times fainter than its bright companion. This makes it very difficult to observe except when the two stars are near apastron (farthest apart) because the fainter star is nearly obliterated by the glare from the brighter star. The spectral type of the faint star was finally determined. This also was a difficult observation, since the light from the bright companion almost covered the spectrum of the fainter star.

By methods previously discussed, the diameters of the two stars can be computed from their absolute magnitudes and spectral types. The data have been collected in Table 18-1. Since Sirius B has an A5 spectrum and its absolute magnitude is +11.5, it is off the main sequence to the left. This classifies it as a white dwarf. It was the first white dwarf to be discovered.

TABLE 18-1   Summary of Data for Sirius

	brighter star Sirius A	fainter star Sirius B
Apparent visual magnitude	−1.43	+8.7
Absolute visual magnitude	+1.4	+11.5
Spectral type	A1V	wd(A5)
Temperature	10,500°K	8700°K
Diameter (sun's diameter = 1)	1.71	$0.024 = \frac{1}{42}$
Mass (sun's mass = 1)	2.47	0.99
Density (sun's density = 1)	0.50	73,000
Density (g/cm³)	0.65	96,000
Period (years)	50.0	
Average separation, a	$7.57'' = 20.2$ a.u.	

The most complete catalog of visual double stars, compiled by Robert G. Aitken (1864–1931), was published by the Lick Observatory in 1932. It lists 17,180 pairs, of which relatively few have yet been observed long enough to show orbital motion. It has been followed by supplementary catalogs from the Lick Observatory and others. Aitken's card catalog of double stars is still kept up to date by the astronomy department of the University of California. Some 23,000 stars are known to be double. A list of double stars for amateurs wishing to work in this field of astronomy will be found in the *Observer's Handbook*, published by the Canadian Astronomical Society. For an example of a curve from Aitken's catalog see Fig. 18-4.

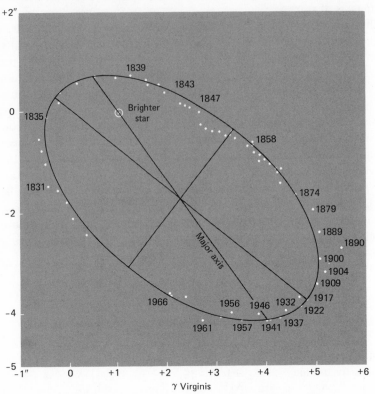

FIG. 18-4   Observations of γ Virginis from Aitken's catalog of visual binary stars. The curve drawn is an ellipse that passes approximately through the observed points. Those points off the curve may be poor observations, although it is remotely possible that there is a third star in the system.

18.3 SPECTROSCOPIC BINARIES

The spectra of most stars are made up of single absorption lines shifted toward the red or blue from their normal positions by the Doppler effect. The shift is caused by the component of the star's space velocity in the line of sight, as discussed in Chapter 16. An exception to this is the bright component of the visual double star, Mizar (ζ Ursae Majoris) in the Handle of the Big Dipper. In 1889 E. C. Pickering (1840–1919) at Harvard found that the lines in the spectrum of this star are alternately single and double (see Fig. 18-5). This was the first spectroscopic binary star to be discovered. The interpretation is as follows:

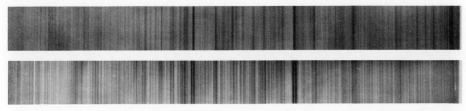

FIG. 18-5 Two spectra of the double-line spectroscopic binary star Mizar. (Yerkes Observatory photographs)

The bright component of Mizar is composed of two stars, which are equally bright. They move around a common center of mass in orbits of approximately equal size and shape ($e = 0.54$). The orbits are in a plane inclined at an angle of about 50° to the "plane of the sky," a plane at right angles to the line from the sun to the star. The period of revolution is 20.54 days and the orbital velocities measure up to 69 km/sec. However, because of the tilt of the orbits, their actual velocities are about 90 km/sec.

When one of the two bodies is approaching the earth, the other is receding. Thus their lines are shifted in opposite directions by amounts large enough to permit them to be seen separately. Five days later, both are moving at right angles to the line of sight and the two sets of lines are superposed. Five days later still, the lines are again separated. Thus they are alternately double and single.

The lines of a spectroscopic binary can be seen double if the velocities are large enough and if the two stars are not too different in brightness. If the magnitude difference is greater than about one magnitude, the lines of the fainter star are covered up by the continuous spectrum of the brighter star. If only one spectrum is visible, the binary nature of the system can still be detected, because the one visible set of lines is seen to shift back and forth in the period of revolution of the brighter star (and also the fainter one) about the center of mass.

Under ordinary circumstances the angle of inclination cannot be determined. A fast-moving star will show the same amount of shift, if the angle is small, as a slower-moving star with large inclination. If the binary happens to be also an eclipsing binary, the angle of inclination can be found.

In Fig. 18-6, upper, two stars are drawn in unequal, concentric circular orbits. The two components remain on opposite sides of their respective orbits, but travel with unequal velocities, 100 and 60 km/sec. When one star approaches, the other star recedes, as shown by the velocity curves. If the two orbits are on edge as seen from the earth, eclipses occur at each conjunction.

In the lower curve of Fig. 18-6, one star is too faint and its spectrum is not visible. The shape of the curve indicates that the orbit is elliptical, with the greatest velocity, recession, occurring at periastron. From the velocity curve, the velocity of the center of mass (K), the average velocity, the eccentricity, and position of the orbit can be found, except for the inclination. The curve at the right is a computed ellipse of eccentricity $e = 0.4$. It appears to the eye to be a

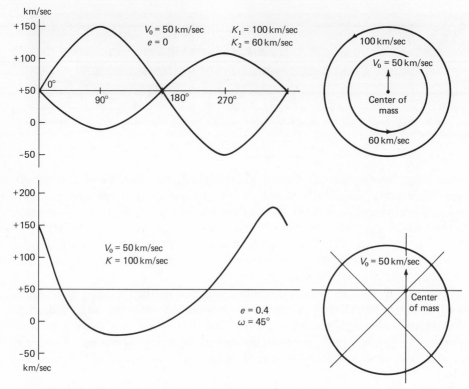

FIG. 18-6 Synthetic velocity curves: a double-line spectroscopic binary (upper) with equally bright stars but unequal masses traveling in concentric circular orbits; a single-line spectroscopic binary (lower) in elliptical orbit.

circle, but careful measurement will show that the axes are unequal. These curves are synthetic, but closely approximate those of some actual binary stars.

18.4 ECLIPSING BINARIES

An eclipsing binary star is always also a spectroscopic binary. The latter is not necessarily an eclipsing binary. The limiting conditions under which eclipses can occur depend on the angle of inclination and the relative size and distance apart of the two components of the system. If the orbital motion is at right angles to the line between the earth and the binary ($i = 0°$), obviously no eclipse can occur. If the orbital plane is lined up exactly with the earth ($i = 90°$), eclipses must occur and will be either total or annular. For intermediate angles of inclination, partial eclipses can occur when neither star completely hides its companion. If the stars are relatively close together and are of the same or nearly the same size, eclipses can occur for smaller angles of inclination than if the stars are smaller and farther apart. Partial eclipses are more frequent than total or annular eclipses.

When two stars in an eclipsing binary are equally bright, during totality the light of one star is prevented from reaching the earth; that is, the total light is diminished by one half. This decrease in brightness, equal to a decrease of 0.75 magnitude, is easily detected visually. Usually the stars are not equally bright, and one eclipse is deeper than the other. The deeper eclipse is called the *primary eclipse;* the shallower is the *secondary eclipse.*

If the two stars are equal in size, a total eclipse is instantaneous. This is not likely to happen, since the orbital inclination must then be exactly 90°. If the stars are unequal in size and the angle is near but not exactly 90°, the total phase can last longer and the light will be constant for that length of time. Between the time when the eclipse begins (first contact) and the beginning of totality (second contact) the farther star is partially eclipsed. The eclipse ends in the reverse order. Between two total eclipses, usually but not always halfway, the smaller star is projected onto the disk of the larger star and the eclipse is annular.

While total eclipses are more rare than partial eclipses, they are more important since they give the astronomer more information about the system. During a partial eclipse the light decreases gradually to a minimum and then returns immediately to maximum at the same rate as during the decline. The light loss can range from a small amount, detectable only by the most sensitive equipment, to deep eclipses of about four magnitudes. No case is known in which the fainter star is completely dark.

The computation of all the elements in an eclipsing binary system is complicated because of the numerous factors involved. For a complete solution, it is necessary that spectroscopic observations be added, that the lines of each component star be visible in the spectrum, and that the eclipses be total and annular alternately. In the most favorable cases the inclination of the orbit, the size, mass, density, and distance apart of the two components can be found, along with other minor data. For partial eclipses, only approximate solutions can be made.

The nature of the eclipses can be determined by the appearance of the light curves (see Fig. 18-7). In the case of AR Cassiopeiae (usually the three-letter

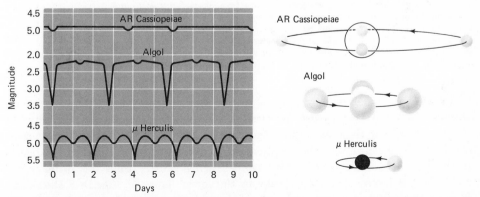

FIG. 18-7 Light curves of three eclipsing binary stars. An explanation of the type of eclipse is shown at the right. The size of the stellar components and their distances apart are drawn to scale.

abbreviation of the constellation is used) the minima are flat at the bottom, indicating total and annular eclipses. But which is which? Only a close inspection of each eclipse can give the answer. The primary eclipse of AR Cas is annular and the secondary is total. This cannot be determined from Fig. 18-7 because of the small scale. A detailed light curve (Fig. 18-8) shows that the secondary is flat and the primary eclipse is rounded at minimum. This is because the smaller star is completely hidden at secondary, while the larger star is darker at the limb than at the center and the loss of light continues during the annular phase as the smaller star moves across its areas of increasing brightness.

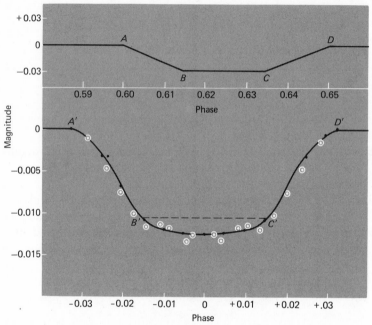

FIG. 18-8 Light curves of AR Cassiopeiae drawn from photoelectric measures and computations by electronic computers. A and A′ mark the beginnings of partial eclipses. B and C are the beginning and end of the total eclipse. B′ and C′ are the beginning and end of the annular eclipse. D and D′ are the ends of partial eclipses. The time scale (phase) at the bottom is given in decimal fractions of the period of the binary. (C. M. Huffer, University of Wisconsin)

In AR Cas the secondary eclipses do not occur midway between the primary eclipses, as would be the case for circular orbits. The unequal interval between eclipses is characteristic of eccentric orbits, as a result of the nonuniform motion (by Kepler's second law). The elliptical orbit also changes the lengths of duration of the two eclipses, the secondary eclipse being shorter than the primary, as shown in the figure. The dashed line in the lower figure of Fig. 18-8 shows the depth of the primary eclipse expected if the larger star were equally bright over its surface. The amount of darkening was calculated from the actual light curve.

The star is about 40 percent dimmer at the limb than at the center. The two stars are spheres, because the maxima are flat, as shown in Fig. 18-7. The observations were made through a yellow filter.

Compare the curve of AR Cas with the lower curve for μ Herculis. In the latter case the maxima are rounded at the top, because the stars are not spheres. They are elongated toward each other by tidal attraction, and the light varies continuously as the stars move in their orbits presenting areas of different size toward the earth. The same effect may be observed in the light-curve of AH Cephei, shown in Fig. 18-9.

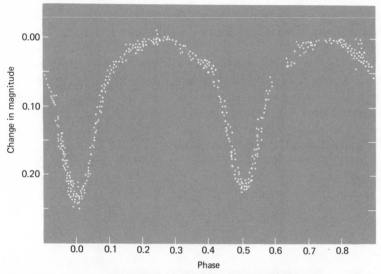

FIG. 18-9 A photoelectric light curve of AH Cephei taken through a yellow filter. (Huffer and Sievers, California State University, San Diego)

In the case of Algol, the light curve rises between primary and secondary. This is caused by reflection. The light from the brighter star is reflected by the fainter star in the direction of the earth, adding to the light from the dimmer star. This may also be a heating effect. The eclipses occur at equal intervals, indicating circular orbits.

The duration of the total or annular phases of the eclipses compared to the duration of the entire eclipse permits the determination of the relative size of the two components of eclipsing binaries. In AR Cas the larger star is about the same size as the sun and the diameter of the smaller star is only 40 percent of that of the larger star. Since the light curves are shallow, an exact value of the ratio is not possible.

There is no evidence of a reflection effect or of tidal distortion in the case of AR Cas. The smaller star is nearly four magnitudes fainter than its companion, so there is no possibility of photographing its spectrum. The mass and density of each cannot be computed accurately.

AH Cep shows very definite tidal distortion. The two components are very nearly of the same brightness and must be relatively close together. Computation shows that these stars are about 5.2 and 5.6×10^6 miles in diameter, with about 3.25×10^6 miles between their surfaces. Their absolute magnitudes are -3.0 and -2.7; the spectral types are B0 and B0.5. The brighter star has the smaller diameter. They are more than 1000 times brighter than the sun.

Algol was the first eclipsing star to be discovered. The name means the "demon star," and it is represented on old star maps as the head of Medusa held in the hand of Perseus. (See the star maps at R.A. 3^h; Decl. $+40°$.) Algol loses nearly two thirds of its light every 2.87 days. The eclipses, which are partial, last about 8 hours. The changes of brightness are observable with the unaided eye, but the observer must know when primary eclipse occurs. The two stars are slightly larger than the sun; the fainter star is a little larger than its brighter companion.

The problem of determining the temperatures of the faint companions of eclipsing stars has been assisted by observations made through colored filters. Figure 18-10 shows light curves made at the University of Wisconsin through the standard yellow and blue filters. Since the telescope used is a refractor, the ultraviolet filter could not be used, since ultraviolet light will not pass through a glass objective. The spectrum of the faint component of RZ Cas is too faint to show, but the depths of the two minima (yellow and blue) show that the fainter star is slightly cooler than the brighter. If the two stars had the same temperature, they would be the same color, there would be no change of color during the eclipses, and the light curves would have the same depth. As the curves show, the eclipses are partial and there is a slight darkening at the limb.

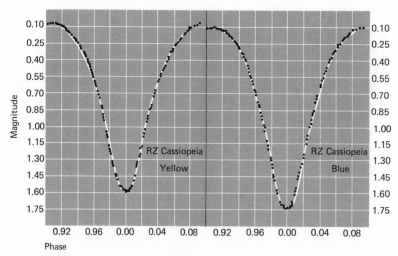

FIG. 18-10   Light curves of the eclipsing binary star RZ Cassiopeiae through a yellow and a blue filter. The light loss is expressed in magnitudes in the vertical scale. The time scale at the bottom is in decimals of the period, which is $1^d 5^h$. The eclipse is partial. (Light curves by Charles M. Huffer)

Another eclipsing star studied at the University of Wisconsin with a reflecting telescope shows light losses of four magnitudes in ultraviolet, three magnitudes in blue, and only two magnitudes in yellow light. In this case, since the fainter star is in front during totality, its spectrum can be obtained and the spectral type determined. It is listed in the catalogs as G5 and is in agreement with the observed change of color. The brighter star is A0.

Methods are available by which the elements of stars with accurate light curves can be determined by the solution of a complicated mathematical formula. Among the elements is the inclination of the orbit, i. This angle cannot be found for spectroscopic binaries that are not also eclipsing binaries. The semimajor axis, orbital velocity and mass were previously uncertain, but they can be accurately determined if the inclination is known.

Therefore, the study of eclipsing stars has been important in the investigation of star diameters, masses, and densities. If the data found from studies of all favorable cases of binary stars are put together, a curve such as Fig. 18-11 can be drawn, from which it can be seen that there is a relation between mass and luminosity of the stars. This curve shows that the brightest stars have the most mass; that is, it is the mass of a star that determines the amount of energy a star can generate. This curve is called the *mass-luminosity relation*. It is assumed that all normal stars conform to this relation; however, some—particularly the white dwarfs—do not, as the figure clearly shows. Most of the main-sequence stars conform, but about 10 percent do not. Fortunately, most of the nonconforming stars can be identified by some peculiarity in their spectra.

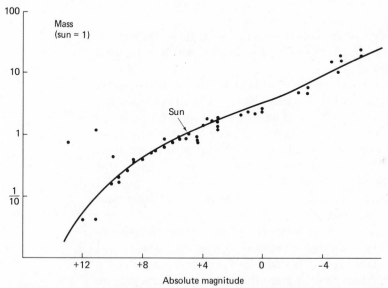

FIG. 18-11   The mass-luminosity relation. The three points to the left of the curve represent white dwarf stars, which have more mass than normal for their luminosities.

18.5 RESULTS OF BINARY STAR STUDIES

From studies of visual binaries and spectroscopic binaries, including eclipsing stars, the following results have been obtained:

1. The shortest period is 4^h 39^m for an eclipsing star in Nova Herculis 1934. To have a short period, the components of a binary system must be close together and must have large orbital velocities to counteract the strong mutual gravitation. The farther apart the stars are, the smaller are their orbital velocities and the longer their periods. The longest periods are as long as 30 years for some spectroscopic and eclipsing binaries to probably thousands of years for visual binaries that are very far apart and very slow moving. There is no essential difference between visual and spectroscopic binaries, except for the amount of separation. All types of binary stars are held together in pairs by their mutual gravitation.

2. Most eclipsing stars have periods less than 10 days, because the probability of an eclipse decreases as the separation increases. However, three or four eclipsing binaries with periods near 30 years have been discovered. The shortest period for a visual binary is about five years, although a few with shorter periods have been reported.

3. The eccentricities vary from zero for the shortest periods to about 0.9 for the longest-period visual binaries, increasing gradually with lengthening periods. Only a few eclipsing binaries with eccentricities other than zero are known; for example, AR Cas, as noted in Section 18.4.

4. The orientation of the orbits in space seem to be completely at random, except for the eclipsing binaries, which must have orbits near 90°.

5. Most binaries have both components on the main sequence. However, some visual binaries have white dwarfs as the fainter components. The two parts of eclipsing binaries are about the same size, mass, and luminosity, although in about 3000 known cases it would be expected that all combinations are possible. At least one case is known where the larger star is 40 times the diameter of the smaller one. Their magnitudes are also nearly the same, but a difference of at least four magnitudes is not uncommon.

6. Triple stars are usually composed of two close companions and one widely separated component. Algol, for instance, has an eclipsing pair with a period less than three days, and a third (and maybe a fourth) star orbiting the other two in more than one year. Another more recent discovery has been that two or three visual binary orbits show short-period motion superposed on the elliptical orbit. In one visual binary case,

61 Cygni, the variations have been attributed to the perturbations of a third body orbiting one of the two stars in the visible pair. This body must have a mass about $\frac{1}{60}$ the mass of the sun. The period is about five years, compared with 720 years for the two bright stars. This mass is too small to generate sufficient energy for the star to be luminous, so it is an invisible companion. It might be called a planet, since its mass is about 16 times that of Jupiter.

It is now possible to compare the sun with the other stars. The data come from many sources—photometry, spectroscopy, and mathematical and physical studies—and mostly from pairs of stars. It is evident that the sun is an average star in almost all respects. To simplify the comparison, the data are collected in Table 18-2. The entries for the maxima and minima of the stars are approximations only.

TABLE 18-2   Comparison of the Sun with Other Stars

	the sun	the stars (maximum)	the stars (minimum)
Apparent magnitude	−26.7	−1.42	?
Absolute visual magnitude	+4.87	−10	+19
Effective temperature	6000°K	85,000°K	1000°K
Color	Yellow	Blue	Infrared
Spectral type	G0	O5	I
Luminosity	1	1,000,000	1/400,000
Diameter	1	3000	1/400
Mass	1	50	1/20
Density (g/cm³)	1.31	10^6	10^{-10}

QUESTIONS AND PROBLEMS

Group A

1. Do both stars of a binary necessarily have the same period? Why?
2. Can the mass of a star with no companion be determined?
3. The total mass of a binary system is six solar masses. Star A is twice as far from the center of motion as Star B. Compute the mass of each star.
4. Do spectroscopic or visual binaries have shorter periods? Why?
5. Show by a diagram why the probability of an eclipse decreases as the separation of a binary increases.
6. From the velocity curve for a single member of a spectroscopic binary, how can it be determined (a) whether the orbit is elliptical and (b) when the star is at periastron?
7. Assume the two components of a binary system to be too close together to be detected visually. Under what conditions

could they be detected (a) spectroscopically only or (b) by light-intensity measurements only?

8. By means of a diagram show why the minima of a light curve have flat bottoms when caused by total or annular eclipses, but are rounded when caused by partial eclipses.

9. If the time interval between the primary and secondary eclipses of a binary is exactly the same, does it follow that their orbits are circles? Explain.

10. What characteristics of the light curve of an eclipsing binary indicate (a) whether an eclipse is total or annular, (b) whether the stars are spheres, (c) the relative surface brightness of the stars, and (d) whether the orbit is elliptical or circular?

11. Interpret the light curves of eclipsing binaries in Fig. 18-7. Compare the apparent magnitudes, depths of eclipses, and size of components. Specify the types of eclipses.

12. What is the value of making light curves of eclipsing binaries with different color filters?

13. Describe how the (a) separation of stars, (b) periods, and (c) eccentricities of orbits vary among visual, spectroscopic, and eclipsing binaries.

Group B

14. The period of the visual binary 70 Ophiuchi is 88 years. Its parallax is 0.199″ and its semimajor axis is 4.6″. Compute the total mass of the system. *Answer:* 1.6 solar masses.

15. What is the orientation of the orbit of a visual binary if, for an observer on the earth, (a) it is impossible to detect a Doppler shift, (b) the maximum space velocity of the components can be measured by the Doppler shift, and (c) the maximum space velocity of each component is equal to the tangential velocity?

16. Star A of an eclipsing binary has a temperature of 12,000°K and half the diameter of star B, whose temperature is 6000°K. What is the percentage decrease in brightness when star A is directly (a) in front and (b) in back of star B? *Answer:* (a) 5 percent.

*17. If, in Fig. 18-8, the partial phase of the eclipse begins at phase −0.030 and totality begins at phase −0.015 (decimals of the period), show that one star is 3 times the diameter of the other. (*Hint:* The stars are externally tangent at the beginning of the partial phase and internally tangent at the beginning of totality.)

*18. Using the method of question 18, Chapter 17, identify Algol. Observe it nightly and estimate its magnitude. Compare β with κ Persei, 4°N of β, 4.0 K0, and π Persei, 2°SW of β, 4.6 A2. Look up the times of minima in *Sky and Telescope* and watch the eclipse for an hour or two before and after minima.

19 star clusters and nebulae

Newton's law of gravitation explains why binary stars stay together; that is, because of their mutual attractions. A binary star may have one or more other stars held to it also by gravitation. These stars are usually quite far away, revolving around the pair in periods several times longer than that of the close pair. In addition to these triple or multiple stars, there are entire groups of stars, called *star clusters*, that are fairly close together in space. Their common bond is gravitation, or they may be moving together with nearly parallel proper motions like the Ursa Major cluster discussed in Chapter 16.

There are several recognized types of clusters, which range in number of stars from loose aggregates of a few stars to vast collections of stars too numerous to count in even the largest telescopes. Five classes are recognized (Table 19-1).

TABLE 19-1   Types of Star Clusters

class	example
1. Moving clusters	Ursa Major cluster
2. Open clusters	Pleiades
3. Globular clusters	M13, the great cluster in Hercules
4. Star clouds	Milky Way in Sagittarius
5. Star associations	Group of O and B stars in Orion

19.1 MOVING CLUSTERS

The first moving cluster to be discovered was a naked-eye group of stars in Taurus, near the first-magnitude star Aldebaran. It is a V-shaped cluster called the Hyades. Like the Ursa Major cluster, it was discovered by the common proper motions of its members, which will eventually carry them to a point east of Betelgeuse. This cluster contains at least 150 members, is some 30 light-years in diameter at its densest part, and is now 130 light-years from the sun. It is also classed as an open cluster.

19.2 OPEN CLUSTERS

Open clusters, as the name suggests, are groups of stars that are far enough apart to be easily resolved in the telescope (Fig. 19-1). They are also called galactic clusters, because they are in our Milky Way galaxy. A moving cluster might be called an open cluster, since the stars are always far apart. However, most open clusters are too far away to have measurable proper motions and are discovered by their appearance, rather than by their motions.

FIG. 19-1   An open cluster in the constellation Libra, photographed with the 200-inch telescope. (Photograph from the Hale Observatories)

The Pleiades (the Seven Sisters) is the best-known example of an open cluster. Six stars (or nine, under the most favorable observing conditions) are visible to the unaided eye. With binoculars perhaps 25 are visible, and with a telescope it is estimated that there are 250 members ranging from third to fourteenth magnitude. This cluster is embedded in nebulosity, not visible to the eye, but seen on a long-exposure photograph, such as Fig. 19-2, or detected with a photocell. This material is not self-luminous, but either reflects light from

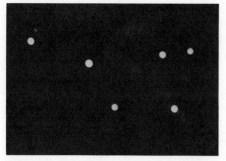

FIG. 19-2 An extrafocal photograph of the Pleiades, an open cluster, showing the brightest stars (left). A long-exposure photograph (right), shows the nebulosity around the Pleiades stars. In an extrafocal photograph the plate is placed near, but not in, the focus of the telescope. The star images are small circles, not points. The brightness of the star is determined by the blackening, not the size, of the image. (Yerkes Observatory photograph)

close stars or radiates by fluorescence. These processes depend on the distance away and the temperature of the stars. The stars in the Pleiades are of spectral class B, and the nebulosity is of the reflection type, probably composed of dust. The Pleiades cluster is also classed as a moving cluster.

There is a double cluster in Perseus, designated as h and χ Persei. It is about 9000 light-years away. Another open cluster is the Praesepe, or Beehive, cluster in Cancer. Still another is in the faint constellation Coma Berenices (Bernice's Hair) near the Handle of the Big Dipper.

The open clusters vary in size from 20 members to more than 1000. Some of the galactic clusters have common motions that are often opposite to the direction of motions of nearby fields of stars. It has been suggested that they will not always stay together, but will be broken up by gravitational attraction as the stars move in different directions. About 900 open clusters are known in our galaxy.

The distances of open clusters can be found, if they are also moving clusters. The diameters can be found when the distances are known. The diameters range from about 10 to 40 light-years, but the stars on the borders of a very populous cluster may be as much as 50 light-years from the center. The star density in the neighborhood of the sun is about one star per 11 cubic parsecs, if the components of double stars are included (see Table 16-1). The density of stars in open clusters range from 1 to 80 stars per cubic parsec.

For clusters that do not show measurable proper motions, it is more difficult to determine the distances. But there is another way even if the clusters are too distant to use the proper-motion method. In the H-R diagram the most conspicuous feature is the main sequence, where the stars are arranged by absolute magnitude and spectral class, or temperature. The spectral classes cannot be obtained if the stars are too faint. But temperature also determines the color of a star. With sensitive photographic and photoelectric photometry, the color index (Section 16.3, Table 16-5) of a star may be determined. When the color index is plotted against visual magnitude, as in Fig. 19-3, a line similar to the main sequence is obtained. Assuming that it is a main sequence, that the absolute magnitude of the sun is +5, and that its color index is 0.60 (see Chapter 16, page 324), the scale of the color-magnitude diagram may be fixed. From this information the distance can be found, as shown in the following example:

EXAMPLE: In Fig. 19-3 the color index of a certain star is +0.6. Its apparent magnitude is +15. Find the distance to the star.

SOLUTION: The most useful information from Fig. 19-3 is the absolute visual magnitude, which is assumed to be correlated with the color-index for all clusters. For color-index +0.6, the absolute visual magnitude, M, is +5.

Substituting in equation (16-7), which reads

$$M = m + 5 - 5 \log d$$

and solving for $\log d$, we have

$$5 \log d = m - M + 5$$
$$5 \log d = 15 - 5 + 5 = 15$$

and

$$\log d = 3$$

whence

$$d = 1000 \text{ parsecs} = 3260 \text{ light-years}$$

It is also possible to solve the problem in the above example without the use of logarithms. From Fig. 19-3, the absolute magnitude of the star is +5, or 10 magnitudes brighter than its apparent magnitude of +15. This difference

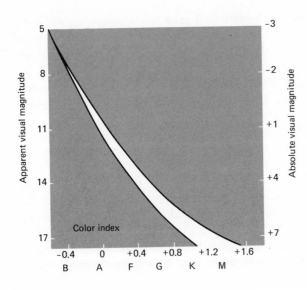

FIG. 19-3 Color-magnitude diagram for a typical hypothetical open star cluster. The main sequence of its component stars should lie inside the band outlined by the curved lines.

corresponds to a brightness ratio of 10,000 (see Section 16.4). The distance ratio is equal to the square root of the brightness ratio, or 100. Thus, the star is 100 times farther away than the standard distance of 10 parsecs, or 1000 parsecs as computed above.

For the general case, compute the distance modulus, $(m - M)$, as defined in Chapter 17. Reduce the distance modulus to the light ratio and take the square root to find the distance ratio.

19.3 GLOBULAR CLUSTERS

The stars in globular clusters are much more numerous and closely packed than in open clusters (see Fig. 19-4). Thirty thousand stars have been counted in the outer regions of M13, the great cluster in Hercules. The density in the center is so great that it cannot be resolved even in the 200-inch telescope. There must be at least 100,000 stars in a typical cluster of this type.

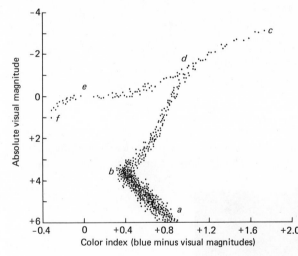

FIG. 19-4 Typical color-magnitude diagram of a hypothetical globular cluster. CD is the giant branch of the H-R diagram; AB is the main sequence. The stars are of Population II. Existing telescopes cannot reach the fainter stars.

There are several ways in which the distances to globular clusters can be determined. In the nearer and larger clusters, individual stars in the outer regions can be photographed. Many of these stars are found to be variables of the RR Lyrae type. Their absolute magnitudes are approximately zero, but there is a possibility that the scale might need further revision, possibly from 0 to +0.5. If this is the case, the distances of these clusters may be increased by a factor of 1.26. Still another correction is necessary. If a cluster is located behind a cloud of dust or gas, the cluster stars are actually brighter than they appear to be. This is because many of them are located near the Milky Way where such clouds are found. (See Section 19.6.)

EXAMPLE: In M13, the cluster in Hercules, the modulus is $m - M = 14.8$. The correction for absorption, K, is introduced into equation (16-7), as follows:

$$M = m + 5 - K - 5 \log d \qquad (19\text{-}1)$$

or

$$5 \log d = m - M - K + 5 \qquad (19\text{-}2)$$

In this case, $K = 0.4$ magnitude. Since the absorption makes the star appear dimmer and therefore more distant, the modulus must be decreased by 0.4 magnitude. That is, the modulus is $m - M = 14.4$, instead of 14.8. We may write this magnitude difference as

$$m - M = 10 + 4 + 0.4$$

and compute the brightness ratio by the use of Table 16-4. Thus,

$$\text{brightness ratio} = 10{,}000 \times 39.81 \times 1.445 = 575{,}000$$
$$\text{distance ratio} = \sqrt{575{,}000} = 760$$
$$\text{star's distance} = 760 \times 10 \text{ parsecs} = 7600 \text{ parsecs}$$
$$= 24{,}800 \text{ light-years}$$

Or, substituting in equation (19-2), we have

$$5 \log d = 14.8 - 0.4 + 5 = 19.4$$
$$\log d = 3.88$$
$$d = 7600 \text{ parsecs} = 24{,}800 \text{ light-years}$$

A second way of determining distances is similar to the color-magnitude method for open clusters (see Fig. 19-5). In this figure, the stars are all of

FIG. 19-5 Globular star cluster Omega Centauri. (Harvard College Observatory)

Population II and there are no young stars. This method requires the matching of the main sequence and also of the giant branches. These techniques lead to an average absolute magnitude for this type of cluster, which can then be assumed and substituted in equation (19-1) or (19-2), and the distance calculated. This will, of course, be an approximate distance. The apparent magnitude can be measured by photography with short-focus cameras, which make the image of a cluster as small as that of a bright star to which its magnitude is compared. Or, even better, the total light of the cluster can be measured with a photocell and compared with that of a nearby bright star. Then the magnitudes, with a correction for space absorption, K, can be used in the equation.

The absolute magnitudes of globular clusters range from about -5 to -9, averaging -7. The modulus is this figure subtracted from the apparent magnitude. The average globular cluster is 12 magnitudes brighter than the sun; that is, the average luminosity is 60,000. The brightest globular cluster is about 400,000 times brighter than the sun.

After the distances have been computed, it is possible to compute the size, if the angular diameters can be measured.

EXAMPLE: M13 has an angular diameter of 18'. Compute its linear diameter.

SOLUTION: Assuming its distance from the previous example as 24,800 light-years,

$$\frac{d}{D} \times 3438' = x'$$

from equation (10-3) adapted to this problem. Here $R = D =$ distance; $d =$ linear diameter; and the angular diameter has been substituted for p, the parallax. Substituting,

$$d = \frac{18' \times 24{,}800 \text{ l-y}}{3438'} = 130 \text{ light-years} = 40 \text{ parsecs}$$

The dense center, where the stars cannot be seen separately, is about 20 light-years in diameter.

Assuming that all globular clusters are about the same size, their distances can be determined by measuring their angular diameters and reversing the formula above. The distances range from about 15,000 to 200,000 light-years, although two clusters are known to be 400,000 light-years distant. They may not be members of our group of globular clusters.

Most of the globular clusters are located alongside the Milky Way. About 120 are known, but there are probably others behind the thick parts of the Milky Way. None is within 4° of the *galactic equator*, the central line (a great circle) of the Milky Way, because the light of any cluster in that direction would be totally absorbed by clouds of gas and dust in space. These clusters usually appear to be spheres, but some are slightly flattened by rotation.

A normal globular cluster has a spectrum much like the spectrum of the sun taken with a low-dispersion spectrograph, such that the lines are crowded together and only broad-lined features are shown. The color of a cluster away from the Milky Way is about the same as the color of the sun. Clouds of gas absorb blue light more strongly than they do red light. Therefore, a cluster behind a gas cloud is redder and also dimmer than one in which there is no material to interfere with the light. Clusters at 30° or 40° or more from the Milky Way are free from absorption and are yellow. Nearer the Milky Way, the clouds of gas and dust increase in thickness and the clusters are redder and fainter until their light is completely absorbed and none gets through. The difficulty in allowing for the absorption correction K is that the clouds are not uniform in density and it is necessary to measure the amount of absorption near the direction of each cluster.

The globular clusters form a nearly spherical system whose center is in the constellation Sagittarius and located at the center of the galaxy (see Chapter 20). Their motions indicate that they are moving independently about the center in highly flattened orbits with periods of hundreds of millions of years. It is thought that they come in close to the center of the galaxy and pass through regions thick with stars, gas, and dust. There is little probability of colliding with stars, but the gas particles in the clusters collide with the gas and dust in space and are swept up in the encounter. This may account for the fact that the globular clusters contain only Population II stars.

With knowledge of the size of globular clusters, it should be possible to calculate their densities—the number of stars per cubic parsec. However, it is impossible to count the stars at the centers. It is known that the number of stars increases greatly toward the center, with the central density something like 100 times the average. That is, there may be as many as 1000 stars per cubic parsec at the center.

The sky would indeed be very spectacular if we could see it from the center of such a cluster. The stars would be 10 times closer together than in our sky; they would, therefore, be 100 times brighter. There would be nearly 6000 first-magnitude and 1 million sixth-magnitude stars! There would be no obscuring material, and no Milky Way. There would be plenty of room for the stars to move around. Collisions would be infrequent, if not impossible. But the perturbations during close approaches would be very difficult to compute!

19.4 STAR CLOUDS AND ASSOCIATIONS

Star clouds are so called because the stars are so numerous and close together that they look like luminous clouds. The brightest clouds are located in the constellations Sagittarius and Scutum in the densest regions of the Milky Way. They are probably visible only because the clouds of gas and dust in the Milky Way, which are dense enough to hide the globular clusters, have open regions that permit us to see the stars behind. It is likely that the Milky Way has other star clouds hidden behind it.

The star clouds apparently have greater densities than the open clusters, but they are far less densely populated than the globular clusters.

It has been known for many years that the stars of classes O and B are not scattered at random, but occur in groups mostly in the Milky Way. The Russian astronomer, V. A. Ambartsumian, proposed that these stars are physically related and gave them the name *associations*. About 80 associations are known, each of which contains between 10 and 100 stars.

The most striking and first-recognized association is in Orion. This association is 1600 light-years distant and 400 light-years in diameter. In the Orion region are large clouds of material faintly visible in long-exposure photographs (see Chapter 1, Fig. 1-1). Since the association stars are all at about the same temperature, they are also about the same age. It is assumed that they were formed from clouds of dust and gas in the region at about the same time. Since they are hot, they convert their hydrogen to helium at a fast rate and must be fairly young—a few million years.

There are other associations—for example, in the Scorpius-Centaurus cluster, in Lacerta, and in Perseus—that are also young and very much like the Orion association.

Most associations lie in the Milky Way, where there is star streaming. That the stars are moving in streams was first discovered by J. C. Kapteyn (1851–1922), a Dutch astronomer, in 1905. It is now known to be a result of the rotation of the galactic system. The relative motions of the association stars and the general field stars in the Milky Way tear any group of stars apart in less than 100 million years. The effect on an association is to stretch it out along the Milky Way. Two or three stars have been found that are moving away from each other and from Orion. If they were originally part of the association, they have been in motion only 2.5 million years, a figure in agreement with recent theories of evolution of stars and the short lives of the O and B stars.

Table 19-2 is a summary of our knowledge of star clusters.

TABLE 19-2   Summary of Star Clusters

	open	globular	associations
Number known	About 500	More than 100	Less than 100
Estimated number	20,000		
Location in galaxy	Arms	Halo, nucleus	Arms
Diameter (parsecs)	$1.5 - 20$	$25 - 125$	$30 - 275$
Mass (sun = 1)	$100 - 1000$	$10^4 - 10^5$	$10 - 1000?$
Number of stars	$50 - 1000$	$10^4 - 10^6$	$10 - 100?$
M (absolute visual)	$0 - -10$	$-5 - -9$	$-6 - -10$
Stars per cubic parsec	$0.1 - 10$	$0.5 - 1000$	Less than 0.1
Age (years)	$10^6 - 10^{10}$	$10^9 - 10^{10}$	$10^6 - 10^8$
Examples	Pleiades, Hyades	Hercules, M13	Orion
	Praesepe	ω Centauri	Perseus
			Scorpius, Centaurus

19.5 NEBULAE

In the 18th century a French astronomer, Charles Messier (1730–1818), was looking for new comets. He was bothered by immovable objects that looked like comets and that he was continually "discovering." He therefore decided to make a list of these objects and in 1781 published a catalog, which is still referred to. Messier's catalog listed over 100 objects, which were called *nebulae*. The word nebula comes from a Latin word meaning clouds, mist, or vapor. But Messier included star clusters and galaxies, as well as true nebulae. We have previously mentioned M13, the globular cluster in Hercules. This is the thirteenth object in Messier's catalog. The Andromeda galaxy is M31. The Pleiades cluster is M45 and the Great Nebula in Orion is M42.

In 1890, Messier's list was combined with several others in a book called the *New General Catalog* (NGC), which was later continued under the name *Index Catalog* (IC). The NGC or IC number is used today in referring to all but the most familiar objects. M31 is also NGC 224. In recent years the number of photographic objects of these types has increased so rapidly that new discoveries are no longer numbered, but located on star charts, such as the *Palomar Sky Atlas* and others, by their right ascensions and declinations.

The spectra of the "nebulae" made it necessary to revise the designations. A star cluster is easily recognized. The spectrum of a cluster is much like the spectrum of the Milky Way, if both are taken with a low-dispersion spectrograph. That is, they have continuous spectra crossed by dark lines. Objects formerly called spiral nebulae also show this kind of spectrum, are therefore recognized as collections of stars, and are now called *galaxies*. The true nebulae, such as M42, either have bright-line spectra or reflect the spectra of nearby stars, possibly with some bright lines. They are now recognized as gaseous objects, not collections of stars. The luminous gas around the stars in the Pleiades cluster satisfies this definition of a nebula.

There are three kinds of true nebulae: bright, dark, and planetary. There is no essential difference between the first two classes. If a gas is near a bright star, it is luminous—a bright nebula. If there is no star near, the nebula is dark and can be "seen" only because it is silhouetted against the bright stars or gas in the background. The best example of a dark nebula is the Horsehead Nebula in Orion (Fig. 19-6). It will be noticed that the upper part of the photograph of this nebula contains bright nebulosity, whereas the lower part seems to be nearly devoid of stars, those that are visible being foreground stars, in front of the gas. The bright part is illuminated by the bright stars in Orion. Both bright and dark gaseous nebulae belong to a larger classification called *diffuse nebulae*.

The ability of a star to illuminate a mass of gas depends on its brightness and its distance from the gas. For example, a blue supergiant of absolute magnitude $M = -5$ can illuminate gas at a distance of about 30 light-years. A star as faint as the sun is effective only to a distance of one fourth of a light-year, much farther than the limits of the solar system, but well short of the distance of the nearest star.

If a star is cooler than class B1 (about 20,000°K), the light of the star is reflected by the gas, and the spectrum of the nebula is that of the spectrum of the star. A good example is the nebulosity around the Pleiades. If the star is hotter than 20,000°K, the nebula shows bright lines typical of a gas under low

FIG. 19-6   The Horsehead Nebula in Orion, an example of a dark nebula. *(Photograph from the Hale Observatories)*

pressure. In this case, we know that the density of the gas is a million million times less dense than the earth's atmosphere at the surface of the earth. Thus the atoms are very far apart and do not interfere much with each other, as they do in the stars. The gas absorbs energy from the nearby hot stars, stores it for a very short time (a microsecond), and then reradiates it in the natural frequency of the gas. This is called fluorescence. A good example is the Orion nebula, which has a bright-line spectrum and is associated with hot stars of class O and B. About 30 elements have been identified in the Great Nebula. (A 3-hour exposure of this nebula taken with the 100-inch telescope is shown in Fig. 19-7.)

At first some unidentified lines in nebular spectra were attributed to an unknown gas, called nebulium. These lines are now known to be similar to the lines in the solar corona and are produced by doubly ionized oxygen and neon and singly ionized oxygen and nitrogen. The strong lines in the green are responsible for the colors of nebulae. There are also strong lines in the blue and in the red, but they do not affect the color very much, since the human eye is more sensitive to green and yellow light than to the other colors. Weak lines due to other elements are also found in the spectra of the brighter nebulae.

There are some dark nebulae in the Milky Way. In the constellations Cygnus, Sagitta, and Aquila, the Milky Way appears to be split into two parts. The western part fades and almost disappears in Ophiuchus and Scutum, but reappears in Scorpius. This apparent split is called the Great Rift. In two areas, dark regions are found to be almost completely devoid of stars. The northern one

FIG. 19-7   A 100-inch telescope photograph of the Orion Nebula, an example of a reflection nebula. *(Photograph from the Hale Observatories)*

is in Cygnus, the Northern Cross, and the southern one is near Crux, the Southern Cross. These dark areas are so black in contrast with the bright areas that they are called Coalsacks. It is a coincidence that they are found near the two crosses.

Barnard was among the first to photograph the Milky Way. In certain areas other dark regions were once thought to be holes between the stars where the observer looked into "the blackness of space." Barnard suggested that these "holes" are regions where the light of distant stars is absorbed by invisible clouds in space. His work was continued and extended by Frank Ross (1874–1960), also of the Yerkes Observatory. The dark areas are dark nebulae. The Great Rift can be traced across Fig. 19-8 and some of Barnard's dark nebulae can be found on Fig. 19-9.

19.6 INTERSTELLAR DUST

Robert Trumpler (1896–1957) at the Lick Observatory investigated the open clusters and noticed that, although they all seem to be the same size, their stars are not all the same brightness. He also noticed that stars are redder in certain areas than in others. This discovery led him to suspect that there are clouds of obscuring material in addition to Barnard's dark nebulae. This work was

FIG. 19-8   A portion of the Milky Way in Cygnus, showing the Great Rift. *(Photograph from the Hale Observatories)*

followed by two-color photometry by Joel Stebbins (1878–1966) and his associates at the Washburn Observatory in Wisconsin and at the Mount Wilson Observatory in California. They later used three- and six-color photoelectric measures. This work was followed by others using similar techniques.

The technique in colorimetry (the accurate measurement of color) is to measure the amount of starlight that passes through colored filters. If two or more filters are used, the color index of a star can be determined (see Chapter 16). If stars of class B are observed (the spectrum determines the class), these stars can be seen and identified to distances of perhaps 1000 parsecs. The nearby stars and those well away from the Milky Way are blue, as is normal for class B, and their color indices are near zero, as listed in Chapter 16. However, as the Milky Way is approached, the colors become redder until, in some regions of Sagittarius, the B-stars are as red as K-stars. This is because the blue light is absorbed and scattered more strongly than the other colors and the red light gets through. This is similar to the reddening of the sun at sunrise and sunset, but of course the stellar reddening takes place out in space. The reddening effect of the earth's atmosphere is allowed for in computing color indices. By this method,

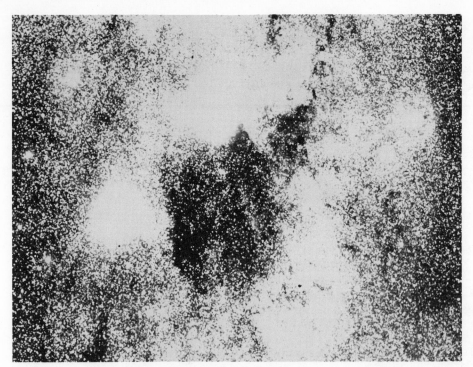

FIG. 19-9   The region of the Coalsack in Cygnus. *(Yerkes Observatory photograph)*

clouds of absorbing material are located by a method different from those already mentioned.

The difference between the measured color index of a star and the normal color index of an unreddened star of the same spectral class is called its *color excess.* The color excess determines whether or not the star's light is selectively absorbed in its journey through space. It has been found that the total absorption is roughly three times the color excess, if both are expressed in magnitudes. This is the correction term K in equation (19-1).

Studies by photography and photoelectric photometry show that clouds of obscuring materials are concentrated in the plane of the Milky Way. The stars near the sun and at some distance from the galactic equator are assumed to have normal color indices, but there is some evidence that the sun itself is in a region not entirely free from obscuring matter. The stars would be about twice as bright as they are now, if the sun were in dust-free space.

The discovery of black globules, which had also been photographed by Barnard, led to the dust-cloud theory of evolution of the solar system discussed in Chapter 15.

Near the beginning of the 20th century, German astronomer J. Hartmann, in his studies of spectroscopic binaries of class B, found that the lines of calcium did not show the changes of velocity as did the other lines in the spectrum. He also found that these lines are sharper (narrower) than the others. It was decided that the calcium atoms absorbing those wavelengths were not in the stellar atmospheres but in clouds of calcium between the stars and the earth. Other similar clouds in space were detected by Otto Struve and others at the Yerkes Observatory. The lines in the violet region of the spectrum are produced by singly ionized calcium. The density is very low, resulting in sharp lines. Other atoms identified are sodium, potassium, and even iron. It is possible that molecules of hydrocarbon (CH) and cyanogen (CN) are present at very low densities and in small quantities. Also some of the stationary lines may be produced by shells of gas around binary stars.

19.7 PLANETARY NEBULAE

There is a subclass of O-type stars called Wolf-Rayet stars for the two men who discovered them. The class designation is W. About 200 of these stars are known. They are among the most luminous of all stars, except novas and supernovas, with absolute magnitudes −4 to −8. They are also very hot, having temperatures estimated between 50,000°K and 100,000°K. Their spectra consist of greatly broadened, diffuse emission lines of highly ionized helium, nitrogen, oxygen, silicon, and carbon, with absorption lines on the violet side of each. A few W-stars are components of binary systems.

One explanation of the physical nature of the W-stars is that there is a central star 1.5 to 2.0 times the diameter of the sun, surrounded by a transparent atmosphere about 3 or 4 times larger. The velocities indicate that this shell is expanding, since the absorption lines produced by the atmosphere between the star and the earth have Doppler shifts, indicating high velocity of approach. It has been suggested that the envelope has been ejected from the star with velocities as high as 1000 km/sec.

Other stars show evidence of similar shells. At times the star Pleione in the Pleiades has shown bright lines, which later disappeared. The shell may have been blown away by the radiation pressure from the intense heat. The bright star Deneb shows similar bright lines.

A few binary stars are known to have envelopes of gas. Probably the best-known example is β Lyrae, an eclipsing star with a period of 12.9 days. The two components are elongated toward each other by tidal attractions and there are streams of gas being exchanged between them. The entire system is surrounded by a disk of diffuse gas that rotates around the center of mass of the binary system. It is estimated that the more massive and brighter star is losing mass at the rate of about one solar mass every 6000 years. Since its total mass is about 50 solar masses, the total mass would be lost in 300,000 years. However, it is probable that the mass loss will continue only until the two stars have equal masses.

It is quite obvious that many stars that appear to the eye to be single, uninteresting stellar bodies are actually double or multiple, with complicated atmospheres in violent motion.

A group of short-period eclipsing stars consists of pairs that are small and close together. They are known as W Ursae Majoris stars, because W UMa was the first of this type to be discovered. Its period is only 8 hours, and the surfaces of the two stars are nearly touching. They are thought to be embedded in nebulosity. The two stars are nearly equal in size and brightness, but not in mass, so there is an exchange of material between the two.

It is evident from this discussion that the size of stellar envelopes varies from the smallest pairs of the W UMa type to the large ones like the W-stars and on to the envelopes around novas and supernovas. The largest shells of gas are found in the *planetary nebulae*. The name is intended to suggest that they have visible detail, but not that they are like planets. The best-known example is the Ring Nebula in Lyra (Fig. 3-5). The largest in angular diameter and the nearest is NGC 7293 in Aquarius (Fig. 19-10). This is the only planetary whose parallax has been determined. It is 85 light-years distant and appears about half as big as the moon. It is very diffuse and difficult to see and is not listed in Messier's catalog, so he must have missed it.

FIG. 19-10 The planetary nebula NGC 7293, photographed in red light with the 200-inch telescope. (Photograph from the Hale Observatories)

About 500 planetary nebulae are known, and there must be thousands of others that have not yet been photographed. They vary in size from NGC 7293 to small, starlike objects recognized as nebulae by the bright lines in their spectra. The greenish color results from the strong lines of doubly ionized oxygen in the green region of the spectrum. These lines are emitted by gas at extremely low density. The density of planetaries is estimated to be only about 10,000 atoms per cubic centimeter.

The planetary nebulae vary from about 20,000 to 200,000 a.u. in diameter— several hundred times the diameter of the orbit of Pluto. Since the parallaxes cannot be measured directly, it is necessary to use indirect methods to determine their distances and diameters. The statistical study of their proper motions and radial velocities show them to be very distant and probably associated with stars of Population II. If the number of atoms in a unit volume can be estimated from the spectrum, and the volume calculated from the diameters measured on photographs and combined with estimated distance, the mass of the atmospheric shell can be calculated. The average is about 0.1 or 0.2 times the mass of the sun.

Inside each nebular shell is probably a star, although the star is sometimes too faint to photograph. The nature of the star in NGC 7293 is fairly well known. Its absolute magnitude is +9.8. Its blue color shows that it is a star of spectral type O. Since it is much fainter (330,000 times) than the usual O-type star, it must be a very hot white dwarf. Its size can therefore be computed by the usual method. Its diameter is only $\frac{1}{60}$ that of the sun and its density 300,000 times the density of water! The ionization of oxygen in the nebular shell is produced by the intense ultraviolet radiation from the star. This nebula is probably a typical one, representing a stage in stellar evolution in which a white dwarf is produced by the explosion of a nova.

From the number of known nebulae of this class it has been estimated that three stars per year in our galaxy become planetaries and three planetaries become white dwarfs. Thus in 5 billion years—the estimated age of the solar system—15 billion white dwarfs have been produced. This is not an unreasonable figure, since the white dwarfs are too faint to be seen at distances of more than a few parsecs, and must be very numerous.

19.8 THE CRAB NEBULA

One final object should be mentioned, since it is receiving a great deal of attention from radio-astronomy observers. The Crab Nebula (M1 in Taurus) has been classed as a planetary nebula, but the motions in its gaseous envelope are much greater than in the typical planetary. It is known to be expanding outward at a rate of some 1300 km/sec. From this rate of expansion it has been identified as the expanding shell around the supernova of 1054, which was brighter than Venus and was visible for two years. Its distance is about 1300 parsecs. Photographs of the Crab Nebula taken in four colors are reproduced in Fig. 17-7.

This nebula is occasionally occulted by the moon and is used to study the possible existence of a lunar atmosphere by its refraction effect. No such effect was found, however.

A pulsar of very short period (0.033 second) was found in the Crab Nebula as noted in Chapter 1. The discovery was made by Staelin and Reifenstein at the U.S. National Radio Observatory in October, 1968. Its position agrees exactly with that of the star in the nebula thought to be the supernova that produced the nebula, the remnant of the explosion of A.D. 1054. This pulsar was the first to be detected photoelectrically in visual light. The discovery was made by W. J. Cocke, M. J. Disney, and Donald J. Taylor at the University of Arizona Observatory on January 15 and 16, 1969. It was confirmed a few days later at the McDonald Observatory in Texas, the Kitt Peak Observatory in Arizona, and the Lick Observatory in California. (See Fig. 19-11.)

The Crab Nebula is one of the strongest sources of radio waves in the sky. It is also a strong source of X rays. The distribution of the nebula's optical,

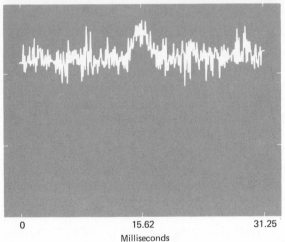

| 0 | 15.62 | 31.25 |

Milliseconds

FIG. 19-11 The Crab Nebula, showing the location of the first optical
pulsar (top). The first recording of a
light pulse from a pulsar, taken on
January 17, 1969, is shown in the lower
illustration. (Upper photograph from
the Hale Observatories; lower photograph courtesy Donald J. Taylor, University of Arizona)

radio, and X-ray wavelengths is the same as would be produced by electrons
spiraling at nearly the speed of light around magnetic field lines. However, the
source of the high-energy electrons was a mystery until the discovery of the
pulsar. Observations made at a time when the nebula was being occulted by the
moon showed that the peak of the X-radiation is at the center of the optical
pulsar. This suggests that the pulsar is the source of the energy. The fact that the
frequency of pulsation is slowly decreasing at a steady rate indicates that the
pulsar is losing energy.

A second pulsar is located only 1.2° from the pulsar just described. Its
period is 3.745^s, the longest known to the date of discovery. Other pulsars have

been discovered in the remnants of other supernova explosions, suggesting that the pulsars are formed as a result of the explosions.

The X rays also vary with the same frequency as that of the pulsar, whose pulsed radiation is approximately 5 or 6 percent of the total X-ray radiation of the entire nebula. The interpretation is that the nebula contains an extremely small neutron star, the pulsar, and that the pulses are produced by the rotation of the star (Fig. 19-12).

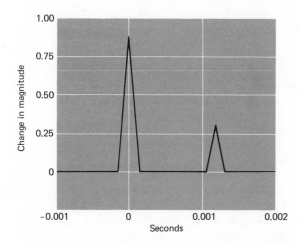

FIG. 19-12 An approximate light curve of the pulsar in the Crab Nebula.

When a star has used up all of its energy-producing material—hydrogen, helium, and other products formed in the evolutionary process—it collapses rapidly. The resulting pressure forces the electrons and protons to combine into neutrons. The neutrons may then become packed to a density of 10^{12} to 10^{15} times the density of water. A mass equal to the mass of the sun would occupy a spherical volume only 10 to 20 miles in diameter. This mass can then rotate at a rate of a few seconds or less—the rate of variation of the pulsar.

QUESTIONS AND PROBLEMS

Group A

1. How would the motions of the stars in a small moving cluster appear to an observer on the earth if the stars were moving in parallel paths (a) directly away from the sun, (b) directly toward the sun, and (c) at right angles to the direction of the sun?

2. List the possible sources of error when determining the distance to a cluster by means of the color-magnitude diagram.

3. In the color-magnitude diagram of an open cluster it is found that the apparent magnitude of main-sequence stars of color index 0 is +16. Assuming these stars to be similar to those on the main-sequence of the H-R diagram in Fig. 16-3, find the

absolute magnitude of the stars and compute the distance to the cluster. *Answer:* 10,000 parsecs.

4. What are three ways of determining the distance to a globular cluster?

5. If the earth were located in the center of a globular cluster, how would the appearance of the sky change from our present view? Compare the apparent (a) number, (b) brightness, (c) color, and (d) distribution of stars.

6. If the average absolute magnitude of the stars in the faintest globular cluster (total $M = -5$) listed in Table 19-2 is $+5$, show that the cluster should contain 10,000 stars.

7. What is the diameter of a globular cluster that is (a) 20,000 parsecs from the earth and has an angular diameter of 15', and (b) 30,000 parsecs from the earth and has an angular diameter of 5"? *Answer:* (a) 280 light-years.

8. Explain in terms of the photon theory why a star must be of spectral class O or B1 to cause fluorescence in a nearby nebula (see Section 7.9).

9. How does the spectrum of nebular light reflected from nearby stars differ from that due to fluorescence?

10. How does the modern definition of the term nebula differ from that used by Messier? Why did Messier include such diverse objects in his catalog under the heading of nebula?

11. The measured color index of a star is $+1.6$, whereas the normal color index of a star of the same class is 0.6. How much brighter would the star appear if there were no loss of light due to absorption?

12. Why do the Wolf-Rayet stars show bright-line spectra, whereas most stars show dark-line spectra?

Group B

13. There are 1000 stars per cubic parsec in the center of a globular cluster. (a) What is their average distance apart? (*Hint:* Assume the stars to be uniformly spaced throughout a cube, 1 parsec on a side.) (b) How many suns could be placed between two adjacent stars, if the suns were touching each other? *Answer:* (b) 2,200,000.

14. Compare the diameter of the central star in NGC 7293 to that of the usual star of its spectral class. *Answer:* 1/570.

15. Calculate the average density of the white dwarf star in NGC 7293, assuming that it has the same mass as the sun.

20 the milky way galaxy

The Milky Way galaxy is the great system of stars of which the sun, the earth, and the solar system are a part. We see this galaxy from the inside. The various kinds of stars, gaseous nebulae, and star clusters it includes have been defined and described in previous chapters. We now examine in more detail their locations and motions, and consider the galaxy in its entirety.

20.1 GALACTIC COORDINATES

In order to describe the location of objects with respect to the Milky Way and to study the distribution of stars and other objects, a great circle has been drawn on the celestial sphere, following as nearly as possible the center line of the Milky Way. This great circle is called the *galactic equator*.

The galactic equator is inclined to the celestial equator at an angle of 62°. It reaches declination $+62°$ at right ascension $0^h 49^m$ in the constellation Cassiopeia, and declination $-62°$ at right ascension $12^h 49^m$ in the Southern Cross. The distance in degrees from the galactic equator to any point on the celestial sphere, measured on a perpendicular great circle, is called the *galactic latitude*.

Galactic longitude is a second coordinate in the galactic system. It is measured northward along the galactic equator from the galactic center, a point that will be defined later in this chapter. The right ascension of the center is $17^h 42.4^m$ and its declination is $-28° 55'$. This point has galactic longitude 0° and galactic latitude 0°. Galactic longitude is measured from this point to the great circle perpendicular to the galactic equator through the point in question. Galactic latitude is measured from the galactic equator along this perpendicular great circle to the point. Both coordinates are measured in degrees. The north and south poles of the galaxy, two points 90° from all points of the galactic equator, are at right ascension 12^h 49^m and declination +27.4° and right ascension 0^h 49^m and declination −27.4°, respectively.

The Milky Way passes through the following constellations, from north to south: Cassiopeia, Cepheus, Cygnus, Sagitta, Aquila, Serpens, and Sagittarius, all of which are visible from the northern hemisphere in the summer. Continuing toward the south and below the horizon are Scorpius, Centaurus, Crux, and Vela. The greatest star density is in Sagittarius and Scutum, where the great star clouds are located. Figure 20-1 is a composite of the Milky Way with the star clouds near the center. The Great Rift is shown extending from Cygnus to Sagittarius.

FIG. 20-1 A map of the Milky Way drawn at the Lund Observatory, Sweden, by Martin and Tatjana Keskūla under the direction of Knut Lundmark. The panorama, which took almost two years to complete, has 7000 stars on it. Although photographs were used as a basis for the map, the panorama makes greater uniformity possible than would a composite photograph. The galactic center is in Sagittarius, with Auriga at either end. *(Lund Observatory, Sweden)*

Continuing along the galactic equator northward from the center into the winter sky, the constellations are, in order: Puppis (part of Argus, the ship), east of Canis Major; Monoceros, east of Orion; Auriga, near Perseus; and back to Cassiopeia. These constellations contain some very bright stars, but the Milky Way itself is faint and difficult to see.

It can also be seen from Fig. 20-1 that the western branch, the upper side of the figure, disappears to the left of center. This is the region of Ophiuchus (above) and Scutum (below). The dark areas are produced by absorption by the dark clouds mentioned in Chapter 19. It will also be seen in the photograph that the number of stars decreases greatly from the equator to the poles, which are at the top and bottom of the figure. Also, the number decreases towards the constellation Auriga at each side. A point 180° from the galactic center is the *anticenter*. The Milky Way is widest in Sagittarius and Scorpius (about 40° wide) and narrowest in Auriga.

The *galactic center* is the point that most nearly represents the center of the Milky Way by star counts. It is also the center of the rotating system of stars, the galaxy, and the system of globular clusters. The International Astronomical Union decided in 1958 that the best representation of this point is in the direction of Sagittarius, where there is heavy concentration of interstellar clouds that prevents the center itself from being observed by optical instruments. It can be observed, however, with radio telescopes.

20.2 GALACTIC CONCENTRATION

The first systematic attempt to investigate the structure of the galaxy was made by William Herschel, who made star counts—he called them star gauges —for 19 years and published his results in 1785. He divided the sky into star fields and counted the stars in each. His "universe" is represented in Fig. 20-2. Herschel did not know any stellar distances. Since he observed in England, he could not see very far below the celestial equator. He located the sun near the center—shown by the large dot slightly to the left of center on the diagram. The split in the Milky Way is shown to the right. He did not explain what he meant by the large dots around the edges, but perhaps intended them to show the positions of the globular clusters that lie along the edges of the Milky Way.

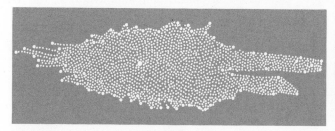 

FIG. 20-2   The universe of Sir William Herschel. The sun is represented by the heavy dot to the left of center. *(Yerkes Observatory photograph)*

Herschel counted stars in 683 selected fields and obtained star densities ranging from one to about 600. These gauges show the galactic concentration, if the galaxy is considered to be a flat disk with the sun near the center. In Herschel's diagram, the number of stars is much greater if one looks to the right and left of the sun on the drawing rather than up and down. This work was continued in the southern hemisphere by his son, Sir John Herschel (1792–1871), who took his father's 20-foot-long telescope to South Africa. These two men could have counted 200 million stars with their 40-foot telescope instead of the smaller instrument. The planet Uranus was discovered with a smaller telescope (8-inch aperture), which Herschel preferred to the larger (48-inch aperture) telescope because it was easier to operate.

The star gauges were made by the Herschels visually, before the invention of photography. Star counts have been continued from photographs made in various parts of the sky. One difficulty with counts is the obscuration by dust and gas clouds in space. As was pointed out in Chapter 19, these clouds increase in density toward low galactic latitudes. The fact that there are no globular clusters visible between 4° north and south galactic latitudes shows that the clouds are so dense they are opaque. There are undoubtedly other globular clusters on the far side of the Milky Way.

Also, the color excesses of stars increase toward the center along the galactic equator, indicating that the clouds increase in density toward the galactic center. The clouds of dust and gas are not uniformly distributed, but are most dense in the Great Rift and near the Sagittarius region. As the galactic latitude increases, the color excess decreases, though not uniformly. Therefore the K-term used in Chapter 19 must be determined for each area separately.

One estimate is that a third of the mass of the galaxy is in the form of gas, which is mixed with dust. Hydrogen is the most abundant element in the clouds, with helium second. For every 20 atoms of hydrogen there is one atom of helium. Because helium is four times heavier than hydrogen, about one fifth of the clouds is helium by mass. The dust, which might be called an impurity in the gas, makes up about 1 percent of the total mass of the clouds.

The ratio of the number of stars in a unit area on the galactic equator to the number in a unit area at the galactic poles is called the *galactic concentration*. This ratio varies with the magnitude of the stars counted and with the galactic longitude. The figures, before correction for absorption and averaged for all longitudes, are as follows:

photographic magnitude	average galactic concentration	visible with
5.0	3.4	naked eye
9.0	3.9	1-inch telescope
15.0	10.4	16-inch telescope
19.0	27.0	100-inch telescope
21.0	44.2	200-inch telescope

It has been estimated that there are 1 billion stars of photovisual magnitude 20; and, because most stars have positive color indices, about half as many have a photographic magnitude of 20. Because of absorption and the fact that the far side of the galaxy cannot be seen, P. J. van Rhijn of Holland estimated the total number of stars in the galaxy as 30 billion. This number is almost certainly too small, as indicated by the computed mass. Van Rhijn estimated the combined light of all stars as equal to the light of 1092 stars of visual magnitude 1.0.

If the stars are uniformly distributed in space, and if there is no absorption, the number of stars should increase by a ratio of 3.98 for an increase of one magnitude. This can be shown from the inverse square law of light and the increase in volume of a sphere, which is proportional to the cube of the radius. However, it has been found that the increase in number of stars per magnitude is not 3.98, but is between 2.5 and 3.0, depending on the galactic latitude. This indicates that either or both of the assumptions—that the stars are equally bright and are uniformly distributed—are false.

Considering the galactic concentration and the increasing absorption towards the galactic equator, the idea grew during the first half of the 20th century that the galaxy is a disk-shaped collection of stars and nebulous material. The sun is not at the center, but is located at some distance from the center. The center must be in the direction of Sagittarius, since the greatest concentration of stars and obscuring material is in that direction, but its distance cannot be determined by star counts alone.

When all corrections have been made as accurately as possible, the result is that the galaxy is considered to be a system of about 100 billion (10^{11}) stars and a nearly equal mass of interstellar gas and dust.

20.3 ROTATION OF THE GALAXY

The rapidly rotating planets, Jupiter and Saturn, and the ring system of Saturn, have been flattened by rotation. In the solar system, the sun and its planets are confined to nearly the same plane. Photographs of the Milky Way, especially those made with a wide-angle camera (Fig. 20-3) show the stars in the galaxy to be flattened along the galactic equator. Several galaxies (Figs. 20-4 and 20-5) are elongated, showing evidence of rotation. These photographs indicate that the galaxy must also be in rotation. When the spiral Andromeda galaxy was found to be a rotating collection of stars, the theory of rotation of our galaxy was strengthened.

Another hint that the galaxy may be rotating was not recognized at first. The sun's motion among the stars was investigated during the first part of the 20th century. Sir Arthur Eddington (1882–1944) in England studied star streaming. W. W. Campbell (1862–1938) in the United States calculated the sun's velocity among the nearest stars. His results gave a velocity of 19.0 km/sec toward a point at right ascension $18^h 0^m$ and declination $+28°$, in the constellation Hercules. As more distant stars (O-stars and giant and supergiant variables) were used in the problem, a greater velocity for the sun's motion was obtained. With respect to the globular clusters, the sun's velocity appeared to be nearly 300 km/sec toward right ascension $20^h 24^m$ and declination $+62°$, in Cygnus. It was found later that the changes of direction and the increase in velocity result from the rotation of the galaxy.

Most globular clusters are located alongside the Milky Way in the region of Scorpius and Sagittarius. Harlow Shapley (1885–1972) had determined their distances by the use of cluster variables (RR Lyrae stars) and by their apparent diameters. He concluded that they form a nearly globular system concentric with the galaxy. Before Shapley's studies of clusters, the galaxy was thought to be a disk-shaped galaxy about 10,000 parsecs (32,600 light-years) in diameter, with

FIG. 20-3   Photograph of the Milky Way taken with a special wide-angle camera. (Courtesy A. D. Code and T. E. Houck, Washburn Observatory, University of Wisconsin)

the sun at the center. Shapley's theoretical galaxy was 75,000 parsecs (245,000 light-years) in diameter, with the center in the direction of Sagittarius 20,000 parsecs (65,000 light-years) from the sun.

Shapley tried to measure absorption in space by studying the times of minima of eclipsing binary stars in red and blue light. He thought that blue light traveled more slowly in an absorbing medium than red light. Since he could find no difference in the times of minima, he concluded that there is no absorption in space.

This work was soon shown to be in error by Stebbins and his associates, who found that stars are reddened by selective absorption and that the globular clusters are reddened also. The correction to the distances of the globular clusters was made and Shapley's galaxy was cut to a diameter of about 35,000 parsecs, since the clusters are not as far away as Shapley had estimated. Stebbins de-

FIG. 20-4 NGC 4565, an edge-on spiral galaxy, photographed in red light with the 200-inch telescope. (*Photograph from the Hale Observatories*)

scribed the galaxy as a "ham sandwich," with the absorbing material in the central plane and with a bulge in the direction of Sagittarius. The sun was placed at about 10,000 parsecs from the center.

The next step was to measure the speed of rotation of the galaxy at the sun's

FIG. 20-5 The spiral galaxy NGC 891. (*Photograph from the Hale Observatories*)

distance. The possibility of a solution to this problem by observations from a moving body inside the galaxy was first pointed out by Bertil Lindblad (1895–1965) in Sweden and Jan Oort in Holland in 1926 and 1927. They knew that the ring of Saturn, the planets in the solar system, and the satellites in orbits around the planets all move in such a way that those closest to the centers of attraction move faster than those on the outside. This is in accordance with the law of gravitation. It was therefore suggested that stars closer to the center of the galaxy than the sun should be moving faster than those toward the anticenter.

The idea of differential motion was followed up by Plaskett and Pearce at the Dominion Astrophysical Observatory in Victoria, British Columbia. They had measured radial velocities of stars of class B, which are visible to great distances and could be used in this problem.

Suppose that in Fig. 20-6 the sun is located as shown with respect to the direction of the galactic center and that its speed is proportional to the length of the white arrow on its orbit. A star on an orbit nearer the center has a greater speed, and one farther away has a slower speed. Consider only radial velocity, the motion of a star toward or away from the sun.

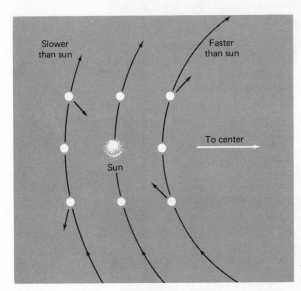

FIG. 20-6   Effects of the rotation of the galaxy on the radial velocities of stars.

Stars on the same orbit as the sun, either preceding or following, have the same speed and therefore no radial velocity. Those in the direction of the center and those away from the center also have no velocity of approach or recession. Their motions are parallel to that of the sun and are therefore tangential motions only, not radial. Stars on different orbits and in other directions should be either receding from the sun or approaching, as shown by the black arrows. These radial velocities are small, but measurable. Not only were stellar velocities used, but also radial velocities of clouds of interstellar gas between the stars and the sun were included.

Since the work of Plaskett and Pearce was published, there have been other solutions by various people and with different results. The following results are probably as accurate as any available and they are given here, although at any time there may be a different solution, based on more complete and more accurate data.

1. Distance from the sun to the center of the galaxy: 10,000 parsecs = 32,000 light-years.
2. Orbital velocity of the sun: 250 km/sec = 155 miles/sec.
3. Period of the sun's revolution: 250×10^6 years. This has been called the *galactic year*.
4. Diameter of the galaxy: 30,000 parsecs = 100,000 light-years (very uncertain).
5. Total mass of the galaxy: 1.5×10^{11} solar masses = 150 billion suns. It has been estimated that nearly half of this mass is in the form of dust and gas.
6. Thickness of the galaxy (disk at center): at least 3000 parsecs = 10,000 light-years.
7. Velocity of escape from the galaxy at the sun's position: 310 km/sec = 193 miles/sec.
8. Density of the sun's neighborhood (all types of matter): 0.15 solar masses/cubic parsec. Equivalent to about 1.5 molecules of hydrogen per cubic centimeter.

20.4 STRUCTURE OF THE GALAXY

Before considering the structure of the galaxy in the light of recent observations and theories, it is desirable to summarize the locations of the objects described previously, starting from the outside.

The globular clusters form a system centered at the center of the galaxy. This system is about 100,000 light-years in diameter, but there may be some stragglers at greater distances. If it is assumed that the globular clusters are stationary or moving at random, they may be used as reference points for determining the relative velocity and direction of the motion of the sun, as was done by using the radial velocities of stars. In fact, it was the apparent high velocity of the sun with respect to these objects that gave the first hint that the galaxy is rotating.

The system of globular clusters makes up what has been called the *halo* or the *corona* of the galaxy (see Fig. 20-7). In addition to clusters there are some stars scattered throughout the halo. And radio observations have detected very diffuse gas molecules, but no conspicuous regions of ionized hydrogen.

The Population II stars are associated with the globular clusters and are located in other regions where the interstellar clouds are lacking or are very diffuse. They are found in the galactic halo and also apparently in the galactic nucleus.

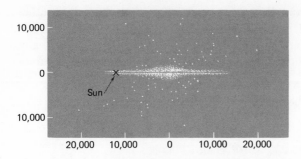

FIG. 20-7 A 1930 diagram of the galaxy, showing the shape of the galactic disk, the central bulge, and the obscuring material in the central plane. The globular clusters, indicated by the dots, are represented as forming the halo. The distances are in parsecs.

The open clusters lie mostly in the plane of the Milky Way. A few, like the Pleiades and Praesepe, appear to be at some distance from the galactic plane only because they are relatively near the sun. The open (galactic) clusters form a flat system, with the sun apparently near the center of those known. Five hundred are known, but there may be many more at greater distances, which are unobservable. Some probably mingle with the dense star clouds and others may be behind the obscuring dust.

The Population I stars are associated with the open clusters and with the interstellar clouds. The idea is growing, and is confirmed by radio observations, that the open clusters, interstellar clouds, and Population I stars lie in spiral arms of the galaxy. This is also in agreement with the structure of other galaxies, particularly the Andromeda galaxy, where the spiral structure is quite obvious. There the distribution of similar groups of stars, nebulae, and clusters is along the spiral arms.

Special wide-angle cameras developed at the Yerkes Observatory can be used to photograph large areas of the sky at one time. When photographs are taken with these cameras on special red-sensitive plates, luminous clouds of ionized hydrogen can be located. These are called HII regions.

The HII regions are very near hot stars, which ionize the gas; that is, each atom is broken down into its proton and electron. This is possible only where there are great amounts of ultraviolet light. The gas shines faintly by fluorescence as a bright, or emission, nebula mentioned in Chapter 19. The density is between 1000 and 10,000 atoms per cubic centimeter, most of which are ionized. A few of the protons recombine with the free electrons and are then capable of radiating the red Hα line during the short interval in which they are neutral and before they are ionized again.

HI regions are areas where the hydrogen is neutral. Hydrogen I cannot be observed on photographs, but fortunately it can be detected by its radio emission. Some of these regions absorb light from hot stars in the background and can be observed by their lines which appear in the stellar spectra, as previously mentioned. There are extensive areas of hydrogen in Orion, Cygnus, and other parts of the Milky Way. They are nearly transparent to light.

The absorption lines of calcium were mentioned in Chapter 19. Other lines have been found that permit the determination of the abundance of elements in the gas clouds. Many of these interstellar lines are double or even triple, having been shifted in position by the Doppler effect. They therefore locate clouds of different velocities and at different distances from the sun.

Gas absorbs selectively in certain wavelengths and produces the reddening of stars. Dust clouds, on the other hand, are efficient absorbers of light at all wavelengths. The Orion nebula is a dust cloud that shines by reflected light. The dark areas in the North America nebula (Fig. 20-8) and in other nebulae are opaque clouds of dust.

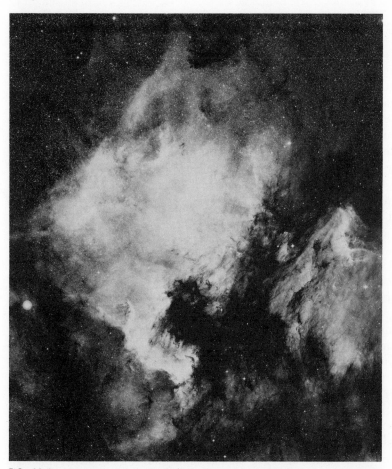

FIG. 20-8 The North America Nebula in Cygnus, photographed in red light with the 48-inch Schmidt telescope on Mount Palomar. (Photograph from the Hale Observatories)

Much about the nature and distribution of the dust particles can be inferred from the way in which they affect light. As was mentioned in Chapter 19, light passing through dust is reddened because the shorter wavelengths are extinguished more efficiently than the longer ones. Another effect was discovered in 1949. The light from certain stars is polarized.[1] It was observed that the polarization is strongly correlated with the amount of reddening, indicating that both phenomena are related to the same particles. To cause the observed polarization, the particles must absorb and scatter the electromagnetic vibrations preferentially in one direction, indicating that they must be unsymmetrical and aligned by the magnetic field of the galaxy. Depending on their composition, they may be shaped like tiny disks or needles, with diameters of 7×10^{-8} to 7×10^{-7} cm, somewhat less than the wavelength of light. This suggests the presence of metallic particles in the dust.

The dust hides the bulge at the center of the galaxy so it cannot be studied by optical methods. But radio waves penetrate any clouds that lie between source and receiver. The discovery of a wave 21 cm long was a tremendous help in radio astronomy. Since it is so much longer than the dimensions of the dust particles, it goes through a dust cloud without interference and can penetrate the thick parts of the galaxy from the direction of the nucleus.

The 21-cm radiation from hydrogen comes from regions of space where there are large amounts of material at very low temperature. The atoms are stimulated to radiate at this frequency by collisions with other atoms. This was predicted theoretically by H. C. van de Hulst in Holland in 1944, but was not found until 1951, by which time radio equipment of sufficient sensitivity had been developed.

Before the use of radio in this problem, Baade had photographed Population II stars of the cluster variable type and had found them in Sagittarius at the very center of the galaxy. He therefore concluded that the nucleus of the galaxy is free of dust.

Radio studies at the 21-cm wavelength have detected a very turbulent layer of hydrogen 10,000 light-years from the center and about 1000 light-years thick. It shows velocities between 50 and 100 km/sec. The density of this turbulent, neutral hydrogen is about one atom per 2 cubic centimeters—half the density in the region of the sun. But it must be mixed with ionized hydrogen, which produces the turbulence.

Two separate absorbing clouds moving with different velocities have been found between this mass of hydrogen and the sun. They have been interpreted as two arms of the galaxy. These arms and others found by both optical and radio investigations were put on a diagram (Fig. 20-9), at first by Morgan and later by others. Our galaxy has been described as a spiral with several arms (possibly five, six, or seven), similar to the galaxy in Andromeda.

Radial velocities of stars and gas clouds show that the velocities increase toward the center of the galaxy, but that there is a point where the velocity is a maximum of 226 km/sec at a distance of 1800 parsecs (nearly 6000 light-years) from the center. From there it drops to 220 km/sec about halfway to the center

---

[1] Light consists of electromagnetic waves (see Chapter 7) that vibrate perpendicularly to the path of the ray. If it were possible for an observer to see the vibrations of a light ray coming toward him, they would appear as spokes radiating out rather symmetrically from the path of the ray. In polarized light the vibrations in one direction are reduced or missing. For example, horizontal vibrations are reflected more strongly from horizontal surfaces than are the vertical vibrations and so the reflected light is partially polarized.

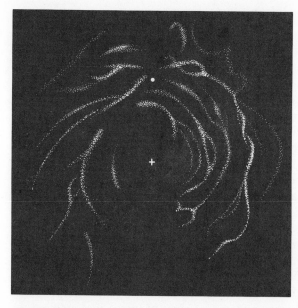

FIG. 20-9 The spiral structure of the galaxy as traced by 21-cm observations in Australia (left side) and in the Netherlands (right side). The galactic center is marked by the cross and the position of the sun by a dot.

and decreases still further until a point is reached, about 1500 parsecs from the center, where the measures are not accurate. The velocity of escape of a star from the galaxy at the sun's distance is about 310 km/sec. Therefore there are no stars at that distance with velocities of $310 - 220 = 90$ km/sec with respect to the sun. All stars at that and greater velocities have been removed from the solar neighborhood. The sun is overtaking all stars moving at slower speeds.

In summary, the galaxy is a spiral with several arms. The sun is situated in or near the edge of one of the arms in a region where the interstellar material has already become rarefied. The density in the solar neighborhood is estimated at 1.5 atoms per cubic centimeter. The density of hydrogen between the arms is almost zero.

The spiral arms are estimated to be 1500 light-years wide and 3000 light-years apart. The density in the nucleus is thought to be 7 times the density in the solar neighborhood, where the total mass of gas is equal to the mass of the sun (2×10^{33} grams) per 12 cubic parsecs. The spiral arms lose their identities at the distance of the turbulent hydrogen clouds, 10,000 light-years from the center. If this distance is considered to be the outer limit of the nucleus, two thirds of the total mass of the galaxy is contained in the nucleus. The direction of rotation is such that the arms are trailing.

20.5 STELLAR EVOLUTION

Nearly 200 years ago William Herschel said that studying the evolution of the sun and the stars would be like walking in the forest for an hour. The observer would see trees in all stages of development without seeing a single leaf form, develop, or die. But he would see sprouting seeds, young saplings, full-grown.

and aged trees, and those that had fallen and were returning to dust. Thus he might be able to form an opinion about the life history of a tree.

In a similar way, if it were possible to observe stars in all stages of formation, development, and aging, and if they could be arranged in a proper sequence, it might be possible to formulate a theory of their evolution. Fortunately, astronomers are now able to do just that, although their theories are still new and in need of testing.

When it was found that the H-R diagram classified the stars in order of temperature and size, it was supposed that stars started as large, giant stars of class M that had formed from nebulae. They were thought to condense, growing hotter by converting potential energy to kinetic energy by the fall of material toward the center. Finally, a stage was reached where they began to cool, ending their careers first as red dwarfs and at last as dark bodies roaming around in space and unobservable. It was thought that there are a great many dark stars.

Then the mass-luminosity relation (Fig. 18-11) was discovered. The most massive stars are the brightest, with a few about 50 times the solar mass. The less massive stars, with masses down to 0.05 the mass of the sun, were the faint red stars at the lower end of the main sequence. Hence the theory of evolution was modified to state that a red giant remained a giant of increasing temperature, but with decreasing diameter, the mass remaining approximately constant until it reached the main sequence at a spectral type depending on the amount of mass. That is, a star might evolve into an A-type star or, if it were more massive, into a star of class B or even O. Then it cooled and became fainter; in modern terms, the star moved slowly down the main sequence, ending its life as a black dwarf.

Later, it was demonstrated that energy from contraction alone would not account for the amount of energy radiated by stars. It was then assumed that there must be a source of heat from the interior, probably from the nuclei of atoms. This led to the theory of production of energy from the conversion of hydrogen to helium discussed in Chapter 11. The possible lifetime of the sun was computed to be 10^{11} (100 billion) years at the present rate of radiation, assuming the sun to be composed entirely of hydrogen. This method of computation was also applied to other stars. It can be shown that the massive stars of spectral type B with mass equal to 25 solar masses, for example, use up their atomic hydrogen in a few million years, and that cool M-stars will last much longer than the sun.

A white dwarf was supposed to be a star that exploded because it ran into a small body, perhaps of asteroid size, or a cloud of dust, which released internal energy from the star. This release of energy produced a gigantic explosion. The hot material from inside the star was blown away into space and the rest collapsed into a star of very high density and temperature at least as high as that of the star before the explosion. The result was a white dwarf, surrounded by a gaseous nebula.

20.6 A LATER MODIFICATION

Progress has been made in the theory of evolution in almost all stages. It is assumed that gas and dust clouds are being compressed by radiation pressure from outside, combined with gravitational attraction from within as the cloud condenses (Fig. 20-10). The heat produced by the conversion of gravitational potential energy into kinetic energy is mostly radiated away, so the system is losing a great deal of its energy. When the condensation of the mass reaches a certain state, the particles are held together by their mutual gravitation, since they do not have velocities high enough to permit them to escape; and a protostar is formed.

FIG. 20-10 The Rosette Nebula in Monoceros, photographed by the 48-inch Schmidt telescope. The black spots on the bright nebulosity are dark globules thought to be forming into stars. *(Photograph from the Hale Observatories)*

When a protostar attains a high enough density, it becomes opaque to radiation and the interior becomes hotter. Any solid grains evaporate and the gaseous atoms lose some of their electrons; that is, they become ionized. This also increases the opacity of the body. When the pressure equals the gravitational attraction, the infall of outside particles is stopped. The body is then in hydrostatic equilibrium and has become a stable star.

It has been shown that at this stage the star radiates from its surface, because energy is circulated by convection; that is, there is a convective volume of gas surrounding the center, and the stellar energy is carried from the center to the outside by the motion of the material itself. As the star radiates, its internal pressure and temperature should decrease; but this is prevented by a contraction of the star, which again raises the internal pressure and temperature. The energy that the star radiates comes from kinetic energy due to the star's contraction. This mechanism was mentioned briefly in the discussion of the production of solar energy in Section 11.9. The process of contraction and the building up of pressure, density, and temperature in the interior is so slow that the gravitational attraction and pressure are continually equalized, and thus the star is maintained in a state of hydrostatic equilibrium.

After a long period of time, perhaps millions of years depending on the mass and composition of the material in the protostar, the convection stops near the center, and radiation is the only process by which energy is carried outward. That is, the central zone of the star is in radiative equilibrium, surrounded by a convection zone. Therefore the central temperature is caused to increase; when it becomes high enough, nuclear reactions begin and hydrogen is converted to helium. This is called *hydrogen burning*.

When the nuclear reactions produce enough energy for the internal heat to equal the energy produced by contraction, the star has arrived on the main sequence. Stars of low mass remain in a state of convective equilibrium throughout, and the central temperature is too low to permit hydrogen burning. The lower limit of mass is about 0.05 times the mass of the sun, below which the production of energy is not great enough for the star to shine. No stars below this limit are known.

At the upper limit of mass the internal pressure becomes too high and the star becomes unstable and breaks up. The upper limit is probably below 100 solar masses, and no star of mass greater than about 50 solar masses is known.

If a star has enough mass to permit hydrogen burning, it builds up a core in the center, which at first is a mixture of hydrogen and helium nuclei and later is composed entirely of helium. The core then heats up, partly by the release of gravitational energy due to the infall of material. As this occurs, the star generates more and more radiative energy and brightens up. Also, the exterior becomes hotter. In other words, the star moves up along the main sequence, increasing its absolute magnitude and changing color toward the blue. This phase of its evolution lasts about 10^{10} (10 billion) years, which is most of its lifetime.

When the hydrogen in the center is completely used up, there is a change from hydrogen burning in the core to hydrogen burning in the radiative shell. The star then becomes larger and cooler on the outside; that is, it moves off the main sequence, up and toward the right on the H-R diagram, and becomes a red giant. The inner core increases its helium content until finally it heats up by contraction sufficiently to permit helium burning to begin. This is followed immediately by a "helium core flash," a sudden combination of helium nuclei that causes a very rapid heating of the helium core. Just the right amount of helium is burned to expand the core from its highly condensed state to a convective helium-burning shell. This state, in which the helium is completely exhausted, lasts about 10^8 (100 million) years.

Most of the helium burning takes place during the helium flashes. The amount of energy released during a flash is equal to 10^5 (100 thousand) times the solar luminosity. It is now thought that the red giant then becomes a Population II

Cepheid. This stage is short—of the order of 1000 years—and may occur several times. That is, the star makes a loop between giant and Cepheid stages several times during the helium-burning stage.

The helium burning produces heavier elements in the star's interior. This results in a carbon-oxygen core, surrounded by a helium-burning shell, then a hydrogen-burning shell, and this is surrounded in turn by the original atmosphere. The production of heavier elements is the last source of nuclear fuel. As successive nuclear fuels are exhausted in the core and the surrounding layers, the star continues to contract. Each contraction increases the internal temperature and brings the next nuclear fuel to the ignition point. Finally, all of the fuel is exhausted.

According to present theory, the fate of the star in the final stage depends on its mass. If its mass is less than roughly 1.4 times the mass of the sun, the material resists further compression when the star has shrunk to approximately the size of the earth. At this point, the density in the core may be as high as 10^8 g/cm³ (a mass the size of a pea weighing more than a truck). For more massive stars, the gravitational pull on the overlying layers is so great that the contraction continues at an increasing rate. Soon the core outstrips the outer layers of the star in the speed of collapse, causing the gravitation to become even more intense. If the mass of the star is less than a critical amount, approximately twice the mass of the sun, the core is able to resist further contraction when the material has been squeezed to a density comparable to that of an atomic nucleus (about 10^{14} g/cm³, equal to all the world's people compressed into a sphere the size of a single raindrop). Electrons and protons are forced together, forming neutrons. The extreme heat generated by the rapid contraction acts as an explosive charge that throws off the outer layers in a supernova explosion, forming an expanding nebula. The remaining core is a *neutron star*, with a diameter of only 10 miles.

If the mass of the original star was greater than the critical amount, the gravitation becomes so intense during the contraction that the material is not able to resist compression and the core continues to collapse. Theory is uncertain as to the extent of the collapse, but it appears that gravitation prevents the escape of light, and so it is called a "black hole."

To summarize:

1. A star begins as a black globule that is being compressed by exterior radiation pressure and finally by internal gravitation. During this stage it may become a variable of the T Tauri class. This theory is strengthened because of the location of Population I stars in the spiral arms of the galaxy in regions where there are known to be clouds (nebulae and hydrogen clouds with other elements mixed in).

2. When a protostar has attained a central temperature sufficiently high to permit the transformation of hydrogen to helium, it becomes a stable star on the main sequence. Its

position depends on the mass of gas it originally had (see Fig. 20-11).

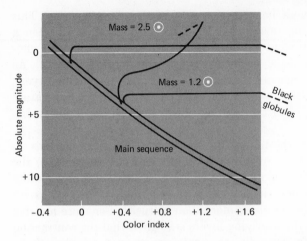

FIG. 20-11 Theoretical tracks in the evolution of stars on the H-R diagram. Numbers above the tracks are masses of the protostars in terms of the sun's mass. Their evolution from black globules is still uncertain.

3. The star remains on the main sequence, increasing in luminosity and temperature for a long time, again depending on its mass, and in the case of stars with the mass of the sun, probably 10 or 12 billion years. During this stage the star is building up a helium core. The energy radiated is produced by the conversion of hydrogen to helium in layers above the core.

4. After 10 or 12 percent of the hydrogen is exhausted, the star begins to expand until it reaches the giant stage. In this stage its luminosity is much greater than that of a star of equal mass on the main sequence. The core contracts until it reaches a temperature of 1.4×10^8 (140 million) degrees K, at which time the helium begins to combine to form still heavier atoms. Carbon can be formed and then combines with helium to form other heavy elements. Higher temperatures can be reached in the more massive stars where iron forms when the temperature reaches 3 billion degrees.

5. The star remains in the red giant stage a relatively short time. It then begins to contract and becomes hotter on the exterior. In a stable star the outward force, resulting from the radiation and gas pressure, equals the inward force, resulting from gravitation on the overlying layers. During the contraction stage, however, the balance between these inward and outward forces may become disturbed, such as by a sudden release of energy. This causes the star to pulsate, becoming a Population II Cepheid.

6. The final stage occurs when the star has exhausted its nuclear fuel. It then becomes a white dwarf, a neutron star, or a black hole, depending on its mass.

The time scale, as has been noted, depends on the mass. Stage 3 for a star with the mass of the sun lasts some 10 billion years. Stage 4 is shorter, perhaps 1 billion years or a little less. This computation is difficult and quite uncertain. Stage 5 is still shorter, from 10 to 100 million years. The white-dwarf stage lasts even longer than all the rest combined—perhaps trillions of years. The galaxy is not old enough yet to produce any black dwarfs.

20.7 OBSERVATIONAL EVIDENCE

As a check on the theory, color-magnitude curves have been drawn for several clusters. It is assumed that all the stars in a given cluster are the same age. When the colors, determined by photography and photoelectric photometry, are plotted against absolute magnitude (assuming the distances can be determined) the H-R diagram of the entire cluster or of several clusters can be studied at the same time, as in Fig. 20-12. It is like looking at all the trees in the forest at once. The stars are seen at various stages of evolution. The point where a curve starts to branch off the main sequence marks the beginning of the expansion of its stars.

The bright clusters (NGC 2362 and h and χ Persei) have blue, hot stars. Since the evolution of these stars is much more rapid than any others, these clusters

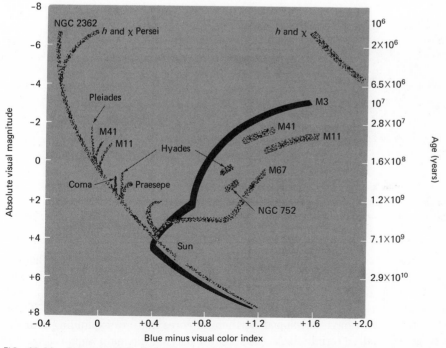

FIG. 20-12   Composite H-R diagram for several star clusters of different ages. *(Adapted from studies made by Alan R. Sandage, Hale Observatories)*

must be only a few million years old. M67 is the oldest. The scale gives the approximate age, corresponding to the absolute magnitudes on the main sequence. According to this scale, the age of the sun is 10 billion years. Some clusters (M11, M41, h and χ Persei) have some stars in the giant stage. Probably there are stars in other stages, but they are less numerous and have not been observed. This diagram appears to confirm the theory, since the stars are found in the various stages that had been predicted. Stars at the lower end of the main sequence are too faint to be observed, as are the white dwarfs.

In Fig. 20-13 the H-R diagram of four hypothetical open clusters and one globular cluster have been plotted. While these plots are hypothetical, they follow closely plots for real clusters. For example, cluster A is similar to the actual diagram of h and χ Persei. As noted in Chapter 16, the H-R diagram is composed of two branches, the main sequence and the giant branch, with a scattering of supergiant stars. In Fig. 20-13 each of the curves A, B, and C shows two similar branches, the main sequence and part of the giant branch. Cluster A shows a short section of supergiants in the upper right. The giant branch of C leaves the main sequence at absolute magnitude $+3$, slightly above the position of the sun at $+5$. The interpretation of these curves is that the stars on the main sequence for the first three clusters have reached the stage of evolution where hydrogen is being converted to helium. In cluster D the faint stars are off the main sequence. It may be that these stars are still contracting from interstellar matter. The stars on this diagram are all Population I stars, except for curve E, which shows a similar branch for Population II stars found in globular clusters.

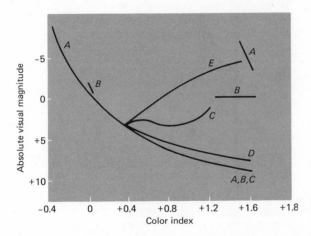

FIG. 20-13 Composite H-R diagram of five hypothetical clusters.

The massive hot giants and supergiants combine their hydrogen into helium at much higher rates than the cooler, less massive stars of classes F, G, and K. The giant branch turns off the main sequence at a point that depends on the mass of the stars and the rate at which hydrogen is being converted to helium. These stars are thought to have developed hot helium cores and are expanding into the giant stage. In modern terminology, the stars are leaving the main sequence.

The giant stage is more rapid than the stage where the stars move up along the main sequence. Hence there are fewer stars in the giant branch and there are corresponding gaps in the H-R diagram.

Cluster A has more stars of spectral class B than the other two clusters. These hot stars are burning their hydrogen at a very rapid rate. Since cluster A does have type-B stars, it must be younger than clusters B and C, which have lost their hot stars. The turnoff point for stars of one solar mass is about where the C-branch leaves the main sequence. The double cluster in Perseus is probably only a few million years old.

It will be recalled that carbon is one of the elements needed for the carbon cycle of transformation of hydrogen to helium. This element is not produced in the interior of the sun at its present stage of evolution. So it must have come from the outside. Carbon is produced at later stages and is thought to be distributed in space by the explosion of a supernova, which apparently blows a star almost completely apart. The same can be said for iron and other heavy elements known to be in the sun.

It is now thought that, since the Population I stars are in dusty regions of space, they must have picked up these elements in the formation stages. Therefore the galaxy must be older than the sun. Another source of heavy elements is from binary stars that are losing mass to each other and to the space around them.

Although the possibility of neutron stars was noted 40 years ago, there was no observational evidence until the discovery of pulsars. The theorists arrived at neutron stars by a process of elimination. In astronomy, periodic behavior has been found in three cases: pulsation of a star, eclipsing binaries, or rotation. The rapid light pulsation (as high as 30 times a second) ruled out the possibility of their being eclipsing binaries because of the forces that would be involved. Also, there is no Doppler effect. No known stars, not even white dwarfs, could pulsate that rapidly. However, a small body could rotate very rapidly, and the neutron stars seem to fit the requirements. White dwarfs would fly apart if spinning that fast, but the smaller size and greater gravitation of the neutron stars would not permit them to do so. To account for the pulsation, it is proposed that the energy radiated by the neutron stars is focused by their strong magnetic fields, so they act as rotating beacons. The slight, but steady, decrease in the rate of pulsation of the Crab pulsar has been related to the energy radiated by the surrounding nebula. It appears that the pulsar supplies energy to the nebula at the expense of its rotational energy.

Observational evidence for black holes is very difficult to obtain because they are invisible. But it is thought that they might be detected indirectly through their gravitational effects on the motions of other stars, or the bending of light from background stars as they pass through the strong gravitational fields.

QUESTIONS AND PROBLEMS

Group A

1. (a) During what month is the densest portion of the Milky Way on the meridian of an observer at 45°N latitude at

9 P.M. local time? (b) Is it the same for your latitude? If not, why not?

2. What changes would have to occur in the earth's rotation and revolution to bring the celestial equator, the galactic equator, and the ecliptic all into the same plane?

3. Draw a cross section of our galaxy as seen edgewise. Label the center and the approximate location of the sun, as in Fig. 20-4. Indicate in light-years the approximate radius of the galaxy and the distance of the sun from the center.

4. Why is the Milky Way least dense in Auriga?

5. Explain how Shapley's assumption that there is no absorption of light in space caused him to overestimate the size of the galaxy.

6. There are only about half as many stars of photographic magnitude 20 as there are of photovisual magnitude 20. Why?

7. If Shapley had been correct in thinking that absorption in space affects the speed of light, what would he have observed about the minima of eclipsing binaries?

8. How can the globular clusters be used as references for determining the sun's motion relative to the galaxy, if they are also in motion?

9. List at least three ways in which the presence of gas and dust in space is detected.

10. Why have HI clouds just recently been discovered?

11. Is it possible that some of the dark nebulae may eventually become bright? Explain.

12. Explain why the stars of a cluster evolve at different rates.

13. Do the most massive, or the least massive, stars have the longer life expectancies? Why?

14. Bethe's theory of nuclear energy production in stars resulted in dramatic changes in thinking concerning stellar evolution. Compare the currently accepted theory with one accepted at the time of Bethe's publication in regard to (a) the changes that occur in a star while it is on the main sequence, (b) the relative number of stars that go through the red-giant stage, and (c) the final evolutionary stage of stars.

15. What evidence indicates that the galaxy is older than the sun?

Group B

16. If all stars were equally bright and uniformly distributed in space, how many more sixth-magnitude stars should be observable than fourth-magnitude stars?

17. Draw a diagram of the celestial sphere showing the galactic equator and poles. Find the inclination of the galactic equator to the celestial equator and to the ecliptic.

18. Using the star maps in this book, estimate the galactic latitude and longitude of several bright stars visible in your evening sky.

21 the universe

An astronomy class was once asked in an examination to write an essay about the expanding universe. Several students wrote compositions about the expanding knowledge of the universe. Although this was not the answer the instructor expected, it did show that the students were impressed by the evolution of man's concept of the universe and the extension of knowledge to greater and greater distances from the earth and the solar system.

21.1 CHANGING CONCEPTS OF THE UNIVERSE

The ancient theories of the structure of the universe were discussed in Chapter 4. The Ptolemaic theory, first published about 150 A.D., lasted until the time of Copernicus in 1543. In this theory the earth was the center, surrounded by its system of sun, moon, and planets. The stars were thought to be attached to an outer sphere and were all considered to be at the same distance. This was somewhat modified in the 14th century by Thomas Digges (Section 4.3), who suggested that the stars are scattered throughout space.

Copernicus put the sun at the center, with the earth as a planet circling it. The stars were at unknown distances from the center. It is interesting to note that William Herschel in 1784 stated the following as his belief:

"The stars, instead of being scattered throughout space, constitute a cluster of definite limits, of which the thickness is small in comparison with its length and breadth and in which the earth (i.e., the solar system) occupies a position somewhat about the middle of its thickness." Herschel thought that he was able to "penetrate to the limits of our stratum" with his telescope. Later these hypotheses were abandoned. He admitted that there is a "sudden concentration" of stars in the neighborhood of

411

the Milky Way. ". . . the stars seem to extend to a distance beyond the reach of the most powerful telescope hitherto constructed and hence the shape of that portion of space which the stars occupy must be entirely unknown to us; that is, the material universe appears to be boundless."

The term *island universes* to designate the spiral nebulae was probably introduced by Immanuel Kant. They were at that time, and later by William Herschel, thought to be resolvable into stars. This belief and the theory of island universes were popular during the latter part of the 18th century. But in the last decade of that century, Herschel changed his mind. In November 1790 Herschel found a planetary nebula (a new class of nebulae he had discovered) that had a prominent central star definitely associated with the nebulosity. This nebula and a few others were apparently near the earth, were small, and could not be considered large clusters of stars. This discovery weakened the island universe theory.

The building of larger and better telescopes helped in the study of the nebulae. John Herschel, observing in the southern hemisphere as well as in the northern hemisphere, came to the conclusion that the spiral M51, in Canes Venatici, was much like the Milky Way system, which he thought to have a spiral form. This led to a revival of the island universe theory. About the middle of the 19th century, some 50 nebulae were "all resolved *without exception*." These included the Orion Nebula, the Crab Nebula, and one with a star cluster in the center "surrounded by spirals."

Without the invention of photography and the perfection of the spectrograph, investigation of the extent and nature of the universe would have been impossible. The first parallaxes were determined in the 1830s (see Chapter 16). Up to the beginning of the 20th century, fairly reliable parallaxes of 55 stars had been measured by several different methods. These parallaxes were limited to those of the nearest stars, which were only a few light-years away. Later, parallaxes of more distant stars were determined by photography with long-focus telescopes that permitted the measurement of very accurate positions of stars and the detection and measurement of the small angles involved in parallax determination. Since distance is inversely proportional to the parallax, more distant stars could be observed and their distances determined. Accurate parallaxes smaller than 0.01 second of arc are still impossible by this method, but distances up to about 100 parsecs are known with acceptable accuracy. Today parallaxes of about 760 stars are known, with parallaxes greater than 0.05 second and an accuracy of about 10 percent.

The next step was the discovery of the relation between the distances of Cepheid variable stars and their absolute magnitudes (see Section 17.4). This method could be used for any cluster of stars which included one or more Cepheids. The spectrum-luminosity diagram (the H-R diagram, Section 16.5) was even more important in the investigation of stellar distances. These methods extended the known distances of stars and star clusters to 1 or 2 million light-years. The H-R diagram method could be used for all stars bright enough for their spectra to be photographed or for their magnitudes and colors to be determined.

The spectroscope was adapted to astronomy about 1860, and in 1864 Sir William Huggins (1824–1910) examined the spectra of "nebulae, shreds and balls of cloudy stuff" that had been observed and sometimes mistaken for comets. Huggins found that there are two classes of these "nebulae." He recognized the gaseous nebulae by their bright-line spectra like those of gases at low pressure. The other type had continuous spectra crossed by a few dark lines, similar to the spectra of the star clusters. Since most of them were spiral in shape, he called them "spiral nebulae." The brightest of these is the Great Spiral in Andromeda, which is dimly visible to the unaided eye.

Some of the spirals contain Cepheid variable stars, and their distances could therefore be determined after the period-luminosity law was established.

There was a spirited debate about 1920 concerning the distances and nature of the spirals. The distance to M31, the Andromeda spiral, was estimated as only 19 light-years and its diameter as about 0.3 light-year. This would put it inside our galaxy, which in 1911 was estimated to be 120 light-years in diameter. During the debate, Harlow Shapley remarked that if the spirals are islands, our galaxy must be a continent, since it was estimated at that time to be much larger than the others.

After the great distances to the spirals had been determined, in about 1923, they were called *extragalactic nebulae*. And since their diameters are roughly the same as that of our galaxy, the name was later changed to *galaxies*, by which name they are known today.

21.2 THE ANDROMEDA SPIRAL, M31

The distance to M31 (Fig. 21-1) was not determined until about 1923. The nova that appeared in M31 in 1885 was thought to be similar to the novas in our galaxy. Its visual magnitude was 7.2. Assuming an absolute magnitude $M = -7$, the average for galactic novas, the distance to M31 is only 7000 parsecs, comparable to the supposed distances of stars inside our galaxy as it was known in 1918.

However, Edwin P. Hubble (1889–1953) photographed Cepheids with the 100-inch telescope over a span of many years and plotted their light curves. Assuming their photographic magnitudes to be +18 and their absolute magnitudes to be −2 (see Chapter 17), the distance to M31 came out to be 100,000 parsecs, well outside our galaxy. At this distance the absolute magnitude of the nova of 1885 would be −12, or 100 times brighter than an ordinary nova. Using a modern distance of 800,000 parsecs, the absolute magnitude would be −16.5 and the star would be classed as a supernova.

More accurate studies of Cepheids in the Andromeda Nebula, as it was called in 1924, placed it at a distance of 270,000 parsecs or 870,000 light-years. At this distance, the diameter of M31 would be about 45,000 light-years, since its measured angular diameter is 3°. At about the same time, Shapley had proposed

FIG. 21-1   The Andromeda galaxy M31 and its two companion ellipsoidal galaxies
M32 (right of center) and NGC 205 (upper left). A 48-inch Schmidt photograph.
(Photograph from the Hale Observatories)

a diameter of 250,000 light-years for our galaxy, which made it some 5 times larger than M31.

These estimates were followed by two observations by Stebbins and Whitford. They showed by the colors of globular clusters that Shapley's galaxy was too large by a factor of two and also that the Andromeda galaxy is about twice as large as measured by Hubble. This was done by scanning M31 with a photo-electric photometer on the 100-inch telescope. M31 and our galaxy were therefore thought to be about the same size. Man was losing his place near the center of the largest collection of stars in the universe!

As mentioned in Chapter 17, using the 200-inch telescope Baade looked for RR Lyrae variables in M31 without finding any. If the faintest photographic magnitude in reach of this telescope is +23, RR Lyrae variables should have been found, if their absolute magnitudes were the same as those of the Cepheids, about $M = -1.5$. Since they were not found, Baade decided that their absolute magnitudes are about zero and that there are two types of Cepheids, which differ by about 1.5 magnitudes, or 4 times in luminosity.

Baade found Population I stars in the spiral arms of M31, where he also found more than 600 emission nebulae. The photographs show dark regions, which he interpreted as absorbing clouds, lying along the arms, of which there are at least five and possibly seven. The Population II stars were found between the arms and in the nucleus. The nucleus is transparent and distant galaxies are visible through it and also between the arms. Bright objects along the edges of the galaxy were thought at first to be globular clusters, but many of them turned out to be HI or HII regions. The entire galaxy is rotating with the arms trailing.

The central part of M31 rotates as a single body, but the outer regions rotate in agreement with Kepler's law. Speeds increase to about 1° (45,000 light-years) from the center, then decrease as in our galaxy. The inclination of the plane of M31 to the line of sight is difficult to measure, but appears to be about 13°. Therefore, the galaxy is not oval but nearly circular. Several faintly luminous extensions indicate that it is perhaps 130,000 light-years in diameter, or 30 percent larger than our galaxy. There is also evidence of a halo around the central spiral.

21.3 THE MAGELLANIC CLOUDS

The Magellanic Clouds are the nearest galaxies to ours. The large cloud is in the constellation Dorado at right ascension $5^h 26^m$, declination $-69°$. It is 33° from the galactic equator and looks like a detached part of the Milky Way, invisible in bright moonlight (Fig. 21-2). The distance is somewhat in doubt, but is probably about 160,000 light-years, and the diameter is 30,000 light-years. It is composed of both populations of stars, bright-line nebulae (probably planetaries), and clouds of gas and dust. The Cepheid variables are normal Type I, or classical. It contains the supergiant star S Doradus, the brightest star known ($M = -10$, approximately), except for supernovas. The radial velocities of about 250 km/sec, discovered about 1915, were surprisingly large. They are, of course, mostly due to the sun's motion in our galaxy.

The Small Magellanic Cloud is in Tucana at right ascension $0^h 50^m$, declination $-73°$. Both clouds are therefore not visible anywhere in the United States except in Hawaii. The small cloud is 44° from the galactic equator. Its distance is 180,000 light-years and its diameter is 25,000 light-years. It is composed entirely of Population II stars, including the Cepheids, which were the first to be used for distance determinations by the period-luminosity relation.

These clouds may be appendages of our galaxy. They apparently are connected by a cloud of hydrogen that has been discovered by radio astronomy, and may therefore be considered to be a single galaxy. There is also a possibility that they are connected to our galaxy, but so far this has not been proved. They may form a pair of galaxies that are in rotation about a common center. Even though these two clouds are the closest galaxies to the sun, there is still a great deal to be learned about them.

21.4 CLASSIFICATION OF GALAXIES

After much discussion, the term *galaxy* is now defined to mean a large collection of stars, dust, and gas clouds held together by their mutual gravitation. The

FIG. 21-2 The Large and Small Magellanic Clouds. Both are classed as irregular galaxies, but the large cloud may be a barred spiral with only one arm. The bright disk at the lower right is a star. (*Harvard College Observatory*)

terms spiral nebulae, extragalactic nebulae, and island universes were suggested but have been abandoned.

Hubble classified the galaxies according to shape into the following three major groups:

1. *Ellipsoidal galaxies* have symmetrical structure ranging from spheres to flattened ellipsoids. Hubble called them E0, E1, . . ., E7, where the number divided by 10 indicates the eccentricity of the visible contour. The nearest ellipsoids are resolvable into stars only with large telescopes and with very careful photography. They are made up entirely of Population II stars, the dust clouds having been swept up, possibly by collisions (Fig. 21-3). Two well-known examples of this type are M32 (E3 type) and NGC 205 (E5 type). They are satellites of M31 and are at the same distance. Their diameters are 8000 and 16,000 light-years, respectively. They can be seen in the photograph of M31 in Fig. 21-1.

FIG. 21-3 NGC 4486, a type E0 galaxy in Virgo. *(Photograph from the Hale Observatories)*

2. *Spiral galaxies* have a distinct nucleus and one or more spiral arms. The arms extend outward from the nucleus and are composed of stars, dust, and gas. These galaxies are made up of Population I stars in the arms and Population II stars in the nucleus, between the arms, and probably in the halo. There are two distinct classes of spirals, the *normal* and the *barred spirals*. These are further subdivided into Sa, Sb, and Sc for the normal spirals and SBa, SBb, and SBc for the barred spirals. The Sa galaxies have prominent nuclei, with small, close-packed arms coming out of the nuclei. The Sb are a little more open, with slightly smaller centers. Our galaxy and M31 are Sb spirals. The Sc galaxies have small nuclei and very prominent open arms. The types of normal spirals are illustrated in Fig. 21-4. The barred spirals have elongated centers, called bars, with arms coming from each end. The letters *a*, *b*, and *c* have the same meaning as for the normal spirals (see Fig. 21-5). There are no well-known objects of this class. The best example is NGC 1300, shown in Fig. 21-6.

3. *Irregular galaxies.* As the name suggests, these galaxies have no regular shape. They are designated by the letters Irr I for those with O and B stars and emission nebulae, and Irr II for those not resolvable into stars. The best-known irregular galaxies are the two Magellanic Clouds (Fig. 21-2), although the large cloud has also been called a barred spiral with one arm.

In Hubble's diagram (Fig. 21-7) the E, S, and SB types

FIG. 21-4 Types of normal spiral galaxies, showing the relation between the
nuclei and the arms. Upper photograph,
Sa; middle, Sb; lower, Sc. (Photographs
from the Hale Observatories)

FIG. 21-5 Types of barred spiral galaxies. Top photograph, SBa; middle, SBb; bottom, SBc. (*Yerkes Observatory photographs*)

FIG. 21-6   NGC 1300, a barred spiral galaxy, photographed with the 200-inch telescope. *(Photograph from the Hale Observatories)*

come together at the point marked S0. This type is flat, like a spiral without arms. Hubble considered it as intermediate between ellipsoids and spirals.

21.5 DISTANCES OF GALAXIES

Accurate distances to galaxies are difficult, if not impossible, to determine. The most accurate method is by the use of Cepheid variable stars, which is possible only if the galaxies are near enough to be resolved into stars with large telescopes.

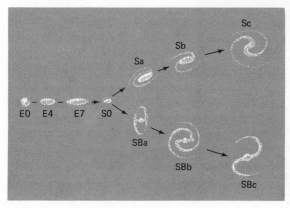

FIG. 21-7   The Hubble classification of galaxies.

The uncertainty in this method has already been discussed (see Chapter 17). The ambiguity resulted in an underestimate of the distances to even the nearest galaxies, the Magellanic Clouds and the Andromeda Galaxy. Even now the zero point of the period-luminosity scale is uncertain to perhaps 0.5 magnitude, leading to an error of 25 percent in distance.

When the brightest stars in a galaxy can be photographed, even though no Cepheids can be found, the absolute magnitudes can be assumed to be about the same as those in our galaxy, and the distances can then be computed by the absolute-magnitude formula. The absolute magnitudes of the RR Lyrae variables are about zero and can be seen to only about 1 million light-years. This distance includes the Magellanic Clouds and only a half-dozen ellipsoidal galaxies. The brighter Cepheids are found in M31, as has been noted, and in other galaxies, thus extending the observable limit to some 15 million light-years. Some observable blue supergiants can be recognized by their color indices and brightness contrasted with other stars in a galaxy. Assuming absolute magnitude $M = -9$, that of the brightest stars in our galaxy, the computation of distances by photometry can be made to perhaps 60 or 70 million light-years. This is relatively close compared to distances of billions of light-years obtained by other methods.

After the distances of galaxies up to the photometric limit were determined and their diameters computed, it was assumed that there is not much variation in diameter and total brightness between the two classes, spirals and ellipsoids. However, there is considerable range in brightness and probably in diameter within each class, with more variation among ellipsoids than in spirals. The ellipsoids range from $M = -10$ to -23, compared with $M = -15$ to -21 for spirals.

There is a possibility that the outer arms of a spiral may be too faint to photograph if a galaxy is too far away. Thus some spirals may be classified as ellipsoids. However, if an average absolute magnitude is assumed, and if the total apparent magnitude can be found with a photoelectric photometer or by some other method, the formula can be used and an approximate distance computed.

EXAMPLE: For the Andromeda galaxy, $m = 3.47$ and $M = -21.2$. Then, substituting in equation 16-7,

$$-21.2 = +3.47 + 5 - 5 \log d$$
$$\log d = \frac{29.67}{5} = 5.934$$
$$d = 860,000 \text{ parsecs}$$

Of course in this example the distance was first determined by the Cepheid period-luminosity method, and the absolute magnitude was computed from the apparent magnitude and the distance.

It is obvious that this method was not accurate. But for statistical investigations, it gave average distances, which were not used after other methods became available. A related method is to assume that all galaxies of a given class are the same size. Then the more distant galaxies appear smaller than those closer to the sun, and the distance is assumed to be inversely proportional to the angular diameter. This is also an approximate method.

The relation between distance and radial velocity of galaxies was discovered by Hubble in 1929. He used the angular diameter method to compute distances. Then he measured the radial velocities with a fast, low-dispersion spectrograph on the 60- and 100-inch telescopes on Mount Wilson. He found that the velocities increase with distance. That is, the lines in the spectra of galaxies are shifted toward the red by an amount that is proportional to the distance of the galaxy. This effect is called the *red shift;* and, if it is a velocity effect, the Doppler formula can be used to determine the radial velocity of the galaxy. This is shown in Fig. 21-8. Written as an equation,

$$\text{Velocity} = V = Hr \tag{21-1}$$

where H is Hubble's constant and r is the distance.

The distances used by Hubble have been multiplied by 3, based on the revision

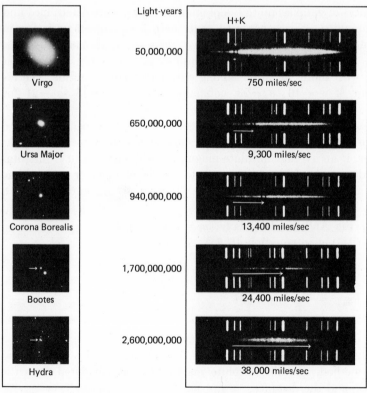

FIG. 21-8 Provisional distances of galaxies estimated from their brightness and their velocities calculated from the red shift. In each spectrum, an arrow indicates the amount of shift of a pair of spectral lines. *(Photograph from the Hale Observatories)*

of the distance scale by Baade. The revised Hubble constant is 75 km/sec per million parsecs or about 14.5 miles/sec per million light-years. This "constant" is subject to change as the distance scale is further revised. Velocities of galaxies up to 41 percent of the speed of light (124,000 km/sec or 76,000 miles/sec) have been measured.

This relation can be inverted and, if the velocity of a galaxy can be measured, its distance can be computed with an accuracy that depends on the value of H. Also, it has been argued for many years that the red shift is not a Doppler effect but may be caused by something other than velocity. One possibility is that light is slowed in a gravitational field, loses some of its energy, and the lines appear to the red of their normal positions. Recent research, however, shows that this velocity effect extends nearly to the speed of light and strengthens the belief that it is a Doppler shift.

21.6 DISTRIBUTION OF GALAXIES

In a large-scale view of the universe the galaxies appear to be scattered at random and to be uniformly spaced throughout the visible universe. Probably a billion are within reach of the 200-inch telescope, which reaches a distance of 2 billion or more light-years. However, there is a tendency for the galaxies to cluster. For example, the Milky Way Galaxy (our galaxy) is one of 18 galaxies within about 3 million light-years. It is near the outside edge of this group called the *local group* of galaxies. These galaxies are listed in Table 21-1.

TABLE 21-1   The Local Group of Galaxies

designation	type[a]	distance (light-years)	diameter (light-years)	solar masses
Milky Way galaxy	Sb		100,000	2×10^{11}
Lg. Mag. Cloud	Irr I	160,000	30,000	2.5×10^{10}
Sm. Mag. Cloud	Irr I	180,000	25,000	
UMi system	dE4	220,000	3,000	
Sculptor system	dE3	270,000	7,000	$3? \times 10^6$
Draco system	dE2	330,000	4,500	
Fornax system	dE3	600,000	22,000	2×10^7
Leo II system	dE0	750,000	5,200	1×10^6
Leo I system	dE4	900,000	5,000	
NGC 6822	Irr I	1,500,000	9,000	
NGC 147	E6	1,900,000	10,000	
NGC 185	E2	1,900,000	8,000	
NGC 205	E3	2,200,000	5,000	
NGC 221 (M32)	E3	2,200,000	8,000	
IC 1613	Irr I	2,200,000	16,000	
NGC 224 (M31)	Sb	2,200,000	130,000	4×10^{11}
NGC 598 (M33)	Sc	2,300,000	60,000	8×10^9
Maffei 1	E3	3,300,000?	6,000?	2×10^{11}?

[a] dE4 = dwarf E4; Irr I = irregular with little or no dust. Other types are discussed in the text.

It will be noticed that the ellipsoids outnumber the rest. If this is true for all of space, there should be about 60 percent ellipsoids and less than 20 percent spirals, with 20 percent irregulars. It seems likely that most of the nearby luminous galaxies have been found and that future discoveries will be limited to the dwarf galaxies. In the local group, our galaxy and M31 are considerably larger than any of the others, with M31 having a calculated mass twice that of ours. There may be other members hidden behind the obscuring clouds in the Milky Way.

In 1971 it was announced that two galaxies had been discovered in a heavily obscured region in Cassiopeia only ½° from the galactic equator. The brighter galaxy, named Maffei I for its discoverer, has been identified as ellipsoidal and in the local cluster. Although its distance is uncertain, it appears to be about 3.3 million light-years, and its diameter about 6000 light-years. Its mass appears to be approximately that of our galaxy. It emits no radio radiation.

The second, Maffei 2, is a spiral which is apparently too far away (about 9 million light-years) to be included in the local group. If its apparent size and distance are accurate, its diameter is of the order of 27,000 light-years. Its mass is still uncertain.

In November 1971, Sidney van den Bergh discovered three new dwarf ellipsoidal galaxies with the 48-inch Schmidt camera on Mount Palomar. They belong to the Local Group of galaxies and are 2 million light-years away. Two of them are probably satellites of M31; the third may be either a satellite of M31 or of the galaxy in Triangulum, M33.

The 48-inch Schmidt camera on Mount Palomar has made a survey and has found more than 600 large clusters of galaxies. The opinion at the Lick Observatory is that the clusters are uniformly scattered throughout space. The Virgo cluster, listed first in Table 21-2, contains more than 3000 members. It is possible that our group is an appendage of this cluster. Perhaps it is the largest in the Coma cluster, which contains some 10,000 members. The names, such as Virgo and Coma, indicate that these clusters are in the direction of the stars in those constellations, but they are of course much farther away. Similarly, the Andromeda galaxy is not in that constellation, but is at a distance of more than 2 million light-years.

Table 21-2 lists ten clusters of galaxies, their distances, and their velocities. A plot is shown in Fig. 21-9.

21.7 THE EXPANDING UNIVERSE

One of the products of Hubble's red shift is the theory of the expanding universe. This theory was first stated in a series of lectures by Abbe Lemaitre (1894–1966) in Belgium and published in a book called *The Primeval Atom*. It was assumed that the entire mass of the universe was originally confined to a volume of space about the size of the earth's orbit. According to this hypothesis, the density at the beginning was very great and the temperature correspondingly high. It was estimated by George Gamow (1904–1968) at 10 billion degrees Kelvin. This was

TABLE 21-2 Clusters of Galaxies

cluster	distance (light-years)	velocity (miles/sec)
Virgo	50,000,000	750
Pegasus	150,000,000	2,300
Perseus	260,000,000	3,100
Coma	390,000,000	4,500
Ursa Major 1	650,000,000	9,300
Leo	1,000,000,000	12,000
Gemini 1	1,200,000,000	14,000
Bootes	1,700,000,000	24,400
Ursa Major 2	1,800,000,000	25,000
Hydra	2,600,000,000	38,000

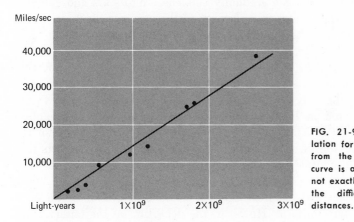

FIG. 21-9 The velocity-distance relation for clusters of galaxies, plotted from the data of Table 21-2. The curve is a straight line. The points do not exactly follow the line because of the difficulty in determining the distances.

the "primeval atom." It was very unstable and exploded, throwing its material out into space at very high velocities. This material cooled rapidly at a rate calculated to be inversely proportional to the square root of the age of the universe, expressed in seconds. Assuming the present computed age and an average of 3°K for the present temperature of the universe, the original temperature was 10^{12} °K or higher. As the material cooled, it had a tendency to collect into separate entities, which became the galaxies and in turn formed into stars.

The fastest-moving galaxies, therefore, reached the greatest distances, their distances from each other being proportional to the speed. The center of this system, the point where the explosion took place, is not known. Its location is not important to the theory, since we seem to be completely surrounded by galaxies and there are more beyond the limits of our telescopes.

In 1965 a new theory was advanced by Allan Sandage (1926–   ) of the Hale Observatories—that the universe is pulsating with a period of 82 billion years and that we are now in the expansion phase.

If the theory of expansion is correct, it should be possible to compute the age—that is, the time since the explosion. Using the data in Table 21-2 for the Hydra cluster, we find that since $s = vt$,

$$t = \frac{s}{v} = \frac{2.6 \times 10^9 \text{ years} \times 186,000 \text{ miles/sec}}{38,000 \text{ miles/sec}}$$
$$= 12.7 \times 10^9 \text{ (or 12.7 billion) years}$$

This is only one method of arriving at the age of the universe. Other methods compute the age as 16 or 24 billion years. In Sandage's words, "The clues indicate that our universe is a finite, closed system originating in a 'big bang,' that the universe is slowing down, and that it probably pulsates once every 82 billion years."

The *steady-state* theory, proposed by a group of British astronomers, assumes that energy is being converted into matter continuously. It also assumes that galaxies are being formed, grow old, and, as their stars evolve to the black-dwarf stage, finally die. In this theory, hydrogen is thought to be created spontaneously, possibly with other atoms. This gas gradually condenses into stars and galaxies. The creation of matter keeps the average density of the universe constant. So the universe as a whole always looks the same. That is, the universe is thought to be infinitely large and infinitely old, with no beginning and no end.

According to this theory, old and new galaxies should be found mixed together in all parts of space. This appears to be the case.

On the other hand, the 3°K radiation indicating a low average temperature of the universe, discovered in 1965, is generally interpreted as the remnant of the original high temperature and constitutes remarkably direct evidence for the "big-bang" theory. Also, the theory of the expansion of the universe predicts that as the galaxies separate, mutual gravitation opposes the expansion and the rate of expansion should decrease. The proof of this theory is difficult; the curve relating speed of recession and distance is not a straight line, as it would be if the rate of expansion were uniform. However, the departure of the observations from a straight line has not yet been determined accurately. The "big-bang" theory seems to be more widely accepted than the steady-state theory.

21.8 EVOLUTION OF GALAXIES

The theory of the development of a galaxy is difficult and has not yet been worked out in satisfactory detail. Suppose that at first the universe was pure hydrogen. If it were spread uniformly throughout space, it would have a tendency to collect into smaller units, where the density would be a little greater than average. These units would grow by attracting other particles to themselves by gravitational pull, and would form galaxies. The *pregalaxy* state might be called a *protogalaxy*.

Smaller collections of gas would similarly form into *protostars*. They would grow by adding particles, as in the dust-cloud theory of the formation of the sun and the solar system (see Chapter 15). At the proper stage of temperature and pressure, a star would begin to convert the hydrogen into helium, develop through various stages into a giant, and decrease in size until the helium core was too hot. Then the star would go through the nova stage. The larger masses would evolve at a relatively more rapid rate, as previously noted.

The important part of this development of stars in a galaxy is that helium, and in some cases carbon and heavier atoms, would have been produced in the stellar interiors. Later they would have been ejected into space by the explosion of novas to be captured by other stars in the process of formation. The importance is that carbon is the element that makes the carbon-nitrogen cycle work in the production of energy. By similar nova and supernova stages, the elements in the periodic table could have been produced. This is, of course, pure speculation.

If this theory is correct, there should be galaxies in various stages of evolution from the young galaxies with blue, hot stars to the older ones where the massive stars have used up their hydrogen and are left with only the older, redder stars. This is indeed the case and has led to a recent revision of the classification of galaxies by Morgan. Morgan's revision of Hubble's system is not intended to replace the older system but rather to add information regarding the constitution of galaxies in view of the discovery of the two stellar populations.

The Morgan classification retains the condensation of the nucleus (such as Sb or SBc) of the Hubble system, then adds a designation based on the colors of the galaxies as produced by stars of different spectral classes within each galaxy. In the sequence irregular galaxies and spirals with very small nuclei are placed at one end. They are designated as "a" galaxies consisting of stars of spectral classes B, A, and F, all of which produce strong radiations at the violet end of the spectrum.

At the red end of the sequence are the giant ellipsoids and spirals with high concentration of light in the nuclei. These are the "k" galaxies, because they are composed mostly of giant stars of spectral class K. Other groups are placed between the two extremes. Thus the classification is based on modern developments in the theory of evolution of stars, since it permits the galaxies composed of young stars of the main sequence (the B, A, and F stars) to be distinguished from those of the old stars of spectral class K.

An irregular galaxy is an unorganized collection of particles illuminated by light from massive, blue stars. It is possibly a young galaxy that has not yet formed into a spiral. The hydrogen atoms by themselves cannot form into dust grains, which must be the result of a sticking together of heavier atoms. The formation of grains makes it possible for the hot gas to radiate away some of its heat, thus helping in the formation of a galaxy under its own gravitation. This shows the importance of the nova stages in the evolution of stars.

As the galaxies develop, the blue stars evolve more rapidly, leaving the less massive stars, like the sun, to go through their cycles at slower rates. Also, the loose particles are used up by falling into the protostars and the older galaxies become nearly free of gas and dust clouds and are eventually composed entirely of dwarf stars.

In his classification of galaxies, Hubble suggested a direction of evolution. He thought that perhaps a galaxy started as a spherical mass of material now thought to have been hydrogen—an E0 galaxy. Because of its rotation it became more flattened. Then it developed arms that extended from the center as in either a normal or a barred spiral. The older galaxies developed more prominent arms. Hence our galaxy, though old, is not in the oldest class.

Shapley suggested that the evolution was in the opposite direction and that the ellipsoids are the remains after the stars in the arms had been thrown out into space. Since spiral galaxies seen edgewise (Fig. 20-4) are much flatter than the ellipsoidal galaxies, it is difficult to see how Shapley's direction of evolution is possible, since a galaxy would tend to become even more flattened, rather than become thicker.

The S0 galaxies are known to be dust-free and are composed of the older Population II stars, which have finished their evolutionary processes because there is no more material on which to grow. The question has come up as to how these galaxies could have become free from dust. The suggestion has been made that, since clusters of galaxies contain large numbers of members of this type, they are produced by collisions. When two galaxies collide (Fig. 21-10), there is so much room between the stars that they pass through without colliding. But the gas and dust particles are swept up in the process and are separated from the stars. Thus the S0 galaxies could have become dust-free during collisions or by the process of star formation from interstellar clouds.

FIG. 21-10   The strong radio source Cygnus A, photographed with the 200-inch telescope. (*Photograph from the Hale Observatories*)

21.9 THE ROLE OF RADIO ASTRONOMY

Radio astronomy developed very slowly after Jansky's discovery of radio waves from space in 1931. Reber began his radio studies in 1936, six years later, and drew a map of the Milky Way (Fig. 8-14) that showed lines of equal intensity similar to the lines of equal elevation on contour maps.

After World War II, development came much faster when radio telescopes were set up in the United States, Holland, and Australia. Radio signals from the sun were first discovered in England in 1942 and by Reber in 1943. These signals were from the quiet sun, and the temperature of the emitting material was estimated at 1 million degrees Kelvin. We now know that the temperature of the corona is at least that high. When the sun is active, more intense radiation is superposed on that from the quiet sun. Still stronger radiation is received from solar flares.

When radio waves are produced by material at various depths in the solar atmosphere, presumably at different temperatures, the heights at which these waves are produced can be studied by radio observations of different wavelengths. For example, the corona produces radiation 15 meters in length. If our eyes were sensitive to that radiation, we should see a sun 10 percent larger at minimum activity than we do with eyes that see only visual light.

A strong source of radio waves from inside our galaxy is from the Crab Nebula (Fig. 17-7). This nebula is the remnant of a supernova and its visual and radio radiations are emitted by electrons accelerated in magnetic fields, as already noted.

In 1948 the first radio source outside the galaxy was discovered. This source, known as Cygnus A (Fig. 21-10), coincides with a distant cluster of galaxies. Since 1948 catalogs from several radio observatories have been published. The third catalog of radio objects from the Cambridge center in England (3C) listed 471 sources, of which 100 have been made observable by improved resolving power of radio telescopes, which permit right ascensions and declinations to be determined with an accuracy of about 10″.

It has been pointed out (Section 20-4) that cold hydrogen radiates at a wavelength of 21 cm. Other waves come from ionized hydrogen in various regions of the Milky Way with wavelengths between 3 and 22 cm, but which do not radiate appreciably at 21 cm. These radiations are called *thermal*. A third class of radio waves, *nonthermal* waves, are probably produced by high-speed electrons moving in magnetic fields.

The nonthermal radio sources are of three or more kinds:

1. *Remnants of supernovas and filamentary nebulae.* The most spectacular and best-known example is the loop or filamentary nebula in Cygnus (Fig. 21-11). This source is nearly circular, with an apparent shell of diameter 50 parsecs in the center. Other sources are the remnants from Tycho's and Kepler's supernovas and the Crab Nebula.

2. *Normal external galaxies.* These include M31, M51, and M81. The radio emission is only about one millionth of the radiation in visible light and is almost a million times the total radiation of the sun. M31 is thought to consist of a disk component about the same as the optical disk of our galaxy, and a halo of much greater extent.

3. *Radio galaxies.* These galaxies differ from the normal galaxies in their high output of radio energy, which ranges up to 100,000 times that from a normal galaxy. Cygnus A and

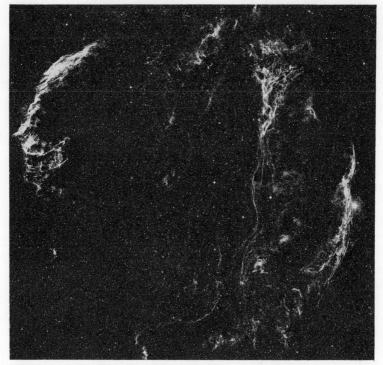

FIG. 21-11   The Filamentary Nebula in Cygnus, photographed with the 48-inch Schmidt telescope in red light. *(Photograph from the Hale Observatories)*

others are composed of two centers seen with a faint bridge between them when observed at 21 cm. The bridge is still more prominent at longer wavelengths. Some double galaxies are very large with components ranging from 10,000 parsecs to more than 100,000 parsecs in diameter, with distances between the centers as great as 300,000 parsecs. They may, therefore, be as much as 10 times larger than our galaxy.

The peculiar elliptical galaxy NGC 5128 (Fig. 21-12) is an unusual example of a radio galaxy.

21.10 QUASI-STELLAR OBJECTS

Some radio sources were originally called *radio stars* because they looked somewhat like stars. However, it was found that they almost certainly are not stars and were called *quasi-stellar radio sources,* or *quasars* for short. Before long, other objects were discovered with similar appearance and spectra that do not emit radio waves. These are also called quasars.

The first object of this type was discovered in 1960. It was listed as 3C48, the forty-eighth object in the third Cambridge catalog of radio sources. The

FIG. 21-12 The peculiar ellipsoidal galaxy NGC 5128 in Centaurus, a strong radio source, photographed with the 200-inch telescope. *(Photograph from the Hale Observatories)*

discovery was made by Thomas Matthews and Allan Sandage at the Hale Observatories. It is a sixteenth-magnitude starlike object in the constellation Triangulum, about 15° from the Andromeda galaxy. Its spectrum shows broad emission lines. Between 1960 and 1963 three others were discovered, and the rate of discovery increased rapidly in the years following. By early 1969 some 10,000 QSO's were known, including about 60 quasars. By 1971 over 100 radio quasars had been identified and there are many more quiet quasars. These discoveries were made from counts of objects down to eighteenth magnitude. It is estimated that perhaps there are 100 million within reach of our largest telescopes.

That the quasi-stellar objects are different from stars is apparent from their optical properties. While they look like stars, on photographs they show small, fuzzy disks (Fig. 21-13). 3C48 is less than 1″ in diameter. 3C273, located in the constellation Virgo, is the brightest (magnitude 13). It is double with a jet coming from the center. The two parts are 30″ apart. This source varies irregularly in brightness by 40 percent in a period of about 10 years. It appears on the Harvard sky-patrol plates, where it has been photographed since 1887, so the variations can be studied over an interval of more than 80 years.

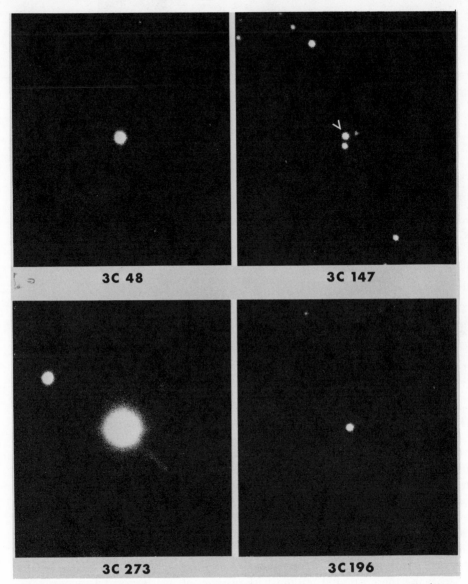

FIG. 21-13 Four quasi-stellar radio sources, photographed with the 200-inch telescope. *(Photographs from the Hale Observatories)*

The fact that these objects also have much more ultraviolet energy than stars, indicates a very high temperature. They show broad emission lines in their spectra. And perhaps most important of all, they exhibit very large red shifts. Two lines of hydrogen and one line of ionized oxygen in 3C273 show a red shift of 16 percent. Hence it cannot be a star, since a star with a velocity of more than 600 km/sec would leave the galaxy. Sixteen percent of the speed of light is 48,000 km/sec, which is comparable to the speed of a galaxy at a distance of almost 2 billion light-years. Other quasars with still higher speeds were discovered in 1965. 3C9 has such a high speed that the ultraviolet lines in the spectrum are shifted into the visible spectrum. The usual lines measured in the red shift of galaxies (Fig. 21-8) are shifted completely out of the visual region of the spectrum. At these high speeds, the linear Doppler shift cannot be used to compute the velocity, and a formula from the theory of relativity must be used.

The usual form of the Doppler equation (7-4) was discussed in Chapter 7. From the theory of relativity, the equation takes the form

$$\sqrt{\frac{1 + (v/c)}{1 - (v/c)}} = \frac{\lambda}{\lambda_0} = 1 + z \qquad (21\text{-}2)$$

where v, c, and λ have the same meanings as in equation (7-4) and

$$z = \frac{\lambda}{\lambda_0} - 1 = \frac{\lambda - \lambda_0}{\lambda_0} \qquad (21\text{-}3)$$

or, solving,

$$\frac{v}{c} = \frac{(1 + z)^2 - 1}{(1 + z)^2 + 1} \qquad (21\text{-}4)$$

For 3C9, Maarten Schmidt of the Hale Observatories found $z = 2.012$. Substituting in the above equation (21-4) and solving, $v = 0.8c$. That is, the velocity of 3C9 is 80 percent of the velocity of light, or 240,000 km/sec. This quasar is an eighteenth-magnitude object near the eastern edge of the Square of Pegasus.

At the beginning of his study, Schmidt, using spectra taken with the 200-inch telescope, measured the red shifts of nine quasars with z varying from 0.158 to 2.012. One beautiful feature of his work was that, starting with 16 percent of the speed of light for 3C273, he could identify lines in successive objects, making them fit ultraviolet lines two or three at a time. As one set disappeared into the invisible part of the spectrum, others came into view from the ultraviolet. In 3C254, where $z = 0.736$, he identified five lines; in 3C287, three; and in 3C9 only two lines were measurable. But this succession of lines fit perfectly when the proper value of z was found.

When the results were tabulated and studied, it was found that all quasi-stellar radio sources have diameters of less than 1" of arc. Their velocities, as measured by the Doppler effect, vary from 12 percent to 82 percent of the velocity of light.

It is, of course, impossible to measure the distances of these very distant objects. Using the Hubble constant of 75 km/sec per million parsecs, as given by Sandage in 1965, the distance of 3C9 is 3.2 billion parsecs or 10.5 billion light-years and is one of the most distant objects yet found. From its visual magnitude of 18.5, the absolute magnitude is computed to be -24.0. This is 2.8 magnitudes brighter than, or 13 times the brightness of, the entire Andromeda galaxy. If its angular diameter is 1", then the linear diameter is 16,000 parsecs or about 40 percent of the diameter of M31.

Schmidt estimated that 3C273 at thirteenth magnitude is 40 times the luminosity of the usual galaxy and 0.1 or less its diameter. Thus the emission of light per unit volume is much greater than usual. Its light variation in 10 years seems to show that it has an even smaller center from which the radiation comes. This center is estimated to be about 5 light-years in diameter, not much more than the distance from the sun to the nearest star. The radiation is 50 trillion (5×10^{13}) times the radiation from the sun. This central portion has two other companion components 3 and 30 times larger.

Various suggestions have been made to account for the production of such an enormous amount of energy. The most easily acceptable seems to be that, in the stages of evolution from a protogalaxy to a normal galaxy, there is a release of energy of this kind. The least plausible may be that of the annihilation of matter by the combination of ordinary matter and antimatter. The proponent of this latter black-hole theory thinks that no other source would provide this amount of energy. In 1965 Sandage announced the discovery of a still different type of object, which he called *quasi-stellar galaxies* (QSG). They are now included among the QSO's, whether or not they are emitting radio energy.

Quasars (QSO's) show very large red shifts, indicating radial velocities between 12 and 82 percent of the velocity of light. They vary in brightness, having periods of the order of one day. Although they are comparatively small, their radiations are enormous—trillions of times greater than the radiation of the sun. Their spectra contain both bright and dark lines—both emission and absorption of energy. There are multiple shifts within any one spectrum. No acceptable theory explains all the characteristics found here. Also, there is disagreement as to the distance of these objects. Most astronomers think the Hubble constant of expansion of galaxies also applies to the quasars and that they are very distant objects. Others think they are inside our galaxy and that the computations of size and amount of radiation are entirely wrong.

As noted earlier, the distance scale is uncertain and hence the Hubble constant is only approximate. Using the 1965 figure of 75 km/sec per million parsecs, the radius of the universe is 4×10^9 parsecs or 13×10^9 light-years. That is, any galaxy at that distance would theoretically be moving away from the center, or from us, at the speed of light. If there are galaxies out to that distance, the age of the universe since the expansion started should be 13 billion years, but may be only one phase of a pulsating universe. There is obviously much left to do in the study of this exciting and rapidly developing field of astronomy. Unfortunately, a 200-inch telescope is required to photograph the spectra of these very distant, faint objects. In this field optical and radio astronomy must work very closely together.

In 1969 a new machine for processing astronomical photographs was built in England and tested in Scotland. It is to be used in connection with a Schmidt camera. The exact coordinates and brightness of stars are determined by scanning the Schmidt plates with a photoelectric scanning device. The rate is 900 stars per hour, which may increase after further improvements are made.

Schmidt plates contain thousands of images. They can be taken in three colors, such as the standard UBV (ultraviolet, blue, and visual) regions of the spectrum. The color indices of all stars, and thus their temperatures, can be determined, as well as their magnitudes in the three colors. Also, the machine can determine the size of star images, as well as their positions, and can detect images not visible to the human eye.

The accuracy of position is about 10 microns and the size of image can be measured to less than 1 micron (10^{-3} mm). This technique will make it possible to detect and measure the angular diameters of such objects as quasars and small, distant galaxies. It will permit the astronomer to discover young stars and even protostars that could not be found otherwise. And these measures can be made very much faster than by human studies of the Schmidt plates.

On the first plate, taken in the constellation Perseus, 1003 young stars were found in a region where only 15 were known previously. The theoretical rate of formation of stars may have to be radically revised.

QUESTIONS AND PROBLEMS

Group A

1. Distinguish between Population I and II stars as to temperature, distribution in galaxies, age, and association with gas and dust.

2. Compare the apparent diameter of the only galaxy visible to the unaided eye in the northern hemisphere, M31, with that of the moon. The angular diameter of the brighter portion of M31 is 3°.

3. Compute the angular size of the (a) Large Magellanic Cloud, and (b) Small Magellanic Cloud. *Answer:* (a) About 11°.

4. How do the spectra of distant galaxies differ from those of emission nebulae?

5. The absolute magnitude of the Andromeda galaxy, M31, is estimated to be -21.2. To how many suns is this equivalent? (The absolute magnitude of the sun, $M = +4.9$.) *Answer:* 2.8×10^{10}.

6. List the various methods of measuring distances to galaxies and give the approximate maximum distance measured by each method.

7. Show that measuring distances to galaxies depends fundamentally on the diameter of the earth.

8. What assumptions are made in measuring distances to galaxies in which individual stars cannot be distinguished?

9. When measuring the distance to a remote cluster of galaxies, astronomers use only the light from the brightest galaxies. Why not use the light from the cluster as a whole, as is done with globular star clusters?

10. Use the velocity-distance relation to find the distance in light-years to galaxies that are receding from ours at (a) 20 percent and (b) 30 percent of the velocity of light. Assume

$H = 75$ km/sec/million parsecs. *Answer:* (a) 2.6×10^9 light-years.

11. The red-shift relation discovered by Hubble indicates that galaxies are receding from ours in all directions. Does this imply that our galaxy is at the center of the universe? Explain.

12. (a) Calculate the velocity with which the Andromeda galaxy should be receding from ours according to the velocity-distance relation. Assume H from Question 10. (b) The Doppler shift of light from M31 indicates a relative velocity of approach of 300 km/sec. Explain.

13. If the distances are actually 5 times larger than those in Fig. 21-8, recompute the Hubble constant.

14. What evidence indicates that quasars are very distant? Explain why the estimates of energy produced by quasars depend on the accuracy of measurements of their distances.

15. Compare the modern concept of the universe with that held before 1920.

16. Compare the resolving power of radio telescopes used in determining locations for the third Cambridge catalog with that of the 200-inch telescope.

17. In what ways do quasi-stellar objects differ from stars and galaxies?

Group B

18. The apparent visual magnitude of the 1885 nova in M31 was 7.2. If it is assumed to be an average nova of absolute magnitude $M = -7$, calculate the brightness ratio and verify the distance quoted in the text.

22   man's conquest of space

Man dreamed of flying and of visiting the moon and planets centuries before the first airplanes and spaceships were invented. Most of these dreams were not scientifically feasible, but many of them were remarkably accurate predictions of things to come. And, feasible or not, they spurred man on to the space efforts that have resulted in the artificial satellites, lunar expeditions, and planetary probes of today.

22.1 FORERUNNERS OF SPACE EXPLORATION

The first imaginary space voyage of which we have a record was written in A.D. 160 by Lucian, a Greek satirist. The name of Lucian's book was *True History*, but he tells his readers at the very beginning that they should not believe it. In the story, a ship is caught in a violent storm and a giant waterspout deposits it on the moon.

In a book appropriately called *Sleep*, published in 1634, Kepler also transports people to the moon. When they reach a certain point in space, the moon's attraction takes over. This was before Newton proposed his law of universal gravitation.

Newton himself introduced the idea of artificial satellites in his *Principia Mathematica* in 1687. To explain the motion of the moon in this book, Newton drew a diagram showing the earth and a hypothetical mountain high enough to reach beyond the atmosphere. "If a cannon ball were fired horizontally from this mountain with sufficient velocity," said Newton, "it would fall continually around the earth." He recognized the importance of avoiding atmospheric friction for such a feat.

Many other writers penned fictional space voyages after the time of Newton. The most familiar are probably those of Jules Verne, a 19th-century French author. For his moon trips, Verne imagined a bullet-shaped spaceship fired out of a large cannon. Made of aluminum, the craft contained a water-filled shock absorber that cushioned the passengers against the pressures of acceleration.

Escape from the earth requires a speed of 7 miles/sec (about 25,000 mph) at the earth's surface. Space flight was not possible until propellants were developed that could produce such speeds. The forerunner of today's chemical rocket propellants was gunpowder. Developed by the Chinese almost 1000 years ago, it consisted of a mixture of sodium nitrate, sulfur, and charcoal, and was soon used for military purposes. Rockets were used on the battlefield as early as 1400, but space flight was not forgotten. About the year 1500, a Chinese experimenter named Wan-Hoo fastened 47 gunpowder rockets to the bottom of a chair and climbed aboard. Servants ignited the space vehicle, but no trace of Wan-Hoo or his ship was ever found.

The "rockets' red glare" over Fort McHenry during the War of 1812 is immortalized in our National Anthem. A few years earlier, in 1807, some 25,000 military rockets had been used in the destruction of the city of Copenhagen. Oddly enough, the full potential of this weapon was not exploited until World War II, more than a century later.

During the early years of the 20th century a Russian school teacher, Konstantin Tsiolkovsky, built the first wind tunnel for the study of friction on the skin of a spacecraft. He also proposed the use of liquid propellants in rockets. It was Tsiolkovsky who first used the term *sputnik*, meaning "fellow traveler."

A book published by the Austrian mathematician, Hermann Oberth, in 1923 is entitled *The Rocket in Interplanetary Space*. It contains detailed specifications for high-altitude rockets and even space stations. Many scientists consider it the book on which the space age is founded.

In 1925 Walter Hohmann, a German scientist, published a book explaining five possible flight paths from the earth to Mars and from the earth to Venus, even though no vehicles or propellants were available to achieve them. At about the same time Robert Esnault-Pelterie, a French inventor and flying enthusiast, wrote books on rockets and rocket fuels. He coined the term *astronautics*.

First to launch a liquid-propelled rocket was the American physicist Robert H. Goddard (1882–1945) in 1926. Goddard devoted his life to the publishing of treatises on rocket propulsion, and he conducted many successful launchings. Strangely enough, he kept his first launchings secret for many years. He developed techniques that are essential in today's space program, including the centrifugal pump for rocket propellants and the gyroscopic guidance method. Unfortunately, he did not live to see the fruition of his work in space research.

In Germany, the Society for Space Travel was organized in 1927. It helped produce a motion picture. *The Woman in the Moon,* in which the concept of the *countdown* was first used. Some of the American space experts of today received their basic training as members of this society. During World War II the society was asked to develop a long-range military rocket. Known as the V-2, it was powered by alcohol and liquid oxygen. It weighed close to 15 tons, had a range of 190 miles, and attained altitudes of 100 miles. Traveling at speeds of almost 2000 miles/hr, the V-2 was virtually impossible to intercept. More than 2000 V-2s were fired during the war.

After the war, 300 carloads of V-2 parts were shipped to the White Sands Proving Grounds in New Mexico for further experimentation (Fig. 22-1). Additional test facilities were built later at Cape Kennedy (formerly called Cape Canaveral) and at other locations. The Russians, who captured some of the German rocket sites, also pursued the development of large rockets after the war.

In 1957 and 1958 both the United States and the U.S.S.R. conducted atmospheric experiments with rockets as part of the IGY (see Section 1.3). These space

FIG. 22-1   A V-2 rocket being readied for launching at the White Sands Proving Ground in 1946. (*U.S. Army photograph*)

efforts soon extended far beyond the original goals. Both military and scientific possibilities became evident. In 1958 the American Congress established the National Aeronautics and Space Administration (NASA) for the purpose of joining all government agencies, the aerospace industry, and the academic and scientific community into a coherent program of space exploration. The well-known John F. Kennedy Space Center in Florida and the mammoth Manned Spacecraft Center in Houston, Texas, are NASA installations.

22.2 ROCKET PROPULSION

Gunpowder will burn in a vacuum, since its ingredients contain oxygen as well as fuel. Hence the gases expelled from a gunpowder-type rocket are produced without the aid of the surrounding air. This makes the rocket ideally suited as a space engine, which must operate in an airless environment.

By definition, then, a *rocket* is an engine that can operate in a vacuum because it is not dependent upon oxygen in the atmosphere for combustion. It differs from a jet engine, which uses the oxygen of the air to combine with its fuel.

Rocket power is dependent on the reaction principle, which is another name for Newton's third law of motion (Chapter 4). Contrary to the notion of some people, a jet or rocket is not propelled by the push of escaping gases against the surrounding atmosphere. If this were true, it would be impossible to use a rocket in the vacuum of space. Instead, the craft is propelled by the unbalanced forces inside the engine directly opposite the nozzle, as shown in Fig. 22-2. The magnitude of these forces depends upon the weight of the expelled gases and the rate at which they leave the rocket. This is called the *exhaust velocity*.

The propellant, the source of power for a rocket, consists of a *fuel* and an *oxidizer*. The fuel is the liquid or solid part of the propellant that combines with the oxidizer to produce the gases that propel the rocket. The oxidizer may also

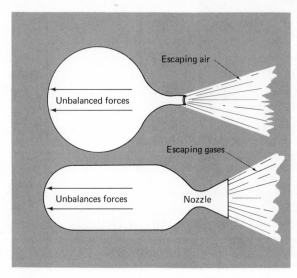

FIG. 22-2 Unbalanced forces inside a balloon and a rocket. The propelling forces are entirely internal and do not depend on any external environment.

be either liquid or solid. Fuels and oxidizers are chosen for maximum *thrust* and *specific impulse*. Thrust, the measure of the unbalanced forces inside the rocket, is usually given in pounds. The thrusts of some of the important American and Soviet rockets are listed in Table 22-1.

Specific impulse combines thrust, burning time, and the weight of the propellant by the formula:

$$\text{specific impulse } I_{sp} = \frac{\text{thrust} \times \text{burning time}}{\text{weight of propellant}} \qquad (22\text{-}1)$$

Specific impulse is therefore given in pound-seconds per pound, or merely in seconds, since the pound units cancel out in the formula.

TABLE 22-1   Thrusts of U.S. and Soviet Rockets

name	use	thrust (pounds)
Sputnik I	First man-made satellite	About 600,000
Redstone	First U.S. satellite	About 95,000
Atlas	Mercury series of flights	360,000–380,000
Titan II	Gemini series of flights	430,000
Titan IIIC	Communications satellites	470,000
Saturn IB	Test flights	1,500,000
Saturn V	Apollo series of flights	7,500,000
Vostok	Later Soviet vehicles	700,000

The specific impulses of several liquid fuel-oxidizer combinations are given in Table 22-2.

TABLE 22-2   Specific Impulses of Some Liquid Propellants

fuel	oxidizer	I_{sp} (sec)
Octane	Nitric Acid	220
Ethyl alcohol	Red fuming nitric acid	220
Octane	Hydrogen peroxide	230
Ethyl alcohol	Liquid oxygen (LOX)	240
Gasoline	Liquid oxygen	242
Octane	Liquid oxygen	248
Boron	Liquid oxygen	330
Liquid hydrogen	Liquid fluorine	370
Liquid hydrogen	Liquid oxygen	375

In a solid-propellant rocket engine, both fuel and oxidizer consist of dry chemicals. The first gunpowder rockets were of this type. The propellant mixture in a modern rocket is still called *grain*, a term that has been borrowed from gunpowder terminology. The advantage of solid propellants is that they can be mixed and placed in the rocket ahead of time and are instantly ready for launching.

Liquid propellants, on the other hand, must be loaded just before launching, since the fuel and oxidizer are usually kept at a very low temperature. The specific impulses of some solid propellants are given in Table 22-3.

TABLE 22-3   Specific Impulses of Some Solid Propellants

propellant	I_{sp} (sec)
Black gunpowder (potassium nitrate, charcoal, sulfur)	80
Galcit (potassium perchlorate, asphalt, oil)	180
NROC (ammonium picrate, sodium nitrate, resin)	180
Cordite (nitrocellulose, diethylphenyl urea)	200
Yellow gunpowder (tetranitro carbazale, potassium nitrate, carbon, wood flour, polyvinyl acetate)	200
JPN (nitrocellulose, nitroglycerine, diethylphthalate, diethylphenyl urea, potassium sulfate, carbon black, candelilla wax)	205

One disadvantage of solid propellants is that cracks may form in the mixture and this may lead to uneven burning and engine failure. Solid propellants also deteriorate with age and are extremely sensitive to temperature changes, so it is important to determine carefully the useful life of a propellant in a rocket that will not be launched immediately after loading.

Two other measures of rocket performance are the *thrust-to-weight ratio* and the *mass ratio*. The ratio of an engine's thrust to the total weight of the fueled rocket before liftoff is called the thrust-to-weight ratio. Thus, the thrust of a rocket weighing 200,000 pounds must be 400,000 pounds if it is to have a thrust-to-weight ratio of 2. Obviously, a rocket with a thrust-to-weight of less than 1 will not leave the ground; and the higher the ratio the greater will be the acceleration.

A more inclusive term is the mass ratio of a rocket. A rocket may have a high thrust-to-weight ratio and still not go very far if the propellant is soon exhausted. The mass ratio, on the other hand, is the relationship between the rocket's total mass and its mass after the propellant is used up. The higher the mass ratio the faster the rocket will travel, even though the specific impulse is the same as in that of a slower vehicle. It is for this reason that rockets are *staged* and that the stages are jettisoned as the propellant in them is used up. Then the mass of the final stage will be small in comparison with the total mass of the rocket before launch.

Space scientists are looking for other methods of propulsion to overcome the limitations of both solid and liquid propellants. One of the most promising is nuclear power. When liquid hydrogen is heated by a nuclear reactor, a specific impulse of over 1000 seconds is possible. And the resulting temperatures are no higher than in some of the high-impulse chemical reactions. Theoretically, specific impulses of 1 million seconds and more are possible by nuclear methods, but the resulting heat problems have not yet been solved.

If nuclear engines are used in manned vehicles, adequate shielding must be provided for the crew. This, plus the weight of the nuclear engine, offsets much of the high specific impulse the engine provides. In other words, the mass ratio is not as high as might be desired.

Plasma and ion engines are also being developed for use in space vehicles. A hot, ionized stream of gas is called *plasma*. The plasma can be heated by electrical means, by nuclear power, or even by solar energy. Once ionized, the gases are expelled from the engine to provide thrust. If liquid hydrogen is used as the propellant, exhaust velocities of 20,000 ft/sec can be attained, compared with 8000 to 10,000 ft/sec for chemical systems.

Even the slight radiation pressure of the sun can be used for rocket propulsion, although the resulting thrust is small indeed. The push of a 100-kilowatt searchlight, for example, is less than 0.0001 pound. But once a vehicle is in space, an extremely small thrust will produce constant acceleration. The speed of light is the theoretical limit of how fast a craft will ultimately travel, although it would take many years of constant acceleration to attain such speeds. Rockets powered in this way have been called *photon rockets*. They may look remarkably like sailboats, since large solar sails would be used to make maximum use of the sun's pressure. These sails will be maneuvered very much like their seagoing counterparts, with no need to worry about squalls or calms.

Why this urgent search for new and greater power? Simply because the power sources of today's rockets are not adequate for the space missions of the future. By way of illustration, an initial specific impulse of 750 seconds is necessary in order to complete a manned round trip to Mars in a year's time. To reduce the time to three months, the specific impulse must be raised to 1400 seconds. And in order to explore the entire solar system, a specific impulse of approximately 5000 seconds is necessary. These requirements are considerably beyond those listed in Tables 22-2 and 22-3 for present solid and liquid propellants.

22.3 ROCKET GUIDANCE

Once a rocket has been launched, an elaborate system of tracking and guidance is used to keep it on course and to assure the successful completion of the mission. Electronic devices are used to measure the vehicle's environment and to transmit the data to a ground station. This is called *telemetry*. Telemetry is also used to monitor the performance of the vehicle and its crew. Telemetered data are broken down into sequential segments known as *bits*. These bits are then decoded in the *readout*. In photographs, like the ones from the moon or Mars, the bits are converted into light impulses and pieced together very much as in the transmission of photographs by radio methods on the earth.

The process of following the movements of a rocket in flight is called *tracking*. Radar, radio, and cameras are used for this purpose (see Fig. 22-3). At

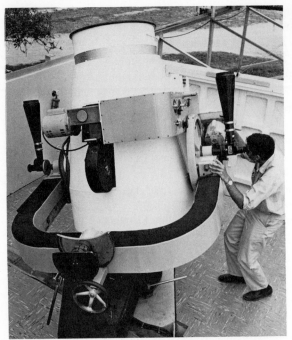

1958 α₁
1 Aug. 1958

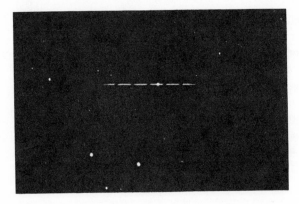

FIG. 22-3 The Baker-Nunn camera (top) on Mt. Haleakala, Hawaii, can detect a 1-foot object thousands of miles in space. To photograph dim satellites, the camera tracks them across the sky, so they appear as sharp dots on the negative (middle). Stars show up on such negatives as long streaks interrupted by the camera's timing device. With the camera stationary, bright satellites become streaks, whereas the stars appear as dots during a short exposure (bottom). This unusual picture shows the Geos satellite on December 20, 1965, with a flash of light reflected from a corner reflector on board the spacecraft (bright dot in fourth streak). (*Smithsonian Astrophysical Observatory photo*)

present, optical tracking is the most precise, but it is severely limited both by weather conditions and by distance. NASA operates a network of tracking stations, telemetry ships, and data-processing centers for satellite tracking. The stations of the Manned Flight Network are located in a belt around the earth (Fig. 22-4).

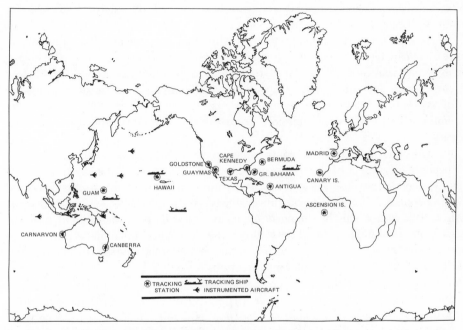

FIG. 22-4 The NASA Manned Space Flight Network is composed of 14 tracking stations, four tracking ships and eight instrumented aircraft that feed the vital flow of information through a half-million miles of communication circuits, including land lines, undersea cables, radio circuits, and communication satellites, to Mission Control Center, Houston, Tex. (NASA)

During flight, space vehicles require light, dependable electric power for the operations of transmitters and other instruments. Nuclear energy, fuel cells, and solar cells have been used for this purpose. The American program for on-board nuclear power is called SNAP (Systems for Nuclear Auxiliary Power). Regular reactors are being tested for this purpose, as well as heat production through the decay of radioisotopes. SNAP units have already worked for several years without breakdown in American satellites.

A fuel cell differs from a dry cell or storage battery in that the ingredients can be fed into it while it is operating. Also, in the fuel cell the electrodes are not changed. Only the fuel and oxidizer are consumed. It is also more efficient than the electrolytic cell. The American spacecraft, Gemini 5, was the first to use fuel cells for on-board power. Liquid hydrogen and liquid oxygen were employed.

For the direct utilization of sunlight in a spacecraft, reflectors more than 30 feet in diameter focus the sun's rays onto a supply of mercury. The mercury vapor is used to run electric generators.

The sun's energy can also be converted into electric power by solar cells, which utilize the same photoelectric effect found in the familiar "electric eye." The current produced by a single cell is very small, so it takes literally thousands of them to operate transmitters on the few watts of power needed. Mariner 4, for example, contained 28,224 solar cells for the production of 10 watts of power. The solar cells must be aimed at the sun, whereas nuclear power and fuel cells do not. The surface of a solar cell can be damaged by micrometeoroids and cosmic radiation.

Because of the large antennas used in tracking, very weak radio signals from spacecraft can be detected over great distances. The 3-watt signal of Mariner 2, for example, was received over a distance of almost 54 million miles. The 85-foot antennas used for this purpose are about 20,000 times more sensitive than the ordinary rooftop television antennas. Because the signals are so weak, they can become garbled with radio emissions from the sun, the stars, and other celestial sources.

Conventional aerodynamic controls can be used for a rocket when it is in the atmosphere of the earth or some other planet. Because of the great speed of rockets, their guide fins are smaller than those on commercial aircraft. Beyond the atmosphere, ballistic controls must be used. With the rocket engines firing, the path of the rocket can be changed by means of exhaust vanes that deflect the gases from the engine. Small jets along the fuselage, called deflection charge controls, are also used. Or the entire engine can be mounted on swivels so that the thrust can be aimed in different directions.

The orientation, called attitude, of a spacecraft may be described by three motions similar to those of a ship or airplane. *Roll* is a rotating motion; *pitch* is an up-and-down nodding motion; and *yaw* is a side-to-side motion. The three axes about which these motions take place are illustrated in Fig. 22-5.

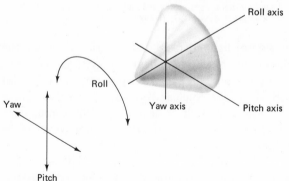

FIG. 22-5 Definition of the three attitude axes of a spaceship.

An instrument called the accelerometer is used to measure the effects of the forces acting on a space vehicle. The inertial resistance of a mass in the accelerometer makes it possible to compute both the velocity and distance traveled. Three are required, one for each of the attitude axes. To keep them properly aligned, rapidly spinning gyroscopes are used. The entire array of gyroscopes and accelerometers, with their associated instruments, is known as the inertial guidance system of the rocket.

Reference stars are used to check the position of a space vehicle, as well as to keep the guidance system oriented. The bright star Canopus is frequently used, since there is no other bright star within 35° that might be confused with the reference star. A sensor on the vehicle is set to respond to light of the intensity of Canopus. If, because of reflection of sunlight from dust particles, or if for some other reason, the sensor loses the star, it can be commanded from the earth to search for the star again.

The orbit of an artificial satellite depends on its velocity. At escape velocity of 7 miles/sec, the orbit is a parabola with the earth at the focus. If the velocity is less than 7 miles/sec, the orbit is an ellipse; if greater, the orbit is a hyperbola. The three types of orbit, all with respect to the earth, are illustrated in Fig. 22-6. The three coincide at the point of insertion into orbit, but the velocity is different for each at that point, and the body goes into the orbit around the earth that is required by its speed.

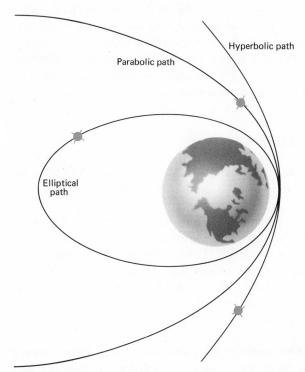

Hyperbolic path

Parabolic path

Elliptical path

FIG. 22-6 Three possible types of orbit of a spaceship. The orbital speeds increase from the ellipse to the hyperbola. The parabolic path is required if the launch speed is equal to the velocity of escape from the earth.

Since the earth's velocity around the sun is 18.5 miles/sec, the spaceship will have this speed plus its velocity with respect to the earth, provided the launch is toward the east, the direction in which the earth rotates on its axis and revolves around the sun. An eastward launch takes advantage of the earth's speed of rotation; a north-south launch is intended for low-altitude vehicles used to study conditions on the earth.

If the total velocity of the ship is 1.414 ($\sqrt{2}$) times the velocity of the earth in its orbit around the sun, its orbit with respect to the sun is a parabola, and the ship leaves the solar system. That is, it reaches the parabolic velocity of 26 miles/sec. If the velocity is less than the parabolic velocity, the orbit is an ellipse with the sun at one focus. Thus the orbital motion of an artificial satellite is governed by the same physical laws that control the motions of other celestial bodies.

During the reentry of a space vehicle into the earth's atmosphere, retro-rockets are fired to reduce the velocity. As altitude is lost, however, the velocity increases according to Kepler's laws of motion. This increased speed sends the vehicle to a slightly higher altitude again. But it also increases the friction on the rocket's surface, so the original altitude is not reached. This cycle is repeated until the vehicle reaches the ground.

During reentry, the kinetic energy of the moving satellite is converted into heat energy. This heat must be dissipated in order to prevent the destruction of the vehicle. A number of methods are used to deal with the heat. In the *ablation* method, a heat shield is constructed of a material that absorbs heat and will melt and fall away during reentry. In *transpiration* cooling, a liquid coolant is forced through holes in the hull and dissipates heat by evaporation. Electro-magnetic fields can also be used to prevent the heated atmosphere from reaching the craft. This method is still in an experimental stage.

22.4 UNMANNED SPACE VEHICLES

A space vehicle designed to obtain information about the environment of some celestial body, including the earth, is called a *probe*. The first artificial satellites were probes.

Artificial satellites can be used as navigation aids that are virtually independent of the weather or the time of day. Successive radio observations of such satellites are used to determine the locations of ships and planes by comparison with special tables. Also, if a satellite emits a signal of its own, the Doppler shift of the signal becomes a valuable navigation aid. A nontransmitting satellite is said to be *passive*, whereas a transmitting one is termed *active*. An active-repeating satellite receives a signal from the earth, amplifies it, and transmits it back to earth.

The first American communications satellites (comsats) were large, passive balloons, called *Echo*. Some of them were more than 150 feet in diameter and were easily visible to the naked eye during favorable viewing conditions. *Telstar I*, launched July 10, 1962, was the first to relay live telecasts between the United States and Europe. *Syncom* orbits at an altitude of 22,400 miles and appears stationary over the rotating earth. Three such synchronous satellites spaced evenly over the equator can provide continuous worldwide communications. The same thing could be accomplished, but not quite as satisfactorily, by orbiting a series of satellites at a lower altitude.

An artificial satellite can also serve as an observing station for the study of the earth's surface, the weather, or some celestial object (Fig. 22-7). An important new series of satellites called ERTS (Earth Resources Technology Satellite) will provide information on the world's food supply, air and water pollution, crop diseases, and natural resources.

FIG. 22-7   An artist's sketch of the OAO-2 in orbit. This man-made satellite is yielding spectral data that are radically changing present ideas about the nature and size of the universe and the composition of interstellar space. (*Grumman Aerospace Corporation*)

Unmanned exploration of the moon was conducted by the Ranger, Orbiter, and Surveyor series of space probes (Fig. 22-8). The Rangers took pictures of the moon before crashing. In the Surveyor series, cameras and other instruments were soft-landed on the lunar surface. The Surveyors sent pictures back to earth and some had mechanical scoops to test the hardness of the lunar soil. Surveyor VI made a small jump and photographed the indentation the footpads made when

FIG. 22-8 The Saturn IB, NASA's launch vehicle for lunar exploration. (NASA)

the craft first landed. A television camera and parts of Surveyor V were recovered by the astronauts of Apollo 12 and brought back to earth, where they were examined for effects of lunar temperature changes and bombardment by particles and radiation after three years on the moon's surface.

American planetary investigations are carried out by the *Mariner* and *Pioneer* family of probes. Pioneer 10, launched on March 2, 1972, will take almost two years to reach Jupiter and will then become the first spacecraft to leave the solar system. It carries a pictorial message for any extraterrestrial beings who might intercept the vehicle in the distant future. (It will take 80,000 years for Pioneer 10 to reach the nearest star.)

22.5 HUMAN SURVIVAL IN SPACE

Some critics of the space program argue that manned vehicles are unnecessary and expensive. They claim that the exploration of the moon and planets can be accomplished by instruments in unmanned vehicles without providing the elaborate equipment required by a crew. But it is impossible to foresee all the conditions that a lunar or planetary space vehicle may encounter. This was dramatically demonstrated during the landing of Apollo 11 on the moon, when astronaut Armstrong manually guided his ship away from rough terrain to a smooth landing some distance from the planned landing site.

The most obvious need of a space crew is a satisfactory air environment. If sufficiently powerful boosters are available, a normal mixture of oxygen and nitrogen can be carried along. But if weight is at a premium, helium can be substituted for the nitrogen. The first U.S. manned flights reduced the weights still further by furnishing the astronauts with pure oxygen at reduced pressure.

When an astronaut leaves his ship, he must take a breathing environment with him. This is accomplished by providing him with a spacesuit, which must also protect him against any radiation or micrometeoroid bombardment. Rigid suits with hinged joints are used for extended stays in space. In effect, the astronauts wear miniature space capsules.

A second consideration for a space crew is to keep the acceleration and deceleration forces within tolerable limits. They are known as *g-forces*, multiples of a person's weight on the earth's surface (Fig. 22-9). Thus a *g*-force of 3 during

FIG. 22-9 A rocket sled in which human reactions to high g-forces can be determined. Here, a fully equipped dummy is shown being ejected during a test run. (*North American Rockwell*)

launch means that the crew experiences three times its normal weight. In the first American manned vehicles, maximum loads of $5g$ were common.

Shortly after enduring the crushing forces of acceleration, the astronaut experiences the completely opposite sensation of *weightlessness*. Actually the term is misleading, because the spacecraft and its occupants are still under the gravitational pull of the earth as well as the sun. It is more nearly correct to say that the vehicle is in a state of free fall, and that under these conditions there is no restraining force on the vehicle or its crew. Weightlessness is, therefore, very much like falling, except that in space the surrounding scenery does not rush by.

While the effects of extremely long periods of weightlessness have not yet been fully determined, it is certain that some type of regular exercise is necessary to prevent the atrophy of muscles and organs. It has also been proposed that slow rotation be used to provide a kind of artificial weight in future space stations. In such an arrangement, down would be toward the outside of the vehicle.

In order to conserve as much weight and space as possible in a capsule, the food supply for the crew is in concentrated form. But liquids must be dispensed in ways that will prevent them from escaping and floating about the capsules as globules.

For longer journeys, all or part of the necessary food and water will be recycled. A number of closed ecological systems are presently under study. The most promising involves the use of a certain species of algae, containing a large percentage of protein. It has been found that a 2-liter container filled with an algal solution can, if illuminated with a powerful light source, supply the food and oxygen needs of an astronaut indefinitely. As the algae are harvested and prepared for food, more are synthesized from the carbon dioxide exhaled by the astronaut, and the water is reclaimed within the capsule. Other recycling schemes involve the use of mushrooms and molds for the conversion of waste products into edibles. Still another plan calls for the chemical conversion of waste materials into rocket propellants. With this plan, the vehicle could be launched with a large food supply instead of a large stock of propellants.

A space crew must also contend with the external hazard of micrometeoroids and cosmic radiation. It is believed that the density of micrometeoroids in the vicinity of the earth is between 100 and 1000 per cubic mile. Fortunately, they are so small that a spacecraft is not seriously damaged by them. In time, however, they can render solar cells inoperative through a sandblasting effect on the transparent shields covering the cells. For larger objects, meteoroid bumpers might be used or spaceships might be equipped with double hulls. Chances of being struck by a large meteoroid are very small, however; and it should even be possible to detect them by radar and to steer around them.

Likewise, cosmic radiation can to some extent be predicted and avoided. The Van Allen belts, for example, have escape corridors over the poles. In addition, the intensity of the Van Allen radiation seems to depend on solar activity. It may become necessary to program launchings for times when solar prominences or flares are expected to be at a minimum.

22.6 MANNED SPACE EXPLORATION

On April 12, 1961, Yuri Gagarin became the first man to orbit the earth, in the vehicle Vostok I. The U.S. manned space program began with the Mercury series. The first American to orbit the earth was Col. John Glenn, Jr., who made the flight on February 20, 1962, in a Mercury capsule.

After a number of successful Mercury flights, the two-man Gemini series was launched. Extended flights of two weeks' duration were included in this program, as well as the intricate maneuver of a rendezvous of two space vehicles. There were 10 Gemini flights.

The three-man Apollo program followed. Project Apollo had the goal of landing astronauts on the moon. The Apollo vehicle accordingly was made up of three sections. The Command Module accommodated three astronauts. It weighed 5 tons and was 12 feet high. It was the only part of the craft that returned to the earth. The Service Module was equipped with rockets so that it could maneuver into and out of orbit. It weighed 25 tons and was 23 feet long.

The Lunar Excursion Module (LM or LEM) was the part of the craft that landed on the moon. Two of the three astronauts rode it down to the lunar surface, where it rested on spiderlike legs. The legs and LM's landing rockets remained on the moon when the astronauts left to rejoin the Command and Service Modules. In later Apollo missions, self-propelled Lunar Rovers were used by the astronauts to explore the moon's terrain. These were also left behind. Obviously, very powerful boosters were required to launch all this equipment at one time (Fig. 22-8). Apollo 17, the last in the series, was launched on December 6, 1972.

Manned American space programs for the remainder of the 20th century include *Skylab* and the *Space Shuttle*. In the former, crews will spend up to 56 days in an orbiting laboratory designed to study the earth's importance to man's well-being and his influence on the ecology of our planet.

The *Space Shuttle* will be a reusable manned spacecraft. Scheduled for launching in the 1980's and 1990's, it will be able to place satellites in orbit or to retrieve them for repair before landing on a conventional runway. Each shuttle may be able to perform as many as 500 missions.

QUESTIONS AND PROBLEMS

Group A

1. Try to find out what types of fuels are used in present space launches.
2. Can a space vehicle go beyond the pull of the earth's gravity? Explain.
3. In which direction should a satellite be launched if it is to be placed in orbit with the least expenditure of energy? List all reasons for your answer.
4. The first manned trip to a planet is planned for Mars, even though Venus is closer. (a) Why? (b) Why might more fuel

be required for a round trip to Venus than to the surface of Mars?

5. If a space vehicle orbiting the earth is in a continual state of free fall, why does it never get any closer to the earth?

Group B

6. Using Kepler's third law and the sidereal period of the moon and its mean distance from the earth, show that the height of a stationary satellite is that given in the text.

7. Find the weights of the following propellants required to supply a thrust of 1000 pounds for 2 minutes: (a) octane and nitric acid, (b) liquid hydrogen and liquid oxygen. *Answer:* (a) 545 pounds.

8. A single-stage rocket has a total weight of 1000 pounds, including 800 pounds of propellant. It has a thrust-to-weight ratio of 2. How many seconds will the propellant burn if it is (a) cordite, or (b) black gunpowder? *Answer:* 80 seconds.

9. If the acceleration of a photon rocket due to the radiation pressure on the sail is 0.001 g, how much will its velocity increase in one year? *Answer:* 192 miles/sec.

10. How many watts would be generated if all the solar energy incident on a mirror mounted on an earth satellite could be converted to electrical energy? Assume that the diameter of the mirror is (a) 30 feet and (b) 50 feet. (The solar constant is 2 cal/min/cm^2 and is equivalent to 130 watts/ft^2.) *Answer:* (a) 92,000.

11. Assuming the average density of micrometeoroids in the neighborhood of the earth to be 100 per cubic mile, find the number a satellite could be expected to encounter in one orbit. The satellite is 10 feet in diameter and is in a circular orbit 1000 miles above the earth's surface. *Answer:* About 9.

23 the development of astronomical thought

23.1 ASTRONOMY BEFORE COPERNICUS (3000 B.C. TO A.D. 1500)

Astronomy has often been called the oldest of the sciences, and it is easy to see why, for man has always been fascinated by the stars. Throughout the ages, from prehistoric to modern times, when man looked at the heavens and saw the stars, he was filled with awe. Of course, we moderns are not as aware of the heavenly bodies as were former generations. Living in the polluted air of our cities, surrounded by high buildings, a person might live a whole lifetime and never actually see the sun rise or set. It is still more difficult to observe that the place of its rising or setting changes as the seasons advance. Is it surprising that our awareness has dulled?

Until fairly recently most people could see the sky unless clouds obscured it; and, being curious creatures, they asked questions. What are these objects in the sky? Where did they come from? How far away are they? What makes them move around in the sky, most of them maintaining the same position in relation to each other? Why do a few wander around by themselves? Do they have any influence on us? If so, what effects do they have?

As people attempted to answer these questions, knowledge of astronomy advanced in many parts of the world. As soon as they associated the change of seasons with the position of the sun and the appearance of certain bright stars and constellations, they began to wonder what other influences the heavenly bodies might have on their lives. Eclipses of the sun and moon, the unexpected appearances of bright meteors and comets—could they be prophetic signs? (See Fig. 23-1.) So they began to observe the sky carefully. Many ancient peoples developed a privileged class whose members devoted all their time to the study

FIG. 23-1   An old engraving showing an eclipse of the sun astounding Peruvian natives. (The Bettman Archive, Inc.)

of the heavens. These observers were held in high esteem. They were considered very wise, and, since it was believed that they could predict future events from a study of their data, they functioned as astrologers as well as astronomers.

In China and Japan their predictions were limited to public affairs, including the welfare of the rulers. Unusual occurrences, such as eclipses of the sun, were considered very bad omens and struck terror into the hearts of the common people as well as the emperor and his court. Calamities that were supposed to occur because of an eclipse or the appearance of some strange object in the sky could be averted if the phenomenon were predicted in time so that proper precautions could be taken. If the unlucky astrologer failed to predict an eclipse, he might be removed from office or even executed. But if he predicted an eclipse that failed to occur, he was congratulated for giving warning in time for steps to be taken to ward off the disaster. Under these circumstances, it is not surprising that many eclipses were listed that were never actually observed.

In addition to predictions affecting the whole country, the Babylonians and Greeks believed that private lives were also influenced by the stars. In order to cast a successful horoscope, it was necessary to know the exact position of the stars when the person was born. Sextus Empiricus described how this could be accomplished. Two astrologers were involved. One stayed outside and observed the sky, while the other watched the progress of the birth and struck a cymbal at the precise moment when the infant was born. Thus the astrologer watching the sky could determine the "ascendant birth star" and cast a horoscope for the infant.

This intensive study of the exact positions of the stars and planets by the astrologers led to advances in the science of astronomy. When astronomers were able to devise measuring instruments, such as the gnomon and the water clock (clepsydra), their careful observations became even more accurate.

The Babylonians, for example, whose records extend from about 2000 B.C. to A.D. 75, were great observational astronomers and calculators. They were interested primarily in the exact time at which celestial objects appeared above and disappeared below the horizon, because they believed these phenomena influenced human affairs. The Babylonians also kept a complete list of eclipses but never tried to explain what they had observed.

The Greeks, on the other hand, tried to find causes and underlying principles and were more critical. Although they proclaimed that they owed a great debt to Babylonia and Phoenicia, their own contributions were superior to those of the Mesopotamian region because they probed more deeply and were more analytical.

The Greeks, Babylonians, and other cultures around the eastern Mediterranean region were able to exchange scientific information. The Greeks borrowed from the Babylonians, and the Alexandrians in turn absorbed knowledge from the Greeks. Thus Western science, profiting by cross-fertilization, made its greatest advances from the 3rd century B.C. to the 4th century A.D. in Alexandria. Scientists from the entire eastern Mediterranean regions were encouraged to come there and exchange ideas while living at the Museum, where they had access to the great Alexandrian Library.

Far Eastern science, on the other hand, did not have the advantage of being able to attract recruits from different regions. Since science makes most progress when it can freely absorb new ideas, this cultural isolation was a distinct handicap. However, the Chinese developed advanced astronomical concepts on their own initiative. For example, a treatise written in the 2nd century B.C. and based on observations going back to the 6th century B.C. placed the north pole in the center of the heavens as the fundamental point of a polar-equatorial coordinate system.

Western astronomy did not develop a similar system until the latter part of the 16th century, nearly 2000 years later. During the period when the Chinese were studying the heavens as a whole, Greek astronomers limited their studies generally to the zodiac. The zodiac includes the paths of the sun, moon, and all the planets known at that time. Investigating the reason the planets wander around the way they do, the Greeks devised the system of circles and epicycles to explain the planetary motions (see Chapter 4).

For our knowledge of astronomy in other parts of the ancient world, we rely on the records we have been able to discover and decipher and on monuments, paintings, and inscriptions that have been preserved, as in the writings of the Babylonians, Greeks, and Chinese.

We have no evidence that the early Egyptians kept records of astronomical observations, although their papyri on mathematical and medical subjects have come down to us. For Egyptian astronomy we rely chiefly on their monuments, such as the Great Pyramid, and on astronomical representations and inscriptions on their tombs (Fig. 23-2). After the Persian conquest at the end of the 6th century B.C., what was called Egyptian astronomy was actually Greek astronomy, differing radically from the true Egyptian astronomy that had preceded it. According to modern scholars, true Egyptian astronomy did not reach the high

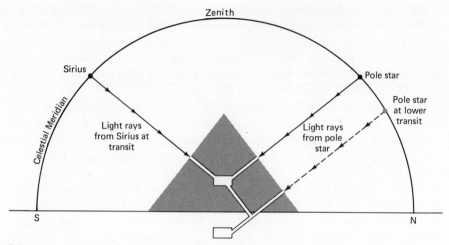

FIG. 23-2   Celestial orientation of the Great Pyramid of Cheops. The four sides of the giant structure face the four cardinal points. During the construction of the pyramid, the rays of Sirius shone down a ventilating shaft into the King's Chamber when the star was a transit. α Draconis, the pole star at that time, shone down the main opening and a secondary shaft when that star was at lower transit. The pyramid dates back to the 25th century B. C.

levels of Greek and Babylonian astronomy, but it was much more accurate than that of most primitive peoples.

In the British Isles the Druids erected monuments of stones, some of which show evidence of having been used for astronomical observations. Stonehenge (see Section 9.1), near Salisbury, is such an ancient monument.

Although the Incas in pre-Columbian America apparently observed the times of the rising and setting of the sun, as evidenced by special markers they set up, they have left no records of their observations.

The Mayans and Aztecs of Central America not only kept records, but their pyramids and other buildings included round towers that served as observatories for observing the equinoxes, solstices, and the setting of the moon when it is farthest north or south (Figs. 23-3 and 23-4). They were interested chiefly in the sun, moon, and Venus. From the motions of these bodies they constructed extremely complicated calendars. A year of 365 days was used for ritualistic purposes, and a 260-day year was used for foretelling future events. These two were combined to produce a 52-year cycle called "a bundle of years." In addition, they had a Venus cycle of 583.92 days, which was so accurate that the error over a thousand years would not be more than one day. The Mayans could handle very large numbers and were not afraid to use them.

Ancient astronomy reached its highest point with Ptolemy's work in Alexandria from A.D. 121 to 151. Among his writings are the *Almagest* and the *Planetary Hypotheses,* in which he presented a new theory of the motions of planets that included epicycles. His contributions were so outstanding that they influenced astronomical thought for the next 14 centuries.

After the 4th century astronomy declined in Alexandria and in the Greek

FIG. 23-3 The Caracol, ruins of a Mayan observatory in the Yucatan Peninsula of Mexico. More than five centuries ago this 75-foot tower was used to make sightings of the rising and setting of the sun, moon, and planets. Stone-lined passages, one oriented due west and another due east, served as lines of sight. The modern name of the observatory is from a Spanish word meaning "snail," because the winding central stairway in the structure resembles a snail shell. (Peabody Museum, Harvard University)

world in general. One reason was the downgrading of theoretical research in favor of practical applications. Another was a growing preoccupation with magic. Barbaric invasions also had their effect. The era of ancient astronomy had ended.

Since Byzantium was relatively safe from invasion, it became the route through which Greek science filtered through to the Arabic world. Arabic science flourished from the 8th to the 12th centuries, and Baghdad became the center of learning. Arabic science preserved and refined Greek science, although it did not make outstanding contributions of its own. However, the Arabic number system (which actually originated in India, not Arabia) greatly facilitated astronomical computations. Its numerals are written according to position to represent units, tens, hundreds, and so on, making most of the systems in use at that time seem clumsy by comparison. Just try multiplying 20 by 14 using the Roman numerals: XX times XIV. How do you go about it?

After the Arabs conquered Spain, the center of learning shifted from Baghdad to Cordoba and Toledo, in Spain, and it is mainly through these centers that the Greek and Arabic science reached the western world in the 11th and 12th centuries.

In the Middle Ages, which extend roughly from the 5th to the 15th century, science was not altogether stagnant (Fig. 23-5). The introduction into Europe of Greek science through Arabic sources stimulated interest, and Greek manuscripts were copied assiduously by the monks in monastaries. The rise of the European universities in the next century and a half advanced scientific knowledge, while the rise of technology from 1350 to 1450 had its influence also, although it was applied chiefly to the solution of practical problems in medicine and navigation.

FIG. 23-4   The Sun Stone, or "Aztec Calendar." Carved more than 100 years before our calendar was adopted, this 25-ton monolith divides the year into 18 months (depicted by the outermost band of figures). The four squares surrounding the sun god depict the four cosmogonic epochs of Aztec mythology: the jaguar, wind, fire, and water. *(Mexican National Tourist Council)*

The invention of printing in 1450 caused a tremendous upsurge of interest in intellectual pursuits in general and in astronomy in particular. By the end of that century over 1000 presses were in operation and a million books had been published. The voyages of the great Spanish and Portugese explorers, which had begun about this time, required astronomical tables. They also proved beyond doubt that the earth is round.

23.2 ASTRONOMY AFTER 1500

The next century was a time of change. With books becoming more available and strange tales coming back from faraway places, it is no wonder that the 16th century brought in new ideas and paved the way for the emergence of modern astronomy. Copernicus shifted the center of the universe from the earth to the sun. Tycho Brahe made thousands of accurate observations, enabling Kepler to formulate his laws of planetary motion. Tycho and Kepler were the

FIG. 23-5 The earliest printed star chart. This woodcut of the northern celestial hemisphere was made by Albrecht Dürer in 1515. A companion chart depicts the stars in the southern sky. *(National Gallery of Art, Washington, D. C., Rosenwald Collection)*

last of the respected astronomers who studied astrology, although Kepler never took it seriously. However, he earned his living by casting horoscopes for the emperor. Tycho probably had more to do with discrediting astrology as a serious science than anyone else. Since that time, astrology has not been generally considered a reputable science.

The 17th century saw the beginning of modern astronomy with the invention of the telescope, which Galileo began to use in 1609. The invention of logarithms soon afterward enabled astronomers to make more computations in less time. It is said that the invention of logarithms lengthened the productive lives of all astronomers. Spectacular advances were made at this time by Galileo and later in the century by Newton.

The body of astronomical knowledge has continued to grow from that time to the present, with the emphasis shifting from the practical to the theoretical and back to the practical. Otto Struve, in *Astronomy of the 20th Century,* calls the period from about 1850 to 1950 the "Golden Age of astronomy," because astronomers during this period were not forced to seek practical applications but "only sought the truth for its own sake."

With the advent of the era of space exploration, astronomy has become practical again, and vast sums of money are poured into projects under pressure to produce results.

Modern technology has grown at an incredible rate. Orbiting observatories, which do not have the turbulent, cloudy atmosphere to contend with, hastened the development of more sophisticated instruments with which to equip them.

Landing on the moon was made possible by a number of highly developed instruments. According to Christopher C. Kraft, director of flight operations at NASA's Manned Space Center at Houston, Texas, the device most responsible for the tremendous advance from earth-orbiting Mercury flights to the Apollo trips to the moon was the high-speed computer.

Communications with the astronauts in the Mercury capsule were sent over low-speed teletype equipment from stations scattered around the world, manned by highly trained technicians, to Cape Canaveral (later called Cape Kennedy). It took 15 minutes to transmit limited data from the distant stations to the Cape.

Seven years later, the spectacular development of high-capacity electronic computers, both on board and on the ground, and the communications satellites made it possible to get almost instantaneously all kinds of data from Apollo to Houston, where it was received on 500 gauges or dials. The ground computer for the first Project Mercury had a storage capacity of 32,000 words; that used on an Apollo flight, 5.5 million words. Such tremendous advances in technology stagger the imagination. They make possible unprecedented advances in science, including astronomy. The technology that made possible the landing of men on the moon may mark the beginning of a new era as significant as the one Galileo initiated in the 17th century.

23.3 BRANCHES OF MODERN ASTRONOMY

Mathematical astronomy deals with motions and orbits. It is based on Newton's laws of motion and his law of universal gravitation. The computation of orbits, based on the attraction by the sun on the planets and by the planets on each other, makes possible accurate prediction of the positions of the planets, comets, and asteroids. Most of the great mathematicians of the early 19th century worked with the problems of mathematical astronomy.

These studies have been extended to double and multiple stars and to motions inside galaxies. Modifications by the theory of relativity lead to the computation of the velocities of very distant objects, and may therefore be applied to the theory of the expanding universe.

Astrometrics is the accurate measurement of positions, and it provides the data needed for mathematical astronomy.

Astrophysics, as the name implies, is concerned with the physical nature of celestial objects, their composition, the types of light and other energy they emit, and the variations in their behavior. The most recent problem of interest to astrophysics is the theory of evolution of stars. The great reflecting optical telescopes of the 20th century were built largely for these studies.

Radio astronomy, as already noted, is the study of radiation from celestial bodies outside the range of the optical spectrum. This newest branch of astronomy is still in its infancy; but it has yielded many important discoveries about the sun and its planets and about distant clusters and galaxies, including quasars and pulsars.

Cosmology is the study of the design and extent of the universe. It includes *cosmogony,* which deals with the origin and density of the universe and of the material of which it is composed. The problem of cosmology is to collect all known data about the universe and its parts and then put together a logical and consistent hypothesis about its structure and history.

23.4 HOW ASTRONOMICAL KNOWLEDGE PROGRESSED

One would think that astronomical knowledge would grow in a steady and orderly fashion, one new idea after another being added to build up a solid body of knowledge. But it does not happen that way. There may be intervals of intense intellectual activity when knowledge seems to grow very rapidly, as in the centuries when Alexandrian science flourished. There may even be a scientific explosion, as in our present era. These periods may be followed by shifts to other regions of the world or by periods when science seems to lie dormant for a while and then reawaken. Often knowledge seems to grow by fits and starts. Sometimes it is lost; sometimes it even goes backward.

For example, in the early part of the 6th century B.C., Anaximander recognized that the earth has a curved surface; but Anaximenes, who followed him, said it wasn't so. He said that the earth was a flat disk, as everybody knew, and proceeded to give proofs that were generally accepted.

Even the greatest astronomers made mistakes and held back the progress of knowledge. In the 3rd century B.C., Aristarchus proposed that the earth and all the other planets travel around the sun. But the great Hipparchus and the other astronomers of the famous Alexandrian school rejected that hypothesis. It was called outrageous and blasphemous. The influence of Hipparchus was so great, not only on his contemporaries but on those who followed him, that it was 1800 years before the hypothesis was proposed again. It appeared in Copernicus' *De Revolutionibus* just before he died, although he had written it 40 years earlier (Fig. 23-6). Even then Andreas Osiander (1498–1552), who wrote the preface, felt obliged to say, "These hypotheses need not be true or even probable," and they were not generally accepted for many years.

Galileo was not always right, either. A book called *Treatise on Comets* by Father Grossi, a Jesuit astronomer, was published about 1623. Galileo attacked the book with such violence that Grassi never attempted a reply, although Galileo was 100 percent wrong. Kepler suggested that the moon might have some

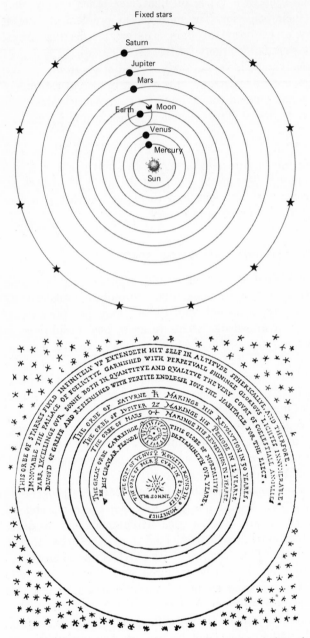

FIG. 23-6 Copernicus placed the sun at the center of the universe (top). This system simplified cosmology and was a major advance in the development of space science. In his conception of the heavens, Thomas Digges made an important change in the Copernican system. Instead of placing the stars on an outer shell, he scattered them at random in space (below), but he seemed to be unaware of the importance of this part of his work. (*Collection of the N. Y. Public Library; Astor, Lenox, and Tilden Foundations*)

influence on the tides and Galileo dismissed the idea with scorn. He advanced a ridiculous theory of his own: that the tides were attributable to the fact that various parts of the earth's surface were moving through space at different velocities!

On the other hand, some astronomers had vision far ahead of their times. Newton anticipated our travels into space when he wrote in the third book of his *Principia Mathematica:* "That the planets can be kept in their orbits by centripetal forces is evident from the motion of a projectile. A thrown stone, propelled by its own weight, is deflected from a straight-line curve, follows a curve in the air, and ultimately falls to earth. If it is thrown with greater velocity, it will go farther. It is conceivable that if the velocity were increased enough, the stone would ultimately shoot beyond the borders of the earth and not fall back again."

It is interesting to speculate on what Newton could have done if modern technology had been available to him in the 17th century.

23.5 CHANGING CONCEPTS OF THE UNIVERSE

As astronomers continue to study the heavens, the idea of the size of the universe changes. Our relatively small, earth-centered world had expanded by the end of the 16th century to a universe with the sun at the center and the earth and the other planets traveling around it, and with the stars scattered at random. In the 17th century, when telescopes showed the Milky Way to be composed of a vast number of stars, the universe had to be extended to include them, and it became a universe with the solar system at the center.

As early as the 18th century, William Herschel began to suspect that the Milky Way was not the whole universe, but he was unable to prove it. He reported to the Royal Society in 1785 that, as the power of a telescope increases, an observer "perceives that the objects that had been called nebulae are evidently nothing but clusters of stars. When he resolves one nebula into stars, he discovers ten new ones which he cannot resolve." But in spite of his suspicions, the Milky Way continued to contain the entire universe until the 20th century.

In addition to the stars, such other smaller systems as the globular clusters and the so-called spiral nebulae were considered part of the Milky Way system.

The discovery that the Andromeda Nebula was nearly 1 million light-years away, and that there are other similar nebulae, was made during the 1920s. And when some systems were found to be composed of stars, as shown by their spectra, and when they were resolved into stars by photography, they were called island universes and were thought to be like our Milky Way system except for size. The name galaxy was finally substituted for the word nebula.

Thereafter, the study of the origin, design, and extent of the universe developed rapidly. The galaxies were found to be nearly uniformly scattered through a universe that Einstein once thought to be infinite in extent. They were found to be moving away from us and from each other. During the 1920s the theory of the expanding universe was a popular astronomical topic. Our galaxy was placed in a small group of galaxies, called the Local Group, and it was estimated that there are 100 million galaxies in reach of the 100-inch telescope. The location of the center of the expanding universe was, and still is, unknown. Observations with the 200-inch telescope helped greatly in these studies. It is estimated that more than a billion galaxies can be observed with the 200-inch telescope.

The discovery of radio waves from space by Jansky in 1931 was not followed by others until Reber built a radio receiver in 1937 and found that the waves come from the Milky Way. This discovery attracted wide attention and led to the building of larger and larger receivers until today this branch of astronomy has resulted in studies of the most distant parts of the presently accepted finite universe. The study of pulsars has become very important as a result of radio studies.

From the relatively small, earth-centered world of former times, the universe has expanded in the past four centuries to include galaxies more than 10 billion light-years away. This figure staggers the imagination. Light reaching us now from the most distant galaxies started billions of years before the earth existed. In former ages men thought they knew the size of their universe. Now we are not so sure.

We are also interested in the state of matter in other regions of the universe. One guess is that the universe is half matter and half antimatter, but there is little evidence of antimatter in our region. We have little hope of discovering life like our own unless we can reach some star with planets at just the right distance to support life.

Progress has been made in recent years with theories of evolution of stars and galaxies, but many points still need to be confirmed and others to be developed. Radio astronomy has been invaluable. Theoretical astronomy is expected to advance rapidly with the help of observations with radio telescopes now being used in many parts of the world. More large optical instruments, such as other 200-inch telescopes, are needed and can be expected in the near future.

High-speed electronic computers are essential in many kinds of problems that would take an impossibly long time to solve with desk calculators. Computers are being used to solve problems concerning the orbits of satellites as well as the flight paths of lunar and planetary space ships. They are also important in theoretical investigations of the internal structure of stars—a vital part of any theory of stellar evolution.

At present our attention is turning toward the theories of evolution of the galaxies and of the universe and to the possibility of life outside the earth and the solar system. Space explorations have almost ruled out any notion of life on Venus or on Mars, once thought to be the most likely planets capable of supporting life. We are waiting for the time to come when we can leave our earth, land on Mars, and even send expeditions beyond our solar system. But even at the speed of light—and according to the theory of relativity, we are limited to this speed—it would take a spaceship more than four years to reach the nearest star. At our present attainable speeds, the time of flight would be prohibitively long.

With the present interest in the theories of evolution, one can speculate on the existence of hundreds of thousands of suns with life-supporting planets like the earth. On the other hand, when one is reminded of the uniqueness of living things—no two animals or even two leaves are exactly alike—one begins to be skeptical. We are also beginning to realize how delicate is the ecological balance that makes it possible for life to exist at all. Now ecologists are warning us that we are in danger of extinction.

Dr. Howard L. Sanders at the Woods Hole Oceanographic Institute warns: "My colleagues and I see irrefutable evidence of the gradual destruction of the planet's environment. Technology is magnificent today and is getting better. But we make these alterations in almost complete ignorance of their long-range consequences. We have now gone so far we face a worldwide emergency."

Technology has enabled us to make advances that seemed impossible only 100 years ago. But unless we take steps to use this technology to preserve our environment, we may find, too late, that it has changed our planet to the extent that it will no longer support life. As we take the first steps to find the conditions that must be met to make life possible at all, we begin to realize how fragile life is, and it would not be surprising if life did not exist in any other place in the universe.

When Galileo looked through his telescope in 1609 and saw for the first time the mountains on the moon, the rings of Saturn, the moons of Jupiter, and found that the Milky Way is composed of stars, neither he nor his contemporaries could possibly have grasped the full significance of his discoveries and how they would influence the development of astronomy during the next several hundred years.

In our own time, the landing of men on the moon may have significance beyond our comprehension. It may be centuries before a true judgment can be formed. We do know that people all over the world were thrilled by this achievement, and it seemed to bring us all closer together. It generated a feeling of hope and confidence in mankind. If man is capable of such a stupendous achievement, there is new hope that he can succeed in making this world a better place to live by improving the world environment and by uniting all men so they can live together in peace.

QUESTIONS AND PROBLEMS

1. List the most important changes in astronomical theories from the time of the most ancient civilization to the present time.
2. List the most important astronomical discoveries since the invention of the telescope.
3. List three inventions that have greatly influenced the course of astronomy.

appendices

A. METRIC-ENGLISH EQUIVALENTS

1 in =	2.540 cm	1 cm =	0.3937 in
1 in^2 =	6.452 cm^2	1 cm^2 =	0.1550 in^2
1 ft =	30.48 cm	1 cm^3 =	0.0610 in^3
1 mi =	1.6093 km	1 m =	39.37 in
1 qt =	0.946 liter	1 l =	1.06 qt
1 oz =	28.35 g	1 g =	0.03527 oz
1 lb =	453.6 g	1 kg =	2.205 lb
1 ton =	0.9065 metric ton	1 metric ton =	1.102 tons
			= 10^6 g

B. SCIENTIFIC NOTATION

In scientific work, very large and very small numbers are usually written as the product of a number from 1 to 9 and a power of 10. This is called **scientific,** or *standard,* **notation.** For example, 316,000,000 is written as 3.16×10^8. The power of 10 indicates the position of the decimal point. If it is positive, the decimal point must be moved to the right. In the example just given, the decimal point must be moved eight places to the right. If the power of 10 is negative, then the decimal point must be moved to the left. For example, 4.73×10^{-6} stands for the number 0.00000473. (See the additional examples below.)

If the number is:

$$6{,}030{,}000{,}000$$
$$7850$$
$$0.0346$$
$$0.00000000902$$

The scientific notation is:

$$6.03 \times 10^9$$
$$7.85 \times 10^3$$
$$3.46 \times 10^{-2}$$
$$9.02 \times 10^{-9}$$

Scientific notation has several significant advantages over the ordinary way of writing numbers. For one thing, it makes long numbers easy to read and to compare with each other. Hence, in lengthy problems, it is possible to estimate (and thus check) the answer quite readily. Another important aspect of this method is the fact that the first part of the number can be used to indicate the precision of a measurement. Thus, 3.724×10^4 centimeters is a more precise measurement than is 3.72×10^4 centimeters.

C. TEMPERATURE CONVERSIONS

To change from Fahrenheit (F) to Celsius (C), use the formula:

$$°C = \frac{5}{9}(F - 32°)$$

For example, $40°F = \frac{5}{9}(40° - 32°) = 4.4°C$.

To change from Celsius to Fahrenheit, use the formula:

$$°F = \frac{9}{5}C + 32°$$

For example, $20°C = \frac{9}{5}(20°) + 32° = 68°F$.

The Kelvin (K) reading is always $273°$ more than the Celsius reading for the same temperature. For example, $250°C = 250° + 273° = 523°K$.

D. RADIAN MEASURE

Calculations involving small angles can be simplified by using a unit of measurement called the **radian** (abbreviated *rad*) instead of degrees. As shown in the diagram below, the radian is related to the radius *r* of a circle. When two radii cut off a segment of a circle which is equal in length to a radius, the angle formed by the two radii equals one radian.

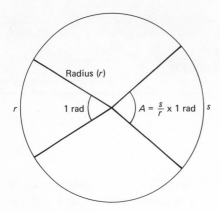

The angle between any two radii is proportional to the segment s of the circle which they cut off. That is, the larger the angle, the larger the segment. Hence, any angle can be expressed in terms of radians. For example, angle A in the above diagram is:

$$A = \frac{s}{r} \times 1 \text{ rad}$$

Since the circumference of a circle equals $2\pi r$, and 2π equals 6.28 (approximately), it follows that there are 6.28 radians in a 360° angle. In order to find the number of degrees in one radian, therefore, we can use the following formula:

$$1 \text{ rad} = \frac{360°}{2\pi} = 57.3° \text{ or } 3438' \text{ or } 206,265''$$

If an angle is very small, as in the case of the slender triangles used in finding the distances to the moon and the sun, the straight line between two points on a circle is almost equal to the corresponding curved segment, as shown in the diagram below. In distance problems, this straight line is called the *base line.*

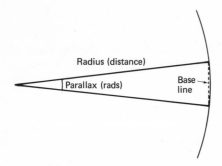

Thus, the formula for the size of the angle becomes:

$$\text{angle (in rads)} = \frac{\text{length of base line}}{\text{length of radius}}$$

In distance problems, the radius is the distance to be measured and the angle is the *parallax.* So the previous formula is solved for the radius as follows:

$$\text{length of radius (distance)} = \frac{\text{length of base line}}{\text{parallax (in rads)}}$$

If the parallax is in degrees, the right side of the formula must be multiplied by 57.3°. It is important to remember, however, that the formula is accurate only when the angle involved is very small.

E. TABLE OF NATURAL SATELLITES

name	discoverer	date	mean distance (miles)	sidereal period (days)	apparent magnitude	diameter (miles)
EARTH						
Moon			238,857	27.32	−12.6	2160
MARS						
Phobos	A. Hall	1877	5,800	0.319	12	13 × 15.5
Deimos	A. Hall	1877	14,600	1.262	13	7.5 × 8.5
JUPITER						
V	Barnard	1892	112,000	0.497	13	100?
Io (I)	Galileo	1610	262,000	1.769	6	2000
Europa (II)	Galileo	1610	417,000	3.552	6	1800
Ganymede (III)	Galileo	1610	665,000	7.155	5	3100
Callisto (IV)	Galileo	1610	1,170,000	16.689	6	2800
VI	Perrine	1904	7,100,000	251	14	70?
X	Nicholson	1938	7,200,000	254	19	10?
VII	Perrine	1905	7,300,000	260	18	0?2
XII	Nicholson	1951	13,000,000	625 (ret)	19	10?
XI	Nicholson	1938	14,300,000	714 (ret)	19	15?
VIII	Melotte	1908	14,600,000	735 (ret)	18	20?
IX	Nicholson	1914	14,700,000	758 (ret)	19	15?
SATURN						
Janus	Dollfus	1966	98,000	0.5	14	200?
Mimas	Herschel	1789	115,000	0.944	12	300?
Enceladus	Herschel	1789	147,000	1.370	12	350?
Tethys	Cassini	1684	183,000	1.888	10	630
Dione	Cassini	1684	234,000	2.737	11	550
Rhea	Cassini	1672	327,000	4.517	10	800
Titan	Huygens	1655	758,000	15.944	8	3000
Hyperion	Bond	1848	919,000	21.276	13	200?
Iapetus	Cassini	1671	2,207,000	79.3	10	700?
Phoebe	Pickering	1898	8,034,000	550 (ret)	15	100?
URANUS						
Miranda	Kuiper	1948	81,000	1.414	17	200?
Ariel	Lassell	1851	119,000	2.520	15	450?
Umbriel	Lassell	1851	166,000	4.144	15	390?
Titania	Herschel	1787	272,000	8.705	14	700?
Oberon	Herschel	1787	364,000	13.463	14	600?
NEPTUNE						
Triton	Lassell	1846	220,000	5.877 (ret)	14	2000?
Nereid	Kuiper	1949	3,500,000	359	19	200?

F. GREEK ALPHABET

A,	α	alpha	I,	ι	iota	P,	ρ	rho
B,	β	beta	K,	κ	kappa	Σ,	σ	sigma
Γ,	γ	gamma	Λ,	λ	lambda	T,	τ	tau
Δ,	δ	delta	M,	μ	mu	Υ,	υ	upsilon
E,	ϵ	epsilon	N,	ν	nu	Φ,	ϕ	phi
Z,	ζ	zeta	Ξ,	ξ	xi	X,	χ	chi
H,	η	eta	O,	o	omicron	Ψ,	ψ	psi
Θ,	θ	theta	Π,	π	pi	Ω,	ω	omega

glossary

ABERRATION OF STARLIGHT. Apparent displacement of a star on the celestial sphere due to the speed of light and the rotation or revolution of the earth.

ABSOLUTE MAGNITUDE. Apparent brightness of stars computed for a distance of 10 parsecs (32.6 light-years), expressed in magnitudes.

ABSOLUTE ZERO. The lowest temperature ($-273°C$), called $0°K$, where almost all molecular motion stops.

ABSORPTION SPECTRUM. Dark lines superposed on a continuous spectrum, due to the absorption by a cool gas in front of a hot source.

ACCELERATION. Rate of change of velocity; may be either positive (an increase) or negative (a decrease).

ACHROMATIC OBJECTIVE. A telescope lens of two or more components to correct for chromatic aberration.

ACTIVE SUN. The sun when producing an unusual number of spots and accompanying phenomena, as during the IGY.

ADVANCE OF THE PERIGEE. The eastward rotation of the orbit of the moon or artificial satellite around the earth.

ADVANCE OF THE PERIHELION. The eastward rotation of the orbit of a planet, particularly Mercury.

ALBEDO. The ratio of the light reflected by an object, especially the moon or a planet, to the total amount of light shining on it; reflective power.

ALPHA PARTICLE. A positively charged particle, consisting of two protons and two neutrons, emitted from the nucleus of a radioactive atom; the nucleus of a helium atom.

ALTITUDE. The angle between the celestial horizon and a celestial object, measured along its vertical circle.

ANGLE OF INCIDENCE. The angle between an incoming ray and the perpendicular to a reflecting or refracting surface.

ANGSTROM.   A unit of length used to measure wavelengths of light; 10^{-8} cm.

ANGULAR DIAMETER.   The angle between two opposite sides of an object.

ANGULAR DISTANCE.   The angle between two objects on the celestial sphere.

ANGULAR MOMENTUM.   The quantity of motion of a body due to its rotation; angular momentum $= mr^2\omega$.

ANNULAR ECLIPSE.   An eclipse of the sun where only a ring of the sun is visible around the moon when the moon is too distant to cover the sun completely.

ANTIMATTER.   Matter made up of antiparticles.

ANTIPARTICLES.   A particle with the same mass as its counterpart, such as antiproton instead of proton, with opposite electrical charge and other basic physical differences.

APHELION.   The point in the orbit of a planet or other body where it is farthest from the sun.

APOGEE.   The point in the orbit of the moon or artificial satellite where it is farthest from the earth.

APPARENT MAGNITUDE.   The brightness of a star or other celestial object as seen from the earth.

APPARENT NOON.   The time when the center of the sun is on the meridian.

APPARENT SOLAR DAY.   The interval between two successive transits of the sun's center across the meridian.

APPARENT SUN.   The real sun that moves along the ecliptic.

APPULSE.   A penumbral eclipse of the moon.

ASCENDING NODE.   The point where the orbit of a celestial body crosses the ecliptic from south to north; the point where the orbit of an artificial satellite crosses the earth's equator from south to north.

ASSOCIATION.   A group of stars with nearly the same physical characteristics, indicating that they probably had the same origin.

ASTEROID.   A small body in orbit around the sun, usually between the orbits of Mars and Jupiter; a minor planet; a planetoid.

ASTROMETRICS.   The branch of astronomy that deals with the accurate measurement of positions and motions of astronomical objects.

ASTRONOMICAL UNIT.   The mean distance between the earth and the sun as fixed by international agreement in 1900; 92,870,000 miles.

ASTRONOMY.   The branch of science that deals with the location, motion, and nature of objects in space.

ASTROPHYSICS.   The branch of astronomy that applies the instruments and theories of physics to the study of the nature of celestial bodies.

ATOM.   The smallest particle of an element that has all the properties of the element.

ATOMIC NUMBER.   The number of protons in the nucleus of an atom; the number of electrons surrounding the nucleus of a neutral atom.

ATOMIC THEORY.   The modern theory of the structure of atoms.

ATOMIC WEIGHT.   The relative mass of an atom compared to carbon 12.

AURORA.   A colorful radiation above the earth's polar regions; northern or southern lights.

AUTUMNAL EQUINOX.   The point on the celestial equator where the sun crosses it on the first day of autumn; R.A. 12^h; Decl. $0°$.

AZIMUTH. The angle measured eastward along the celestial horizon between the north point and the vertical circle through a given object.

BALMER LINES. A series of emission or absorption lines of hydrogen in the visible part of the spectrum.

BARRED SPIRAL. A spiral galaxy having an elongated nucleus with an arm extending from each end.

BETA PARTICLE. A negatively charged particle emitted from the nucleus of a radioactive atom; an electron.

"BIG BANG" THEORY. A theory of the evolution of the universe holding that the universe resulted from an initial explosion.

BINARY STAR. A double star with components close enough together to be affected by their mutual attraction; a visual, spectroscopic, or eclipsing double.

BLACKBODY. A hypothetically perfect radiator that absorbs and re-emits all radiation falling on it.

BLACK DWARF. The supposed final stage in the evolution of a star when it has no more energy to radiate.

BLACK HOLE. A hypothetical volume whose intense gravitation prevents light from escaping.

BODE'S LAW. A sequence of numbers that represent approximately the mean distances of the planets from the sun.

BOHR THEORY. A proposed model of an atom with electrons in orbits around the nucleus.

BOILING POINT. The temperature of a liquid at which the molecules have enough energy to break the attractive forces between them.

BOLOMETER. A physical instrument for measuring the total amount of energy emitted by a radiating body, or reflected from a body; especially well adapted to infrared studies.

BOLOMETRIC MAGNITUDE. A kind of stellar magnitude based on the total radiation from a body.

BRIGHT-LINE SPECTRUM. A pattern of bright spectral lines emitted by an incandescent gas under low pressure.

BRIGHT NEBULA. A visible gas illuminated by a nearby hot star.

BRIGHTNESS. A measure of the luminosity of a body.

CALORIE. The amount of heat energy needed to raise 1 gram of water by 1° Celsius; 4.186×10^7 ergs.

CANALI. Narrow markings on Mars; Italian for channels or canals.

CARBON CYCLE. A series of nuclear reactions involving carbon by which four atoms of hydrogen combine to form one atom of helium.

CARDINAL POINTS. The four principal directions: N, S, E, W.

CASSEGRAIN REFLECTOR. A reflecting telescope that uses a concave primary mirror with a convex secondary; light is brought to a focus through a hole in the primary at the lower end.

CASSINI'S DIVISION. A 3000-mile-wide gap in the ring system of Saturn.

CELESTIAL EQUATOR. The great circle on the celestial sphere 90° from each of the two celestial poles.

CELESTIAL HORIZON. The great circle 90° from zenith and nadir.

CELESTIAL MECHANICS. A mathematical branch of astronomy that deals with gravitation and the motions of bodies in space.

CELESTIAL MERIDIAN. The great circle through the celestial poles and the zenith.

CELESTIAL POLES. Two points on the celestial sphere on the axis of the earth's rotation about which the sky seems to rotate.

CELESTIAL SPHERE. A sphere of infinite radius with center at the eye of the observer.

CENTRIFUGAL REACTION FORCE. The inertial reaction to a centripetal force.

CENTRIPETAL FORCE. A force, directed toward a center of curvature, required to change a body from a straight path into a curved path.

CEPHEID VARIABLE. A star whose light output changes because of periodic expansions and contractions; a pulsating star.

CERES. The first minor planet to be discovered; also the largest.

CHANGE OF STATE. A change of substance from solid to liquid or from liquid to gas, or the reverse.

CHROMATIC ABERRATION. The failure of a lens to bring to one focus all the wavelengths of light that pass through it.

CHROMOSPHERE. The layer of the sun's atmosphere directly above the photosphere.

CIRCUMPOLAR STARS. Those stars near a celestial pole that are always above the horizon.

CLOCK DRIVE. The mechanism on a telescope that turns it at the proper rate to counteract the rotation of the earth.

CLUSTER VARIABLE. A type of variable star with a period less than one day and absolute magnitude about zero; an RR Lyrae variable.

COALSACK. One of two conspicuous dark areas in the Milky Way.

COLLIMATOR. A lens in a spectroscope that changes a diverging beam of light into a parallel beam.

COLOR EXCESS. The difference between the color index of a star and that of an unreddened star of the same spectral class; measured in magnitudes.

COLOR INDEX. The difference between the visual and photographic magnitudes of a star; expressed in magnitudes.

COMA. The hazy shell-like structure around the nucleus of a comet; an aberration in a telescope whereby images formed off the axis of the telescope are not in focus.

COMET. A diffuse astronomical body in orbit around the sun; usually in an elongated orbit.

COMPARISON SPECTRUM. A spectrum placed alongside the spectrum of a star for the purpose of comparing wavelengths.

COMPOUND. A substance composed of two or more elements.

CONCAVE LENS. A lens with one or both surfaces curved like the inside of a sphere; it is thicker at the edges than at the center.

CONCAVE MIRROR. A reflecting surface curved like the inside of a sphere.

CONIC SECTION. The curve formed by the intersection of a circular cone and a plane; a circle, ellipse, parabola, or hyperbola.

CONJUNCTION. A lining up of celestial bodies so that they have the same right ascension or celestial longitude.

CONSTELLATION. An apparent grouping of stars named for a mythical figure, animal, or inanimate object; there are 88 constellations.

CONTINUOUS SPECTRUM. An uninterrupted band of color emitted by an incandescent solid, liquid, or gas under pressure, as observed by a spectroscope or spectrograph.

CONVEX LENS. A lens with one or both surfaces curved like the outside of a sphere and thicker at the center than at the edges.

COPERNICAN SYSTEM. The solar system as described by Copernicus; the sun at the center and its family of planets, their satellites, and other bodies, associated by gravitation.

CORE. The central portion of the earth, the sun, or a star.

CORONA. The outermost layer of the sun's atmosphere.

CORONAGRAPH. An instrument for photographing the sun's prominences and corona at times other than during eclipses.

COSMIC RAYS. High-energy radiations and particles from space.

COSMOGONY. The branch of cosmology that deals with the origin and future evolution of all matter in the universe.

COSMOLOGY. The study of the structure and extent of the universe and its evolution.

CRAPE RING. The faint, innermost ring of Saturn.

CRATER. A depression on the earth, the moon, or Mars.

CRESCENT. The phase of the moon or a planet between new and quarter; bow shaped.

CROWN GLASS. A high-quality window glass used in telescope objectives.

CRUST. The topmost layer of the solid earth; 3 to 30 miles thick.

DARK NEBULA. A nonluminous gas in space, detected because it obscures the light of stars behind it.

DAYLIGHT SAVING TIME. Time obtained by advancing the clocks, usually during the summer; standard time plus one hour.

DECELERATION. The rate of decrease of velocity; negative acceleration.

DECLINATION. The angular distance north or south of the celestial equator, measured in degrees of arc along an hour circle.

DECLINATION AXIS. The axis of a telescope about which the telescope can be turned north or south.

DEFERENT. A circle in the Ptolemaic system centered in the earth, about which a planet or its epicycle was supposed to move in direct motion.

DEFLECTION OF STARLIGHT. The bending of light as it passes near a massive celestial body; predicted by the theory of relativity.

DENSITY. Mass per unit volume, usually grams per cubic centimeter (g/cm^3); also called *mass density*.

DESCENDING NODE. A point where the orbit of a celestial body crosses the ecliptic from north to south; a point where the orbit of an artificial satellite crosses the earth's equator from north to south.

DEUTERIUM. "Heavy hydrogen" with a nucleus composed of one proton and one neutron.

DIFFRACTION. The bending of light rays after passing through a narrow opening or openings between ruled lines on a grating; also produced by a reflection grating.

DIFFRACTION GRATING. A system of finely ruled lines used to produce a spectrum.

DIFFRACTION PATTERN. A system of alternate bright and dark lines produced by interference or diffraction.

DIFFUSE NEBULA. A bright or dark nebula, but not a planetary nebula.

DIFFUSION. The tendency of gas molecules to fill a space uniformly.

DISK OF GALAXY. The central, flat part of a galaxy superposed on the spiral structure.

DIFFUSION OF LIGHT. Scattering of light from an irregular surface.

DISPERSION. The separation of colors of light in a spectroscope.

DIRECT-FOCUS TELESCOPE. A reflecting telescope where observations are made at the focus without the use of a second mirror.

DIRECT MOTION. The usual west-to-east motion of a planet or other body.

DISTANCE. The measure of separation between the centers of two objects.

DISTANCE MODULUS. The difference between the apparent and absolute magnitudes of an object; used to compute distances.

DIURNAL CIRCLE. A circle described by a celestial object on the sky due to the earth's rotation; a small circle parallel to the celestial equator.

DIURNAL LIBRATION. An apparent motion of the moon produced by the rotation of the earth, which permits us to see a little of the moon's limb that is alternately visible and invisible.

DIURNAL MOTION. The apparent motion of objects in the sky due to the earth's rotation.

DOPPLER EFFECT. The apparent change in wavelength or frequency of sound, light, or other radiation produced by the relative motion of source and observer.

DOUBLE STAR. A star that appears single to the unaided eye but seen as two stars in the telescope or detected by the spectroscope.

DRAPER CLASSIFICATION. A sequence of classes of stellar spectra.

DUST-CLOUD HYPOTHESIS. A theory that parts of the solar system came from a mass of gas and dust which condensed in space.

DWARF STAR. A star of moderate luminosity and mass; generally a main-sequence star.

DYNE. A unit of force in the metric system; the force required to give a mass of 1 gram an acceleration of 1 cm/sec/sec.

$E = mc^2$. A fundamental equation from the theory of relativity; relates mass and energy.

EARTHLIGHT. Sunlight reflected from the dark part of the crescent moon, having previously been reflected from the earth's atmosphere.

EARTHQUAKE. A movement of the earth's crust.

EAST POINT. The point on the celestial horizon whose azimuth is 90°.

ECCENTRICITY. The ratio of the distance between the two foci of an ellipse to the length of the major axis; also defines a parabola ($e = 1$) and an hyperbola ($e > 1$).

ECLIPSE. The phenomenon of one body shutting off all or part of the light of another, or by an opaque body passing into the shadow of another opaque body.

ECLIPSE LIMIT. The distance in degrees from a node within which an eclipse of the sun or moon is possible.

ECLIPSE PATH. The track on the earth followed by the shadow of the moon during a solar eclipse.

ECLIPSE SEASON. A time of year when the sun or the earth's shadow is inside an eclipse limit.

ECLIPSE YEAR.   The time between successive passages of the sun or the earth's shadow through the same node; 346.6 days.

ECLIPSING BINARY.   A binary star whose orbital plane is so placed that the two components pass in front of each other as seen from the earth, thus producing eclipses.

ECLIPTIC.   The apparent path of the sun among the stars.

EINSTEIN EFFECT.   The deflection of light in the region of a celestial body of large mass; a revision of the Doppler effect.

ELECTROMAGNETIC SPECTRUM.   The family of radiations that includes light, radio waves, infrared and ultraviolet light, X rays, and gamma rays.

ELECTRON.   A fundamental subatomic particle with negative charge and small mass; revolves in orbit around the nucleus of the atom, unless free to move in space.

ELECTRON VOLT.   The energy required to accelerate an electron through a potential of 1 volt. 1.60×10^{-5} joule.

ELEMENT.   A substance that cannot be divided into simpler substances by ordinary chemical means.

ELEMENT OF AN ORBIT.   One of several quantities that are used to compute the size, shape, and orientation of an orbit; used to determine the position of a body in the orbit.

ELLIPSE.   A closed plane curve in which the sum of the distances from any point on the curve to two internal points (foci) is always the same; a conic section $(e < 1)$.

ELLIPSOIDAL GALAXY.   A galaxy without arms, ellipsoidal in shape.

ELONGATION.   The angle between the centers of two objects, measured along a great circle; usually the angle between the sun and the moon or a planet, or a planet and its satellite.

EMISSION LINE.   A bright line in the spectrum.

EMISSION NEBULA.   A bright nebula.

EMISSION SPECTRUM.   A spectrum consisting of bright (emission) lines, produced by an incandescent gas at low pressure.

ENERGY.   The capacity to do work; the part of the universe that is not matter.

EPHEMERIS.   A table or book that gives the positions of celestial bodies at various times, usually at regular intervals.

EPICYCLE.   A circle that moves along the deferent of a planet according to the Ptolemaic system; a circle on a circle.

EQUATION OF TIME.   Apparent minus mean solar time.

EQUATOR.   A great circle on the earth or on the celestial sphere 90° from the poles.

EQUATORIAL MOUNTING.   A telescope set on two axes, one of which is directed toward the celestial poles, the other at right angles; compensates easily for the rotation of the earth.

EQUATORIAL SYSTEM.   A system of coordinates with the celestial equator as the fundamental plane.

EQUINOX.   An intersection of ecliptic and celestial equator.

ERG.   A unit of energy in the metric system; amount of work done by a force of one dyne acting through a distance of 1 cm.

ESCAPE VELOCITY.   The speed at which a body overcomes the gravitational pull of another and moves off into space.

Ether. A hypothetical medium through which light is transmitted; once thought to permeate all space.

Evolutionary theory. A theoretical explanation of the changes in the universe from its beginning.

Exosphere. The outermost layer of the earth's atmosphere, where molecules escape the earth's gravitational pull.

Expanding universe. The theory that the universe started with an explosion; all parts are moving away from a central point.

Eyepiece. A lens combination in a telescope by which the eye examines the image formed by the objective; magnifies the image.

Facula. A bright region on the sun near the limb or near a sunspot.

Filar micrometer. A device for measuring angles; used at the eye end of a telescope.

Fission. The separation of an atomic nucleus into smaller parts.

Flare. The sudden, temporary brightening of a region on the sun.

Flare star. A star that increases suddenly, unexpectedly, and temporarily in brightness.

Flash spectrum. The spectrum of the "reversing layer" of the sun, seen very briefly as a bright-line spectrum just before and after the total phase of an eclipse.

Flint glass. A hard glass used in one component of an achromatic lens.

Flocculus. A bright region near a sunspot seen in a spectroheliogram.

Fluorescence. The absorption of light of one wavelength and its re-emission in another; usually ultraviolet to visible light.

Focal length. The distance from the center of a lens or mirror to the focus.

Focal ratio. The ratio of focal length to aperture of a camera or telescope.

Focus. The point where converging rays of light meet.

Focus of a conic section. A point that, associated with the directrix, defines a conic section.

Forbidden lines. Any lines in the spectrum of a gas formed by highly improbable changes in the orbits of electrons.

Force. That which can overcome the inertia of a body.

Frame of reference. A set of axes to which positions and motions in a system can be referred.

Fraunhofer lines. Absorption lines in the spectrum of the sun or stars.

Frequency. Number of waves passing a given point in a unit of time.

Fringes. Successive bright and dark areas produced by the interference of light.

Full moon. The phase of the moon when the visible disk is fully illuminated; moon is at opposition.

Fusion. The building up of atoms from lighter ones.

Galactic center. A region in Sagittarius about which the sun and stars of the galaxy rotate.

Galactic cluster. A moving or open cluster of stars in the spiral arms or disk of the galaxy.

Galactic concentration. The ratio of the number of stars on the galactic equator to the number in an equal area at the poles.

GALACTIC EQUATOR. An imaginary great circle that divides the Milky Way into two approximately equal parts; fundamental circle for galactic coordinates.

GALACTIC LATITUDE. The angular distance from the galactic equator to any point on the celestial sphere, measured along a perpendicular circle.

GALACTIC LONGITUDE. The angular distance from the galactic center northward along the galactic equator to a perpendicular from a given point.

GALACTIC POLES. Two points 90° from the galactic equator.

GALAXY. A large collection of stars, dust, and gas in space; a system of millions or billions of stars held together by gravitation.

GAMMA RAY. A quantity of high-energy radiation emitted from the nucleus of a radio-active atom; a photon.

GEGENSCHEIN (COUNTERGLOW). The area of faint, reflected sunlight from a large collection of small particles opposite the sun.

GIANT STAR. A large star of about zero absolute magnitude.

GIBBOUS. The phase of the moon or a planet between quarter and full.

GLOBULAR CLUSTERS. Large systems of stars packed into nearly spherical shape, located mostly in the halo of the galaxy.

GLOBULE. A small, dense, dark nebula; possibly a protostar.

GRANULES. Small areas on the sun, formerly called rice grains, which give the sun a mottled appearance through a telescope.

GRAVITATION. The attractive force which masses exert on each other.

GRAVITATIONAL CONSTANT. G, the constant of proportionality in Newton's law of gravitation; 6.668×10^{-8} dyne-cm^2/g^2.

GRAVITY. The gravitation of the earth.

GREAT CIRCLE. The largest circle that can be drawn on a sphere; its center is the center of the sphere.

GREATEST ELONGATION. The largest angular separation between Mercury or Venus and the sun.

GREAT RIFT. An apparent split in the Milky Way between Cygnus and Crux, due to heavy absorption of light in space.

GREENWICH MERIDIAN. The prime meridian through Greenwich, England.

GREGORIAN CALENDAR. The present calendar introduced by Pope Gregory XIII.

HI REGION. A region of neutral hydrogen in space.

HII REGION. A region of ionized hydrogen in space.

HALF-LIFE. The time required for half of the atoms in a radioactive element to disintegrate.

HALO OF GALAXY. The system of globular clusters, some stars, and diffuse hydrogen gas surrounding the galaxy.

HARMONIC LAW. Kepler's third law of planetary motion; $A^3 = P^2$.

HARVEST MOON. The full moon nearest the autumnal equinox.

HEAD OF A COMET. The nucleus and coma of a comet.

HELIOCENTRIC UNIVERSE. A theoretical universe with the sun at the center.

HORIZON (CELESTIAL). A great circle 90° from zenith and nadir.

HORIZON SYSTEM. A system of coordinates with the celestial horizon as the fundamental circle.

HORIZONTAL PARALLAX. The angle at the center of a celestial object in the solar system, such as the moon or a planet, between the two ends of a radius of the earth.

HOUR ANGLE. The angle between the celestial meridian and an hour circle.

HOUR CIRCLE. A great circle on the celestial sphere through the poles and a celestial object; perpendicular to the celestial equator.

H-R DIAGRAM. A diagram relating spectral types and absolute magnitudes; spectrum-luminosity or Hertzsprung-Russell diagram.

HUBBLE CONSTANT. A number relating distance and velocity of galaxies; about 75 km/sec/10^6 parsecs; 14 miles/sec/10^6 light-years; value uncertain.

HUNTER'S MOON. The full moon following the harvest moon.

HYPERBOLA. A conic section; eccentricity greater than 1.0.

IMAGE. The optical representation of an object, usually seen as a luminous point or area at the focus of a telescope.

IMAGE TUBE. A device that combines a photoelectric sensitive surface and a photographic plate for measuring stellar brightness.

INCLINATION. The angle between the orbital plane of a moving body and a fundamental plane; one of the elements of an orbit.

INDEX OF REFRACTION. The ratio of the velocity of light in a vacuum to its velocity in a given transparent substance.

INERTIA. The property of matter that resists a change in its motion.

INFERIOR CONJUNCTION. Conjunction of Mercury or Venus when between the earth and the sun.

INFERIOR PLANET. A planet with orbit between that of the earth and the sun; Mercury and Venus.

INFRARED RADIATION. Radiation of wavelength longer than visible red light and shorter than radio waves.

INNER PLANET. One of the first four planets: Mercury, Venus, earth, Mars.

INTENSITY. The energy of a light source per second per unit area.

INTERFERENCE. The reinforcing or canceling of light waves.

INTERNATIONAL DATE LINE. An arbitrary, irregular line near longitude 180°, where the date changes by one day.

INTERNATIONAL GEOPHYSICAL YEAR (IGY). An 18-month period in 1957 and 1958 set aside for international cooperation in research of the sun and also the earth and its environment.

INTERSTELLAR DUST. Microscopic dust grains in space.

INTERSTELLAR GAS. Diffuse gas in space.

INTERSTELLAR LINES. Absorption lines in the spectrum produced by interstellar gas.

INTERSTELLAR MATTER. Dust and gas in space; literally "between the stars."

ION. An electrically charged atom formed by the loss or gain of one or more electrons without a change in the nucleus.

IONIZATION. The process by which an ion is formed.

IONOSPHERE. The layer of the earth's atmosphere between 20 and 500 miles above the earth; contains electrically charged particles.

IRREGULAR GALAXY. A galaxy without regular shape.

IRREGULAR VARIABLE. A star whose brightness changes without a regular period; its amplitude may also vary.

ISLAND UNIVERSE. A former name for a galaxy.

ISOSTASY. The theory that the crust of the earth is kept in equilibrium by each part being subject to equal pressure from all sides.

ISOTOPE. One of two or more forms of atoms that have the same atomic number but different atomic weights.

JET STREAMS. Fast-moving streams of air in the upper troposphere.

JOULE'S LAW. A given amount of mechanical energy is always equivalent to a definite quantity of heat energy.

JULIAN CALENDAR. A solar calendar introduced by Julius Caesar.

JULIAN DAY CALENDAR. A consecutive numbering of days beginning on January 1, 4713 B.C.

JUPITER. The fifth planet from the sun.

KEPLER'S LAWS. Three laws of planetary motion stated by Kepler.

KINETIC ENERGY. The energy of motion: K.E. $= \frac{1}{2}mv^2$.

KINETIC THEORY. The assumption that matter is composed of molecules in constant motion.

KIRCHHOFF'S LAWS. Three laws of the production of spectra.

KIRKWOOD'S GAPS. Gaps in the spacing of asteroids, or in the rings of Saturn.

LATITUDE. The angle from the terrestrial equator to a point on the earth, measured along its meridian; also similar angles from the ecliptic and the galactic equator.

LATITUDINAL LIBRATIONS. Librations due to the tilt of the equator of the moon or Mercury to its orbital plane.

LAW OF AREAS. Kepler's second law: The radius vector sweeps out equal areas in equal intervals of time.

LAW OF CONSERVATION OF ENERGY. Under ordinary conditions, energy can be neither created nor destroyed.

LAW OF CONSERVATION OF MASS-ENERGY. Mass and energy are related; $E = mc^2$; the total mass and energy in the universe remains constant.

LAW OF CONSERVATION OF MATTER. Matter cannot be created or destroyed by ordinary chemical processes; it can only be changed from one form to another.

LAW OF ILLUMINATION. The amount of light reaching a point varies inversely as the square of the distance from the source.

LAW OF RECESSION OF GALAXIES. The radial velocity of a galaxy is proportional to its distance; $v = Hr$.

LAW OF REFLECTION. The angle of reflection is equal to the angle of incidence.

LAW OF REFRACTION. A beam of light is bent towards the perpendicular to the surface when passing from one medium to another of higher index of refraction.

LAW OF UNIVERSAL GRAVITATION. Every particle in the universe attracts every other particle with a force that is proportional to the product of their masses and inversely proportional to the square of the distance between them; Newton's law:

$$F = G \frac{m_1 m_2}{d^2}$$

LEAD SULFIDE CELL. A photoelectric device for measuring infrared radiation.

LEAP YEAR. A calendar year with 366 days.

LIBRATIONS. Apparent oscillations of a body, which permit more than half of its surface to be seen from another body.

LIGHT. Electromagnetic radiation visible to the eye.

LIGHT CURVE. A graph showing the relation between time and the brightness of an object; usually the curve of changing brightness of a variable star.

LIGHT-GATHERING POWER. The amount of light a telescope can collect; proportional to the area of the objective.

LIGHT-YEAR. The distance light travels in one year; about 6 trillion miles.

LIMB. The edge of the moon, a planet, the sun, or a star.

LIMB DARKENING. The ratio of brightness at the edge of the sun or a star to its brightness at the center.

LIMITING MAGNITUDE. The faintest magnitude observable in a telescope under given conditions.

LINE OF APSIDES. The major axis of an ellipse; the line joining the points farthest from and nearest to a focus.

LINE OF NODES. The line connecting the two nodes of an orbit; a line common to two planes; located by one of the elements of an orbit.

LOCAL APPARENT NOON. The time at which the center of the apparent (true) sun is on the local meridian.

LOCAL APPARENT TIME. The hour angle of the apparent sun.

LOCAL GROUP. The cluster of galaxies that includes our galaxy.

LOCAL MEAN NOON. The time at which the mean sun is on the local meridian.

LOCAL MEAN TIME. The hour angle of the mean sun.

LONGITUDE. The angle between the prime meridian and the meridian through a point on the earth.

LONGITUDE OF THE ASCENDING NODE. The angle measured eastward along the ecliptic from the vernal equinox to the ascending node of the moon or a planet; an orbital element.

LONGITUDINAL LIBRATION. A libration due to the variable speed of the moon or Mercury in its orbit.

LONG-PERIOD VARIABLE. A variable star whose changes of brightness are usually not strictly periodic, but repeat in about 100 days.

LUMINOSITY. The brightness of a star compared to that of the sun.

LUNAR ECLIPSE. An eclipse of the moon.

LYMAN SERIES. An array of lines in the ultraviolet spectrum of hydrogen.

MAGELLANIC CLOUDS. The two nearest galaxies.

MAGNETIC FIELD. The region of space near a planet or other body where magnetic forces have been detected.

MAGNETIC POLE. One of two points on the earth toward which the ends of a compass needle point.

MAGNIFYING POWER. The apparent size of an object seen through a telescope compared to its size as seen with the unaided eye.

MAGNITUDE. An arbitrary number to indicate the brightness of an object.

MAIN SEQUENCE. A line on the H-R diagram.

MAJOR AXIS. The longest diameter of an ellipse; line of apsides.

MAJOR PLANET. One of the four largest planets: Jupiter, Saturn, Uranus, and Neptune.

MANTLE. The layer of the earth between the crust and the core.

MARIA. Latin name for lunar seas; singular, mare.

MARS. The fourth planet from the sun.

MASCON. A mass concentration below the surface of the moon, mostly under the maria.

MASS. The total amount of matter in an object.

MASS-LUMINOSITY RELATION. The principle that says that the brightness of a star depends on the amount of its mass.

MATTER. Anything that occupies space; the part of the universe that is not energy.

MEAN DISTANCE. The length of the semimajor axis of an ellipse; average of perihelion and aphelion distances, or perigee and apogee, or periastron and apastron.

MEAN SOLAR DAY. Average length of the solar day during the year; interval between successive passages of the mean sun across a meridian.

MEAN SOLAR TIME. Time kept by the mean sun; the interval of time since the mean sun crossed a meridian.

MEAN SUN. An imaginary sun that moves uniformly along the celestial equator, making a complete circuit in one year.

MELTING POINT. The temperature at which the particles in a solid overcome the forces that hold them together.

MERCURY. The first planet from the sun.

MERIDIAN. A great circle through the zenith and a celestial pole; a great circle on the earth through the terrestrial poles.

MESON. A subatomic particle with the charge of an electron, but of greater mass.

MESOSPHERE. The layer of the ionosphere immediately above the stratosphere; the "middle sphere."

MESSIER CATALOG. A catalog of "nebulae" compiled by Charles Messier.

METEOR. The term used to describe the phenomenon associated with the collision of a particle from space and the earth's atmosphere.

METEORITE. A meteoroid of density higher than average, which has struck the earth's surface.

METEORITE THEORY. The theory that lunar craters were formed by meteoric impact.

METEOROID. The particle involved in the meteor phenomenon.

METEOR SHOWER. Many meteors which seem to radiate from a small area on the celestial sphere.

MICROMETEORITE. A microscopic particle in space; an exceptionally small meteoroid.

MILKY WAY. A faint band of light around the sky; composed of a vast number of stars and interstellar matter.

MINOR AXIS. The smallest diameter of an ellipse, perpendicular to the major axis and passing through the center of the ellipse.

MINOR PLANET. An asteroid or planetoid.

MOHOROVIČIĆ DISCONTINUITY (MOHO). The boundary between the earth's crust and mantle.

MOLECULE. A combination of atoms; the smallest particle of a substance that has all the properties of the substance.

MONOCHROMATIC. A word meaning of one color or wavelength.

MORGAN CLASSIFICATION. A modern classification of stellar spectra; a suggested classification of galaxies; both by W. W. Morgan.

MOVING CLUSTER. A group of stars moving in nearly parallel paths and with the same velocity.

NADIR. The point opposite the zenith; altitude $-90°$.

n-BODY PROBLEM. The problem of determining the positions and motions of more than two bodies from their mutual gravitation; not solvable by ordinary mathematics.

NEAP TIDE. A tide at quarter moon when the solar and lunar tides partially cancel each other.

NEBULA. A gas or dust cloud in space.

NEBULAR HYPOTHESIS. The theory of Laplace that the solar system was formed from a nebula.

NEPTUNE. The eighth planet from the sun.

NEUTRINO. A subatomic particle of very small mass and zero charge.

NEUTRON. A subatomic particle of zero charge and mass similar to that of the proton.

NEUTRON STAR. A star composed entirely of neutrons after it has lost its other material following the explosion of a supernova.

NEW GENERAL CATALOG (NGC). A catalog of "nebulae," successor to the Messier catalog.

NEW MOON. The phase of the moon when it is in conjunction with the sun.

NEWTONIAN REFLECTOR. A reflecting telescope that uses a plane mirror to deflect the beam of light to one side of the telescope tube; devised by Newton.

NEWTON'S LAW OF GRAVITATION. The law of universal gravitation.

NEWTON'S LAWS OF MOTION. Three laws of mechanics formulated by Newton.

NODE. One of two points where the orbit of a celestial body crosses a reference plane; for example, a point where a planet or the moon crosses the ecliptic.

NORMAL SPIRAL. A spiral galaxy with a nucleus from which arms extend; not a barred spiral.

NORTH CELESTIAL POLE. The point on the celestial sphere 90° from the celestial equator at one end of the earth's axis of rotation; now located in Ursa Minor.

NORTH GALACTIC POLE. A point in the northern hemisphere of the celestial sphere 90° from the galactic equator; in constellation Coma Berenices.

NORTH POINT. The point of intersection of the celestial meridian with the celestial horizon under the north celestial pole.

Nova. A "new" star; a temporary star that shows an unexpected outburst of light.

Nuclear fission. The splitting of the nucleus of an atom.

Nuclear fusion. The combination of the nuclei of two atoms to form an atom of larger mass.

Nucleus. The central part of an atom, a comet, or a galaxy.

Nutation. A "nodding" of the earth's pole; a variation of precession.

Objective. The large lens or mirror of a telescope; forms an image of a luminous source.

Objective prism. A small-angle prism placed in front of a telescope objective to photograph the spectra of a field of stars.

Oblateness. A measure of the amount of flattening of a body, such as the earth, due to its rotation.

Obliquity of the ecliptic. The angle between the ecliptic and the celestial equator.

Obscuration. Absorption of starlight by interstellar material.

Occultation. The passage of a celestial body behind a larger one.

Open cluster. A loosely formed group of stars; a galactic cluster.

Opposition. The aspect of the moon or a planet when opposite the sun as seen from the earth; elongation approximately 180°.

Optical double star. Two stars that appear close together, but are actually too far apart to be gravitationally held together.

Orbit. The path of a body in revolution around another body or bodies.

Orbital plane. The plane in which a planet, satellite, or star revolves around a central attracting body.

Outer planet. A planet beyond Mars; Jupiter, Saturn, Uranus, Neptune, or Pluto.

Ozone layer. A layer of the earth's atmosphere; the chemosphere.

Parabola. A conic section, every point of which is equidistant from a given point (a focus) and a straight line (the directrix).

Paraboloid. A curved surface formed by rotating a parabola about its axis.

Parallax. The apparent displacement of an object when viewed from two different points.

Parsec. A unit of distance; 3.26 light-years; 206,265 a.u.

Partial eclipse. An eclipse that is not total.

Penumbra. That part of the shadow of an opaque body which only partly cuts off the light from a luminous source.

Penumbral eclipse. The passage of the moon, during a lunar eclipse, through the penumbra of the earth's shadow without going through the umbra.

Perigee. The point in the orbit of a satellite where it is nearest the earth.

Perihelion. The point in the orbit of a planet or comet where it is nearest the sun.

Perihelion advance. See Advance of the perihelion.

Period. The time required for a single revolution of one body around another; also used for rotation.

Period-luminosity law. The relation between the period and absolute magnitude of Cepheid variable stars.

PERTURBATION. The gravitational disturbance of a celestial body which pulls it from its regular orbit.

PHASES OF THE MOON. Changes in the apparent shape of the illuminated portion of the moon or a planet.

PHOTOELECTRIC EFFECT. The emission of electrons from a substance when light strikes it.

PHOTOELECTRIC PHOTOMETER. A device for measuring light intensity by the use of a photoelectric cell.

PHOTOELECTRIC MAGNITUDE. The brightness (magnitude) of a star as measured with a photoelectric photometer.

PHOTOGRAPHIC MAGNITUDE. The brightness (magnitude) of an object as measured on a photographic plate.

PHOTOMETER. An instrument for measuring the amount or color of light from a luminous source; visual, photographic or photoelectric.

PHOTON. A unit of electromagnetic energy; a quantum of radiant energy.

PHOTOSPHERE. The visible layer of the sun; the "surface" or "light-giving sphere."

PHOTOVISUAL MAGNITUDE. The magnitude of an object as measured through a filter, such that it approximates the visual magnitude.

PLAGE. A bright region of the sun observed in monochromatic light; a flocculus.

PLANE MIRROR. A mirror with a flat (plane) reflecting surface.

PLANET. A "wanderer"; one of nine bodies in orbit around the sun.

PLANETARIUM. A projector that throws a reproduction of the sky onto the inside of a spherical dome to show the positions and motions of stars, planets, and other bright objects.

PLANETARY NEBULA. A shell of gas surrounding a hot star.

PLANETESIMAL HYPOTHESIS. The theory of Chamberlin and Moulton that the planetary system was formed from "bolts" of hot material pulled from the sun.

PLANETOID. A minor planet; an asteroid.

PLUTO. The ninth planet from the sun.

POLAR AXIS. The axis of a telescope parallel to the earth's axis and about which the telescope turns to compensate for the earth's rotation.

POLARIZATION. A state in which light waves are confined to a single plane.

POPULATION I AND II. Two classes of stars based on location in the galaxy and state of evolutionary growth.

POSITRON. A particle with positive electrical charge, but mass equal to that of an electron; a positive electron.

POTENTIAL ENERGY. Stored energy; can be converted to kinetic energy.

POUND. The standard unit of force in the English system.

PRECESSION. The conical motion of the axis of a rotating, oblate body.

PRECESSION OF THE EQUINOXES. The slow westward motion of the equinoxes along the ecliptic due to precession.

PRIMARY MINIMUM. The deeper minimum in the light curve of an eclipsing binary star.

PRIME FOCUS. The point in a telescope where the image is formed by the objective, when no second mirror is used; the focus of an objective.

PRIME MERIDIAN. The terrestrial meridian through Greenwich.

PRIMEVAL ATOM. The single mass into which all matter in the universe was originally collected, according to the Lemaitre theory of the expanding universe.

PRINCIPIA. Newton's great book on motion and gravitation; *Philosophiae Naturalis Principia Mathematica*.

PRINCIPLE OF THE PENDULUM. The law that the time of one complete swing of the pendulum depends on its length.

PRISM. A piece of glass or other transparent material with plane sides and triangular ends; used in spectroscopes for refraction.

PROMINENCE. A flamelike projection seen on the limb of the sun; associated with sunspot activity.

PROPER MOTION. The apparent rate of change of direction of a star, measured in seconds of arc per year.

PROTON. A fundamental subatomic particle of positive electrical charge; the nucleus of a hydrogen atom.

PROTOPLANET, PROTOSTAR. The original material that condensed to form a planet or star.

PROTON-PROTON REACTION. One cycle by which four hydrogen nuclei (protons) combine to form a helium nucleus.

PULSAR. A variable star with extremely short period, probably resulting from the rotation of a neutron star.

PULSATING STAR. See CEPHEID VARIABLE.

PYRHELIOMETER. An instrument for measuring radiant solar energy.

QUADRATURE. The relative position of two bodies as seen from the earth when they are 90° apart; especially the moon or a planet when at 90° elongation from the sun.

QUARTER-MOON. The moon in quadrature; also called half-moon.

QUASAR. A starlike object with very high red shift; thought to be a small, intensely bright galaxy.

QUIET SUN. The sun when there is little activity; see ACTIVE SUN.

RADAR TELESCOPE. A radio telescope equipped also with a transmitter.

RADIAL VELOCITY. The component of velocity of an object toward or away from the sun; measured by the Doppler effect in km/sec.

RADIANT POINT. The area of the sky from which meteors radiate.

RADIATION. Energy that can be transmitted through space.

RADIATION PRESSURE. The slight pressure exerted by light or other radiation.

RADIOACTIVITY. The natural disintegration of the nucleus of an atom, in which subatomic particles and gamma rays are emitted.

RADIO ASTRONOMY. The study of astronomy by means of radio waves, which are beyond the infrared part of the spectrum.

RADIO SEXTANT. A sextant in which the lenses and mirrors are replaced by a radio receiver.

RADIO TELESCOPE. A large receiver for collecting and measuring radio radiation from space.

RADIUS VECTOR. A line joining any point on an orbit with a focus; for example, the line between the sun and a planet.

RAYS.   Systems of bright, elongated streaks on the moon.

REACTION.   The equal and opposite force that accompanies every force; see NEWTON's THIRD LAW OF MOTION.

REAL IMAGE.   An actual image at the focus of a lens or mirror.

RED GIANT.   A large, cool star of absolute magnitude brighter than about 0; plotted on the upper right of the H-R diagram.

RED SHIFT.   A Doppler shift toward the red end of the spectrum of most galaxies; basis for the theory of the expanding universe.

REDDENING.   The interstellar reddening of starlight by absorption.

REFLECTING TELESCOPE.   A telescope with mirrors instead of lenses.

REFLECTIVITY.   See ALBEDO.

REFRACTING TELESCOPE.   A telescope with lenses instead of mirrors.

REFRACTION.   The bending of light as it passes from one transparent medium (or a vacuum) to another medium of different density.

REGRESSION OF THE NODES.   The westward motion along the ecliptic (or equator) of the nodes of a planet (or satellite).

RELATIVITY.   The theory formulated by Einstein; a modification of the Newtonian theory of gravitation.

RESOLVING POWER.   The ability of a telescope to separate objects apparently close together; expressed in seconds of arc.

RETROGRADE MOTION.   The apparent east-to-west motion of a planet or comet; opposite to direct motion.

REVOLUTION.   The motion of one body around another.

RIGHT ASCENSION.   The angle on the celestial equator eastward from the vernal equinox to the hour circle through a body; the angle at the celestial poles between the hour circles of the vernal equinox and any other point.

RILL.   A crack in the floor of the moon.

ROCHE'S LIMIT.   The distance from a body inside which a satellite would break up under the body's gravitation.

ROTATION.   The turning of a body about an axis.

SAROS.   A cycle of similar eclipses that recur at intervals of about 18 years.

SATELLITE.   A body that revolves around a larger body; a moon.

SATURN.   The sixth planet from the sun.

SCHMIDT CAMERA.   A type of reflecting telescope that uses a spherical primary mirror and a correcting plate; has a small focal ratio.

SCIENCE.   A branch of learning concerned with observation and classification of data concerning natural phenomena.

SECONDARY MINIMUM.   The shallower minimum in the light curve of an eclipsing binary star.

SECONDARY MIRROR.   A small mirror used in addition to the primary mirror in reflecting telescopes.

SEISMIC WAVES.   Earthquake waves.

SEISMOGRAPH.   An instrument for recording seismic waves.

SEISMOLOGIST.   A scientist who studies earthquakes.

SEISMOLOGY. The science of origin and transmission of seismic waves.

SEMIMAJOR AXIS. One half the major axis of an ellipse.

SEPARATION. The angular distance between the two components of a visual binary star.

SHELL STAR. A star surrounded by a shell or ring of gas.

SHOWER OF METEORS. See METEOR SHOWER.

SIDEREAL DAY. The time interval between two successive passages of the vernal equinox or a star across the meridian.

SIDEREAL MONTH. The period of revolution of the moon around the earth with respect to the stars; about 27⅓ days.

SIDEREAL PERIOD. The period of revolution of one body around another.

SIDEREAL TIME. The hour angle of the vernal equinox; the right ascension of the meridian; star time.

SIDEREAL YEAR. The period of the earth's revolution around the sun.

SIDERITE. A meteorite composed largely of iron and nickel.

SMALL CIRCLE. A circle on the surface of a sphere that is not a great circle.

SOLAR ACTIVITY. Sunspots, prominences, and so forth, on the sun.

SOLAR APEX. The point toward which the sun is moving with respect to the nearest stars.

SOLAR CONSTANT. The amount of solar radiation received at a distance of 1 a.u.; 1.96 cal/min/cm² = 1.39×10^{-1} j/sec/cm².

SOLAR DAY. The time interval between two successive transits of the sun across a given meridian.

SOLAR ECLIPSE. An eclipse of the sun.

SOLAR FLARE. A short-lived, sudden outburst of energy from the sun, usually from a sunspot area.

SOLAR INTERIOR. The mass of hot gas below the solar photosphere.

SOLAR MOTION. The motion of the sun in space with respect to the nearest stars.

SOLAR PARALLAX. The angle at the sun between two ends of a radius of the earth; about 8.8″.

SOLAR SYSTEM. The assemblage of planets, satellites, and other bodies under the gravitational influence of the sun.

SOLAR WIND. A radial flow of particles and radiation from the sun.

SOLSTICES. Two points on the ecliptic where the sun's declination is greatest; +23½° and −23½° declination.

SOUTH CELESTIAL POLE. The point 180° from the north celestial pole; declination, −90°.

SOUTH GALACTIC POLE. The point 180° from the north galactic pole; galactic latitude −90°.

SOUTH POINT. The point on the celestial horizon 180° from the north point; an intersection of the meridian and the celestial horizon.

SPACE MOTION. The velocity of a star with respect to the sun.

SPECTRAL CLASS. The classification of a star from the appearance of the lines in its spectrum; the Draper or Morgan classification.

SPECTROGRAPH. An instrument for photographing the spectrum.

SPECTROHELIOGRAM. A photograph of the sun's atmosphere in the light of a single element.

SPECTROSCOPE. Arrangement of prisms or grating to produce a spectrum for examination.

SPECTROSCOPIC BINARY. A binary star with components too close together to be separated visually, but detected and measured with a spectrograph.

SPECTROSCOPIC PARALLAX. A means of determining the parallax of a star (and therefore its distance) by observing its spectral characteristics, such as the width of the lines.

SPECTRUM. The band of colors produced when light is separated into its component parts.

SPECTRUM-LUMINOSITY DIAGRAM. See H-R DIAGRAM.

SPEED. The rate of change of distance without regard to direction.

SPHERICAL ABERRATION. The failure of spherical lenses and mirrors to bring all the rays from a point source to a single focus.

SPICULE. A narrow jet of hot material rising in the solar atmosphere.

SPIRAL ARM. A curved formation that extends outward from the nucleus of some galaxies; composed of stars, dust, and gas.

SPIRAL GALAXY. A galaxy with a nucleus and spiral arms.

SPRING TIDE. An exceptionally high tide produced by the reinforcement of solar and lunar tides.

STANDARD TIME. Time used by a region on earth, based on mean solar time with respect to a standard meridian.

STAR. A self-luminous, nearly spherical mass of gas.

STAR CLOUD. A portion of the Milky Way where the stars are so close together they appear as a luminous cloud.

STAR CLUSTER. A group of stars held together by mutual gravitation or by their common motion.

STAR MAP. A chart showing positions and magnitudes of stars.

STEADY-STATE THEORY. The belief that the density and shape of the universe are always the same, and that stars are being formed at the same rate at which their mass is being converted to energy and radiated into space.

STEFAN'S LAW. The amount of energy radiated by a body varies as the fourth power of its absolute temperature; $E = aT^4$.

STELLAR EVOLUTION. The continual change in size, mass, luminosity, structure, and other properties, of a star with time.

STELLAR PARALLAX. The angle in seconds of arc subtended by one astronomical unit at the distance of a star; annual parallax.

STRATOSPHERE. The layer of the earth's atmosphere between the troposphere and the ionosphere.

SUBDWARF. A star of lower luminosity than that of a main-sequence star of the same spectral class.

SUBGIANT. A star of luminosity between those of main-sequence stars and normal giants of the same spectral class.

SUMMER SOLSTICE. The point on the ecliptic where the sun reaches its greatest northern declination, $+23\frac{1}{2}°$.

Sun. The star that is the gravitational center of the solar system.

Sunspot. A dark spot seen on the disk of the sun; dark by contrast with the hotter photosphere.

Sunspot cycle. The interval of time from one sunspot maximum to the next; roughly, an 11-year period.

Supergiant. A star of absolute magnitude brighter than about −1.

Superior conjunction. Conjunction of a body on the far side of the sun as seen from the earth; syzygy, with the sun between the object and the earth.

Superior planet. A planet farther from the sun than the earth is.

Supernova. An exploding star much more luminous than the normal nova.

Synchrotron radiation. Radiation emitted by charged particles that are accelerated in a magnetic field and move with velocity near that of light.

Synodic month. The interval between consecutive new or full moons; about 29½ days.

Synodic period. The time interval between successive planetary oppositions or conjunctions of the same kind.

Syzygy. The lining up of three celestial bodies; conjunction or opposition.

Tail. The finest particles of a comet, which extend in a band away from the head.

Tangential velocity. The component of a star's space motion perpendicular to the radial velocity;

$$T = 4.47 \frac{\mu}{p} \text{ km/sec}$$

Telescope. An instrument for observing objects at a distance either visually or by other means.

Temporary star. A star that suddenly brightens by several magnitudes and eventually returns to near normal brightness; a nova; an exploding star.

Terminator. The line of sunrise or sunset on the moon or a planet.

Terrestrial planet. A planet about the size and mass of the earth.

Theory. A set of hypotheses and well-demonstrated laws to explain a phenomenon.

Theory of relativity. The theory advanced by Einstein as a modification of Newton's theory of gravitation.

Theory of the expanding universe. The proposal, based on the red shift in the spectra of galaxies, that the galaxies in the universe are moving outward from an unknown center; the Big Bang Theory.

Thermosphere. A hot layer of the earth's atmosphere.

Tidal hypothesis. A theory, proposed by Jeans and Jeffreys, that the planetary system was formed by a single bolt of hot material pulled from the sun by a passing star.

Tide. The deformation of a body by external forces.

Time zone. A zone 7.5° each side of a standard time meridian.

Ton. A measure of weight; 2000 pounds in the English system; 2200 pounds or 1000 kilograms or 1 million grams in the metric system.

Total eclipse. An eclipse of the sun, moon, or star where the light is completely cut off.

Train or trail. A temporary, luminous streak left by a meteoroid.

Transit. The passage of a body across the meridian or the face of a larger body; an instrument for observing the passage of a body across the meridian; an engineering instrument for measuring vertical and horizontal angles.

TRANSMUTATION. The process of changing from one form to another; a reaction by which one element is changed into another.

TRANSVERSE WAVE. A wave perpendicular to the direction of propagation.

TROPICAL YEAR. The year of the seasons; the interval of time between successive passages of the sun through the vernal equinox; 365.2422 days.

TROPOPAUSE. The boundary between the troposphere and stratosphere.

TROPOSPHERE. The layer of the earth's atmosphere immediately above the crust; extends to between 5 and 10 miles.

TWINKLE. The changing brightness and color of stars due to the earth's atmosphere.

TYCHONIC SYSTEM. A model of the universe proposed by Tycho Brahe.

ULTRAVIOLET. The part of the spectrum with wavelengths shorter than the visible spectrum.

UMBRA. That part of a geometrical shadow in which the direct light from a luminous body is completely shut off.

UNIT MASS. A mass that, when acted on by a unit force, is given a unit acceleration.

UNIVERSAL GRAVITATION. See LAW OF UNIVERSAL GRAVITATION.

UNIVERSAL TIME. The local mean time of the prime meridian; Greenwich Mean Time.

URANUS. The seventh planet from the sun.

VAN ALLEN RADIATION BELTS. Regions of electrically charged particles surrounding the earth.

VARIABLE STAR. A star that changes in light or energy with possible changes in its spectrum.

VECTOR. A directed quantity; may be represented by a straight line with an arrow at one end.

VELOCITY. Rate of change of position along a line; speed and direction.

VELOCITY OF ESCAPE. The speed which a body must have to leave the gravitational field of another body.

VENUS. The second planet from the sun.

VERNAL EQUINOX. The point on the celestial equator where the sun crosses it from south to north at the beginning of spring.

VERTICAL CIRCLE. A great circle on the celestial sphere through a body and the zenith; perpendicular to the horizon.

VISUAL BINARY. A binary star of sufficient separation to permit both components to be seen through a telescope.

VOLCANIC THEORY. The theory that the craters on the moon (or a planet) were formed by the action of volcanoes.

WATT. A unit of power; 10^7 ergs/sec.

WAVELENGTH. The distance from one point on a wave to the corresponding point on the next wave.

WAVE THEORY OF LIGHT. The theory that light travels as a wave.

WEIGHT. A measure of the attraction between the earth (or any other massive body) and a given mass.

WEIGHTLESSNESS.   The absence of weight.

WEST POINT.   A point on the celestial horizon of azimuth 270°.

WHITE DWARF.   A star that has about the same mass as, but much smaller diameter than, an ordinary star; a white star several magnitudes below the main sequence.

WIEN'S LAW.   The wavelength of maximum energy radiated by a body is inversely proportional to its absolute temperature; $\lambda_{max} = 0.289/T$.

WINTER SOLSTICE.   The point on the ecliptic where the sun is farthest south; declination, $-23\frac{1}{2}°$.

WORLD CALENDAR.   A proposed 12-month calendar of equal quarters so designed that any day of the month will fall on the same day of the week each year.

X RAYS.   Electromagnetic radiations of wavelength about 1 angstrom.

X-RAY STAR.   A star (not the sun) that emits a measurable amount of X rays.

YEAR.   Period of revolution of the earth around the sun.

ZENITH.   The point overhead; altitude 90°.

ZENITH DISTANCE.   The angle between the zenith and any point on the celestial sphere measured on a vertical circle; $z = 90° - h$.

ZODIAC.   A band on the celestial sphere extending about 8° on each side of the ecliptic; twelve constellations are in this band.

ZONE OF AVOIDANCE.   An irregular band on the celestial sphere near the galactic equator where practically no galaxies have been found.

index

Page numbers in italics refer to illustrations.

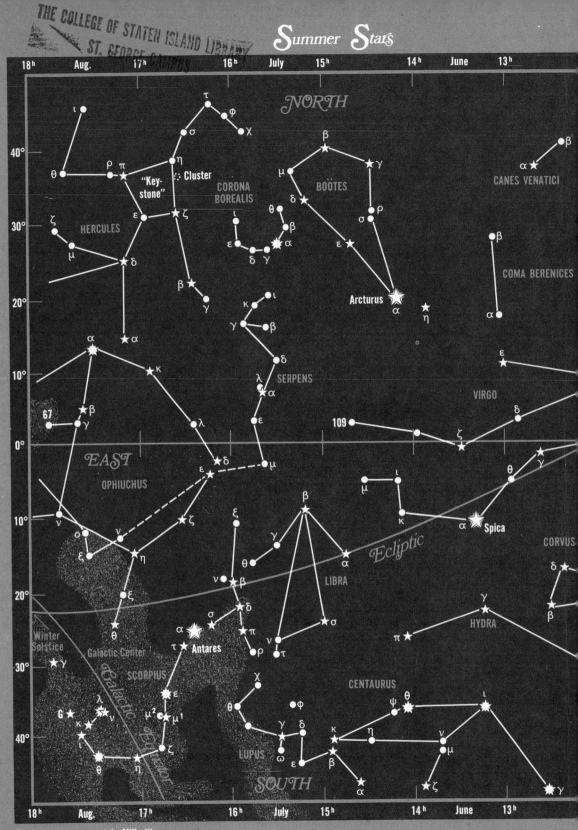

Summer Stars

NORTH

18ʰ Aug. 17ʰ 16ʰ July 15ʰ 14ʰ June 13ʰ

40°

τ
φ
σ
χ

ι

ρ π
θ   "Key-
stone"   ∴ Cluster
30°
ζ   HERCULES   ε   ζ
μ

δ

β

γ

CORONA
BOREALIS

ι
ε
δ γ α
θ δ
β

β
μ
BOÖTES γ

θ δ
σ
ρ

ε

α
Arcturus
α

η

β
β

α

CANES VENATICI

COMA BERENICES

α

20°

κ
γ
β
ι

δ

10°
α
κ

SERPENS
λ
α

ε

ε
VIRGO
δ

67
β
γ
λ

109
ζ

0°
EAST   δ
ε   μ
OPHIUCHUS

θ γ

ν
ο
ξ
ε
ν
η
ζ

ε

β
γ

μ ι
κ

α
Spica

CORVUS
δ
10°
β

ε

ν β
θ
ν

α

LIBRA

γ
β
20°
ε
δ
σ
π
ν
τ
ρ

σ

γ

HYDRA
π

Winter
Solstice
Galactic Center
θ
α
Antares
τ

χ
θ φ

CENTAURUS
ψ θ
ι

30°
γ
SCORPIUS

ε
λ ν
G κ μ²μ¹
ι   ζ
θ η

γ δ
ω
ε
LUPUS
β
α

κ
η
ν
μ

ζ
γ

40°

SOUTH

18ʰ Aug. 17ʰ 16ʰ July 15ʰ 14ʰ June 13ʰ

Ecliptic

Galactic Equator

Shaded areas represent the Milky Way